Biological sequence analysis
Probabilistic models of proteins and nucleic acids

The face of biology has been changed by the emergence of modern molecular genetics. Among the most exciting advances are large-scale DNA sequencing efforts such as the Human Genome Project which are producing an immense amount of data. The need to understand the data is becoming ever more pressing. Demands for sophisticated analyses of biological sequences are driving forward the newly-created and explosively expanding research area of computational molecular biology, or bioinformatics.

Many of the most powerful sequence analysis methods are now based on principles of probabilistic modelling. Examples of such methods include the use of probabilistically derived score matrices to determine the significance of sequence alignments, the use of hidden Markov models as the basis for profile searches to identify distant members of sequence families, and the inference of phylogenetic trees using maximum likelihood approaches.

This book provides the first unified, up-to-date, and tutorial-level overview of sequence analysis methods, with particular emphasis on probabilistic modelling. Pairwise alignment, hidden Markov models, multiple alignment, profile searches, RNA secondary structure analysis, and phylogenetic inference are treated at length.

Written by an interdisciplinary team of authors, the book is accessible to molecular biologists, computer scientists and mathematicians with no formal knowledge of each others' fields. It presents the state-of-the-art in this important, new and rapidly developing discipline.

Richard Durbin is Head of the Informatics Division at the Sanger Centre in Cambridge, England.

Sean Eddy is Assistant Professor at Washington University's School of Medicine and also one of the Principle Investigators at the Washington University Genome Sequencing Center.

Anders Krogh is a Research Associate Professor in the Center for Biological Sequence Analysis at the Technical University of Denmark.

Graeme Mitchison is at the Medical Research Council's Laboratory for Molecular Biology in Cambridge, England.

Biological sequence analysis

Probabilistic models of proteins and nucleic acids

Richard Durbin

Sean R. Eddy

Anders Krogh

Graeme Mitchison

CAMBRIDGE
UNIVERSITY PRESS

PUBLISHED BY THE PRESS SYNDICATE OF THE UNIVERSITY OF CAMBRIDGE
The Pitt Building, Trumpington Street, Cambridge CB2 1RP, United Kingdom

CAMBRIDGE UNIVERSITY PRESS
The Edinburgh Building, Cambridge CB2 2RU, UK http://www.cup.cam.ac.uk
40 West 20th Street, New York, NY 10011-4211, USA http://www.cup.org
10 Stamford Road, Oakleigh, Melbourne 3166, Australia

First published 1998

Printed in the United Kingdom at the University Press, Cambridge

Typeset in Times 11/14pt

A catalogue record for this book is available from the British Library

Library of Congress Cataloguing in Publication Data

Biological sequence analysis: probabilistic models of proteins and nucleic
acids/Richard Durbin … [*et al.*].
  p. cm.
Includes bibliographical references and index.
ISBN 0 521 62041 4 (hardcover). – ISBN 0 521 62971 3 (pbk.)
1. Nucleotide sequence – Statistical methods. 2. Amino acid sequence – Statistical
methods. 3. Numerical analysis. 4. Probabilities. I. Durbin, Richard.
QP620.B576 1998
572.8'633 – dc21 97–46769 CIP

ISBN 0 521 62041 4 hardback
ISBN 0 521 62971 3 paperback

Contents

Preface

At a Snowbird conference on neural nets in 1992, David Haussler and his colleagues at UC Santa Cruz (including one of us, AK) described preliminary results on modelling protein sequence multiple alignments with probabilistic models called 'hidden Markov models' (HMMs). Copies of their technical report were widely circulated. Some of them found their way to the MRC Laboratory of Molecular Biology in Cambridge, where RD and GJM were just switching research interests from neural modelling to computational genome sequence analysis, and where SRE had arrived as a new postdoctoral student with a background in experimental molecular genetics and an interest in computational analysis. AK later also came to Cambridge for a year.

All of us quickly adopted the ideas of probabilistic modelling. We were persuaded that hidden Markov models and their stochastic grammar analogues are beautiful mathematical objects, well fitted to capturing the information buried in biological sequences. The Santa Cruz group and the Cambridge group independently developed two freely available HMM software packages for sequence analysis, and independently extended HMM methods to stochastic context-free grammar analysis of RNA secondary structures. Another group led by Pierre Baldi at JPL/Caltech was also inspired by the work presented at the Snowbird conference to work on HMM-based approaches at about the same time.

By late 1995, we thought that we had acquired a reasonable amount of experience in probabilistic modelling techniques. On the other hand, we also felt that relatively little of the work had been communicated effectively to the community. HMMs had stirred widespread interest, but they were still viewed by many as mathematical black boxes instead of natural models of sequence alignment problems. Many of the best papers that described HMM ideas and methods in detail were in the speech recognition literature, effectively inaccessible to many computational biologists. Furthermore, it had become clear to us and several other groups that the same ideas could be applied to a much broader class of problems, including protein structure modelling, genefinding, and phylogenetic analysis. Over the Christmas break in 1995–96, perhaps somewhat deluded by ambition, naiveté, and holiday relaxation, we decided to write a book on biological sequence analysis emphasizing probabilistic modelling. In the past two years, our original grand plans have been distilled into what we hope is a practical book.

This is a subjective book written by opinionated authors. It is not a tutorial on practical sequence analysis. Our main goal is to give an accessible introduction to the foundations of sequence analysis, and to show why we think the probabilistic modelling approach is useful. We try to avoid discussing specific computer programs, and instead focus on the algorithms and principles behind them.

We have carefully cited the work of the many authors whose work has influenced our thinking. However, we are sure we have failed to cite others whom we *should* have read, and for this we apologise. Also, in a book that necessarily touches on fields ranging from evolutionary biology through probability theory to biophysics, we have been forced by limitations of time, energy, and our own imperfect understanding to deal with a number of issues in a superficial manner.

Computational biology is an interdisciplinary field. Its practitioners, including us, come from diverse backgrounds, including molecular biology, mathematics, computer science, and physics. Our intended audience is any graduate or advanced undergraduate student with a background in one of these fields. We aim for a concise and intuitive presentation that is neither forbiddingly mathematical nor too technically biological.

We assume that readers are already familiar with the basic principles of molecular genetics, such as the Central Dogma that DNA makes RNA makes protein, and that nucleic acids are sequences composed of four nucleotide subunits and proteins are sequences composed of twenty amino acid subunits. More detailed molecular genetics is introduced where necessary. We also assume a basic proficiency in mathematics. However, there are sections that are more mathematically detailed. We have tried to place these towards the end of each chapter, and in general towards the end of the book. In particular, the final chapter, Chapter 11, covers some topics in probability theory that are relevant to much of the earlier material.

We are grateful to several people who kindly checked parts of the manuscript for us at rather short notice. We thank Ewan Birney, Bill Bruno, David MacKay, Cathy Eddy, Jotun Hein, and Søren Riis especially. Bret Larget and Robert Mau gave us very helpful information about the sampling methods they have been using for phylogeny. David Haussler bravely used an embarrassingly early draft of the manuscript in a course at UC Santa Cruz in the autumn of 1996, and we thank David and his entire class for the very useful feedback we received. We are also grateful to David for inspiring us to work in this field in the first place. It has been a pleasure to work with David Tranah and Maria Murphy of Cambridge University Press and Sue Glover of SG Publishing in producing the book; they demonstrated remarkable expertise in the editing and LaTeX typesetting of a book laden with equations, algorithms, and pseudocode, and also remarkable tolerance of our wildly optimistic and inaccurate target dates. We are sure that some of our errors remain, but their number would be far greater without the help of all these people.

We also wish to thank those who supported our research and our work on this book: the Wellcome Trust, the NIH National Human Genome Research Institute, NATO, Eli Lilly & Co., the Human Frontiers Science Program Organisation, and the Danish National Research Foundation. We also thank our home institutions: the Sanger Centre (RD), Washington University School of Medicine (SRE), the Center for Biological Sequence Analysis (AK), and the MRC Laboratory of Molecular Biology (GJM). Jim and Anne Durbin graciously lent us the use of their house in London in February 1997, where an almost final draft of the book coalesced in a burst of writing and criticism. We thank our friends, families, and research groups for tolerating the writing process and SRE's and AK's long trips to England. We promise to take on no new grand projects, at least not immediately.

1

Introduction

Astronomy began when the Babylonians mapped the heavens. Our descendants will certainly not say that biology began with today's genome projects, but they may well recognise that a great acceleration in the accumulation of biological knowledge began in our era. To make sense of this knowledge is a challenge, and will require increased understanding of the biology of cells and organisms. But part of the challenge is simply to organise, classify and parse the immense richness of sequence data. This is more than an abstract task of string parsing, for behind the string of bases or amino acids is the whole complexity of molecular biology. This book is about methods which are in principle capable of capturing some of this complexity, by integrating diverse sources of biological information into clean, general, and tractable probabilistic models for sequence analysis.

Though this book is about computational biology, let us be clear about one thing from the start: the most reliable way to determine a biological molecule's structure or function is by direct experimentation. However, it is far easier to obtain the DNA sequence of the gene corresponding to an RNA or protein than it is to experimentally determine its function or its structure. This provides strong motivation for developing computational methods that can infer biological information from sequence alone. Computational methods have become especially important since the advent of genome projects. The Human Genome Project alone will give us the raw sequences of an estimated 70 000 to 100 000 human genes, only a small fraction of which have been studied experimentally.

Most of the problems in computational sequence analysis are essentially statistical. Stochastic evolutionary forces act on genomes. Discerning significant similarities between anciently diverged sequences amidst a chaos of random mutation, natural selection, and genetic drift presents serious signal to noise problems. Many of the most powerful analysis methods available make use of probability theory. In this book we emphasise the use of probabilistic models, particularly *hidden Markov models* (HMMs), to provide a general structure for statistical analysis of a wide variety of sequence analysis problems.

1.1   Sequence similarity, homology, and alignment

Nature is a tinkerer and not an inventor [Jacob 1977]. New sequences are adapted from pre-existing sequences rather than invented *de novo*. This is very fortunate for computational sequence analysis. We can often recognise a significant similarity between a new sequence and a sequence about which something is already known; when we do this we can transfer information about structure and/or function to the new sequence. We say that the two related sequences are *homologous* and that we are transfering information *by homology*.

At first glance, deciding that two biological sequences are similar is no different from deciding that two text strings are similar. One set of methods for biological sequence analysis is therefore rooted in computer science, where there is an extensive literature on string comparison methods. The concept of an *alignment* is crucial. Evolving sequences accumulate insertions and deletions as well as substitutions, so before the similarity of two sequences can be evaluated, one typically begins by finding a plausible alignment between them.

Almost all alignment methods find the best alignment between two strings under some scoring scheme. These scoring schemes can be as simple as '+1 for a match, −1 for a mismatch'. Indeed, many early sequence alignment algorithms were described in these terms. However, since we want a scoring scheme to give the biologically most likely alignment the highest score, we want to take into account the fact that biological molecules have evolutionary histories, three-dimensional folded structures, and other features which constrain their primary sequence evolution. Therefore, in addition to the mechanics of alignment and comparison algorithms, the scoring system itself requires careful thought, and can be very complex.

Developing more sensitive scoring schemes and evaluating the significance of alignment scores is more the realm of statistics than computer science. An early step forward was the introduction of probabilistic matrices for scoring pairwise amino acid alignments [Dayhoff, Eck & Park 1972; Dayhoff, Schwartz & Orcutt 1978]; these serve to quantify evolutionary preferences for certain substitutions over others. More sophisticated probabilistic modelling approaches have been brought gradually into computational biology by many routes. Probabilistic modelling methods greatly extend the range of applications that can be underpinned by useful and consistent theory, by providing a natural framework in which to address complex inference problems in computational sequence analysis.

1.2   Overview of the book

The book is loosely structured into four parts covering problems in pairwise alignment, multiple alignment, phylogenetic trees, and RNA structure. Figure 1.1

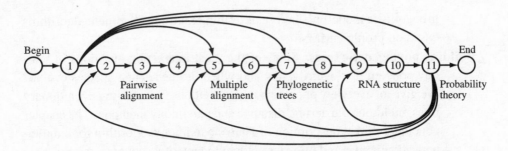

Figure 1.1 *Overview of the book, and suggested paths through it.*

shows suggested paths through the chapters in the form of a *state machine*, one sort of model we will use throughout the book.

The individual chapters cover topics as follows:

2 Pairwise alignment. We start with the problem of deciding if a pair of sequences are evolutionarily related or not. We examine traditional pairwise sequence alignment and comparison algorithms which use dynamic programming to find optimal gapped alignments. We give some probabilistic analysis of scoring parameters, and some discussion of the statistical significance of matches.

3 Markov chains and hidden Markov models. We introduce hidden Markov models (HMMs) and show how they are used to model a sequence or a family of sequences. The chapter gives all the basic HMM algorithms and theory, using simple examples.

4 Pairwise alignment using HMMs. Newly equipped with HMM theory, we revisit pairwise alignment. We develop a special sort of HMM that models aligned pairs of sequences. We show how the HMM-based approach provides some nice ways of estimating accuracy of an alignment, and scoring similarity without committing to any particular alignment.

5 Profile HMMs for sequence families. We consider the problem of finding sequences which are homologous to a known evolutionary family or superfamily. One standard approach to this problem has been the use of 'profiles' of position-specific scoring parameters derived from a multiple sequence alignment. We describe a standard form of HMM, called a profile HMM, for modelling protein and DNA sequence families based on multiple alignments. Particular attention is given to parameter estimation for optimal searching for new family members, including a discussion of sequence weighting schemes.

6 Multiple sequence alignment methods. A closely related problem is that of constructing a multiple sequence alignment of a family. We examine existing multiple sequence alignment algorithms from the standpoint of probabilistic modelling, before describing multiple alignment algorithms based on profile HMMs.

7 Building phylogenetic trees. Some of the most interesting questions in biology concern phylogeny. How and when did genes and species evolve? We give an overview of some popular methods for inferring evolutionary trees, including clustering, distance and parsimony methods. The chapter concludes with a description of Hein's parsimony algorithm for simultaneously aligning and inferring the phylogeny of a sequence family.

8 A probabilistic approach to phylogeny. We describe the application of probabilistic modelling to phylogeny, including maximum likelihood estimation of tree scores and methods for sampling the posterior probability distribution over the space of trees. We also give a probabilistic interpretation of the methods described in the preceding chapter.

9 Transformational grammars. We describe how hidden Markov models are just the lowest level in the Chomsky hierarchy of transformational grammars. We discuss the use of more complex transformational grammars as probabilistic models of biological sequences, and give an introduction to the stochastic context-free grammars, the next level in the Chomsky hierarchy.

10 RNA structure analysis. Using stochastic context-free grammar theory, we tackle questions of RNA secondary structure analysis that cannot be handled with HMMs or other primary sequence-based approaches. These include RNA secondary structure prediction, structure-based alignment of RNAs, and structure-based database search for homologous RNAs.

11 Background on probability. Finally, we give more formal details for the mathematical and statistical toolkit that we use in a fairly informal tutorial-style fashion throughout the rest of the book.

1.3 Probabilities and probabilistic models

Some basic results in using probabilities are necessary for understanding almost any part of this book, so before we get going with sequences, we give a brief primer here on the key ideas and methods. For many readers, this will be familiar territory. However, it may be wise to at least skim though this section to get a grasp of the notation and some of the ideas that we will develop later in the book. Aside from this very basic introduction, we have tried to minimise the discussion of abstract probability theory in the main body of the text, and have instead concentrated the mathematical derivations and methods into Chapter 11, which contains a more thorough presentation of the relevant theory.

What do we mean by a probabilistic model? When we talk about a *model* normally we mean a system that simulates the object under consideration. A probabilistic model is one that produces different outcomes with different probabilities. A probabilistic model can therefore simulate a whole class of objects, assigning each an associated probability. In our case the objects will normally be sequences, and a model might describe a family of related sequences.

Let us consider a very simple example. A familiar probabilistic system with a set of discrete outcomes is the roll of a six-sided die. A model of a roll of a (possibly loaded) die would have six parameters $p_1 \ldots p_6$; the probability of rolling i is p_i. To be probabilities, the parameters p_i must satisfy the conditions that $p_i \geq 0$ and $\sum_{i=1}^{6} p_i = 1$. A model of a sequence of three consecutive rolls of a die might be that they were all independent, so that the probability of sequence [1,6,3] would be the product of the individual probabilities, $p_1 p_6 p_3$. We will use dice throughout the early part of the book for giving intuitive simple examples of probabilistic modelling.

Consider a second example closer to our biological subject matter, which is an extremely simple model of any protein or DNA sequence. Biological sequences are strings from a finite *alphabet* of residues, generally either four nucleotides or twenty amino acids. Assume that a residue a occurs at random with probability q_a, independent of all other residues in the sequence. If the protein or DNA sequence is denoted $x_1 \ldots x_n$, the probability of the whole sequence is then the product $q_{x_1} q_{x_2} \cdots q_{x_n} = \prod_{i=1}^{n} q_{x_i}$.[1] We will use this 'random sequence model' throughout the book as a base-level model, or null hypothesis, to compare other models against.

Maximum likelihood estimation

The parameters for a probabilistic model are typically *estimated* from large sets of trusted examples, often called a *training set*. For instance, the probability q_a for amino acid a can be estimated as the observed frequency of residues in a database of known protein sequences, such as SWISS-PROT [Bairoch & Apweiler 1997].We obtain the twenty frequencies from counting up some twenty million individual residues in the database, and thus we have so much data that as long as the training sequences are not systematically biased towards a peculiar residue composition, we expect the frequencies to be reasonable estimates of the underlying probabilities of our model. This way of estimating models is called *maximum likelihood estimation*,because it can be shown that using the frequencies with which the amino acids occur in the database as the probabilities q_a maximises the total probability of all the sequences given the model (the likelihood). In general, given a model with parameters θ and a set of data D, the maximum likelihood estimate for θ is that value which maximises $P(D|\theta)$. This is discussed more formally in Chapter 11.

When estimating parameters for a model from a limited amount of data, there

---

[1] Strictly speaking this is only a correct model if all sequences have the same length, because then the sum of the probability over all possible sequences is 1; see Chapter 3.

is a danger of *overfitting*, which means that the model becomes very well adapted to the training data, but it will not *generalise* well to new data. Observing for instance the three flips of a coin [*tail, tail, tail*] would lead to the maximum likelihood estimate that the probability of *head* is 0 and that of *tail* is 1. We will return shortly to methods for preventing overfitting.

Conditional, joint, and marginal probabilities

Suppose we have two dice, D_1 and D_2. The probability of rolling an i with die D_1 is called $P(i|D_1)$. This is the *conditional probability* of rolling i given die D_1. If we pick a die at random with probability $P(D_j)$, $j = 1$ or 2, the probability for picking die j and rolling an i is the product of the two probabilities, $P(i, D_j) = P(D_j)P(i|D_j)$. The term $P(i, D_j)$ is called the *joint probability*. The statement

$$P(X, Y) = P(X|Y)P(Y) \qquad (1.1)$$

applies universally to any events X and Y.

When conditional or joint probabilities are known, we can calculate a *marginal* probability that removes one of the variables by using

$$P(X) = \sum_Y P(X, Y) = \sum_Y P(X|Y)P(Y),$$

where the sums are over all possible events Y.

Exercise

1.1   Consider an occasionally dishonest casino that uses two kinds of dice. Of the dice 99% are fair but 1% are loaded so that a six comes up 50% of the time. We pick up a die from a table at random. What are $P(\text{six}|D_{\text{loaded}})$ and $P(\text{six}|D_{\text{fair}})$? What are $P(\text{six}, D_{\text{loaded}})$ and $P(\text{six}, D_{\text{fair}})$? What is the probability of rolling a six from the die we picked up?

Bayes' theorem and model comparison

In the same occasionally dishonest casino as in Exercise 1.1, we pick a die at random and roll it three times, getting three consecutive sixes. We are suspicious that this is a loaded die. How can we evaluate whether that is the case? What we want to know is $P(D_{\text{loaded}}|3 \text{ sixes})$; i.e. the *posterior probability* of the hypothesis that the die is loaded given the observed data, but what we can directly calculate is the probability of the data given the hypothesis, $P(3 \text{ sixes}|D_{\text{loaded}})$, which is called the *likelihood* of the hypothesis. We can calculate posterior probabilities using Bayes' theorem,

$$P(X|Y) = \frac{P(Y|X)P(X)}{P(Y)}. \qquad (1.2)$$

The event 'the die is loaded' corresponds to X in (1.2) and '3 sixes' corresponds to Y, so

$$P(D_{\text{loaded}}|3\text{ sixes}) = \frac{P(3\text{ sixes}|D_{\text{loaded}})P(D_{\text{loaded}})}{P(3\text{ sixes})}.$$

We were given (see Exercise 1.1) that the probability $P(D_{\text{loaded}})$ of picking a loaded die is 0.01, and we know that the probability $P(3\text{ sixes}|D_{\text{loaded}})$ of three sixes given it is loaded is $0.5^3 = 0.125$. The total probability of three sixes, $P(3\text{ sixes})$, is just $P(3\text{ sixes}|D_{\text{loaded}})P(D_{\text{loaded}}) + P(3\text{ sixes}|D_{\text{fair}})P(D_{\text{fair}})$. Now

$$
\begin{aligned}
P(D_{\text{loaded}}|3\text{ sixes}) &= \frac{(0.5^3)(0.01)}{(0.5^3)(0.01) + (\frac{1}{6}^3)(0.99)} \\
&= 0.21.
\end{aligned}
$$

So in fact, it is still more likely that we picked up a fair die, despite seeing three successive sixes.

As a second, more biological example, let us assume we believe that, on average, extracellular proteins have a slightly different amino acid composition than intracellular proteins. For example, we might think that cysteine is more common in extracellular than intracellular proteins. Let us try to use this information to judge whether a new protein sequence $x = x_1 \ldots x_n$ is intracellular or extracellular. To do this, we first split our training examples from SWISS-PROT into intracellular and extracellular proteins (we can leave aside unclassifiable cases).

We can now estimate a set of frequencies q_a^{int} for intracellular proteins, and a corresponding set of extracellular frequencies q_a^{ext}. To provide all the necessary information for Bayes' theorem, we also need to estimate the probability that any new sequence is extracellular, p^{ext}, and the corresponding probability of being intracellular, p^{int}. We will assume for now that every sequence must be either entirely intracellular or entirely extracellular, so $p^{\text{int}} = 1 - p^{\text{ext}}$. The values p^{ext} and p^{int} are called the *prior* probabilities, because they represent the best guess that we can make about a sequence *before* we have seen any information about the sequence itself.

We can now write $P(x|\text{ext}) = \prod_i q_{x_i}^{\text{ext}}$ and $P(x|\text{int}) = \prod_i q_{x_i}^{\text{int}}$. Because we are assuming that every sequence must be extracellular or intracellular, $p(x) = p^{\text{ext}}P(x|\text{ext}) + p^{\text{int}}P(x|\text{int})$. By Bayes' theorem,

$$P(\text{ext}|x) = \frac{p^{\text{ext}}\prod_i q_{x_i}^{\text{ext}}}{p^{\text{ext}}\prod_i q_{x_i}^{\text{ext}} + p^{\text{int}}\prod_i q_{x_i}^{\text{int}}}.$$

$P(\text{ext}|x)$ is the number we want. It is called the *posterior* probability that a sequence is extracellular because it is our best guess *after* we have seen the data.

Of course, this example is confounded by the fact that many transmembrane proteins have intracellular and extracellular components. We really want to be able to switch from one assignment to the other while in the sequence. That requires a more complex probabilistic model which we will see later in the book (Chapter 3).

Exercises

1.2   How many sixes in a row would we need to see in the above example before it was most likely that we had picked a loaded die?

1.3   Use equation (1.1) to prove Bayes' theorem.

1.4   A rare genetic disease is discovered. Although only one in a million people carry it, you consider getting screened. You are told that the genetic test is extremely good; it is 100% sensitive (it is always correct if you have the disease) and 99.99% specific (it gives a false positive result only 0.01% of the time). Having recently learned Bayes' theorem, you decide not to take the test. Why?

Bayesian parameter estimation

The concept of overfitting was mentioned earlier. Rather than giving up on a model, if we do not have enough data to reliably estimate the parameters, we can use *prior knowledge* to constrain the estimates. This can be done conveniently with Bayesian parameter estimation.

As well as using Bayes' theorem for comparing models, we can use it to estimate parameters. We can calculate the posterior probability of any particular set of parameters θ given some data D using Bayes' theorem as

$$P(\theta|D) = \frac{P(\theta)P(D|\theta)}{\int_{\theta'} P(\theta')P(D|\theta')}. \tag{1.3}$$

Note that since our parameters are usually continuous rather than discrete quantities, the denominator is now an integral rather than a sum:

$$P(\theta) = \int_{\theta'} P(\theta')P(D|\theta').$$

There are a number of issues that arise concerning (1.3). One problem is 'what is meant by $P(\theta)$?' Where do we obtain a prior distribution over parameters? Sometimes there is no good rationale for any specific choice, in which case *flat* (uniform) or *uninformative* priors are normally chosen, i.e. ones that are as innocuous as possible. In other cases, we will wish to use an informative $P(\theta)$. For instance, we know *a priori* that the amino acids phenylalanine, tyrosine, and tryptophan are structurally similar and often evolutionarily interchangeable. We would want to use a $P(\theta)$ that tends to favour parameter sets that give similar probabilities to these three amino acids over other parameter sets that assign them very different probabilities. These issues are examined in detail in Chapter 5.

Another issue is how to use (1.3) to estimate good parameters. One approach is to choose the parameter values for θ that maximise $P(\theta|D)$. This is called maximum *a posteriori* or MAP estimation. Note that the denominator of (1.3) is independent of the specific value of θ, and so MAP estimation corresponds to maximising the likelihood times the prior. If the prior is flat, then MAP estimation is the same as maximum likelihood estimation.

Another approach to parameter estimation is to choose the mean of the posterior distribution as the estimate, rather than the maximum value. This can be a more complicated operation, requiring that the posterior probability can either be calculated analytically or can be sampled. A related approach is not to choose a specific set of parameters at all, but instead to evaluate the quantity of interest based on the model at many or all different parameter values by integration, weighting the results according to the posterior probabilities of the respective parameter values. This approach is most attractive when the evaluation and weighting can be done analytically – otherwise it can be hard to obtain a valid result unless the parameter space is very small.

These approaches are part of a field of statistics called *Bayesian statistics* [Box & Tiao 1992]. The subjectiveness of issues like the choice of prior leads some people to be wary of Bayesian methods, though the validity of Bayes' theorem *per se* for manipulating conditional probabilities is not in question. We do not have a rigid attitude; we use both maximum likelihood and Bayesian methods at different points in the book. However, when estimating large parameter sets from small amounts of data, we believe that Bayesian methods provide a consistent formalism for bringing in additional information from previous experience with the same type of data.

Example: Estimating probabilities for a loaded die

To illustrate, let us return to our examples with dice. Assume we are given a die that we expect will be loaded, but we don't know in what way. We are allowed to roll it ten times, and we have to give our best estimates for the parameters p_i. We roll 1, 3, 4, 2, 4, 6, 2, 1, 2, 2. The maximum likelihood estimate for $\hat{p}_5$, based on the observed frequency, is 0. If this were used in a model, then a single observed 5 would rule out the dataset from coming from this die. That seems too harsh. Intuitively, we have not seen enough data to be sure that this die never rolls a five.

One well-known approach to this problem is to adjust the observed frequencies used to derive the probabilities by adding some fake extra counts to the true counts observed for each outcome. An example would be to add one to each observed number of counts, so that the estimated probability $\hat{p}_5$ of rolling a five is now $\frac{1}{16}$. The extra count for each class is called a *pseudocount*. Using pseudocounts corresponds to a posterior mean approach using Bayes' theorem and a prior from the Dirichlet family of distributions (see Chapter 11 for more details). Different sets of pseudocounts correspond to different prior assumptions about what sort of probabilities a die will have. If in our previous experience most dice were close to being fair, then we might add a lot of pseudocounts; if we had previously seen many very biased dice in this particular casino, we would believe

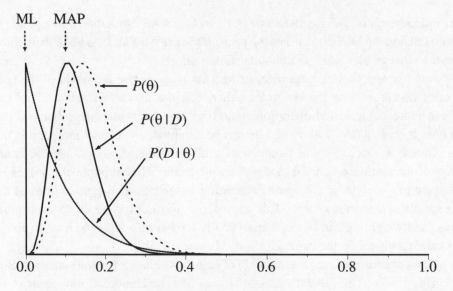

Figure 1.2 *Maximum likelihood estimation (ML) versus maximum a posteriori (MAP) estimation of the probability p_5 (x axis) in Example 1.1 with five pseudocounts per category. The three curves are artificially normalised to have the same maximum value.*

more strongly the data that we collected on this particular example, and weight the pseudocounts less. Of course, if we collect enough data, the true counts will always dominate the pseudocounts.

In Figure 1.2 the likelihood $P(D|\theta)$ is shown as a function of p_5, and the maximum at 0 is evident. In the same figure we show the prior and posterior distributions with five pseudocounts per category. The prior distribution of p_5 implied by the pseudocounts, $P(\theta)$, is a Dirichlet distribution. Note that the posterior $P(\theta|D)$ is asymmetric; the posterior mean estimate of p_5 is slightly more than the MAP estimate.   □

Exercise

1.5   In the above example, what is our maximum likelihood estimate for p_2, the probability of rolling a two? What is the Bayesian estimate if we add one pseudocount per category? What if we add five pseudocounts per category?

1.4 Further reading

Available textbooks on computational molecular biology include *Introduction to Computational Biology* by Waterman [1995], *Bioinformatics – The Machine Learning Approach* by Baldi & Brunak [1998] and Sankoff & Kruskal's *Time*

Warps, String Edits, and Macromolecules [1983]. For readers with no molecular biology background, we recommend *Molecular Biology of the Gene* by Watson *et al.* [1987] as a readable, though encyclopedic, undergraduate-level introduction to molecular genetics. *Introduction to Protein Structure* by Branden & Tooze [1991] is a beautifully illustrated guide to the three-dimensional structures of proteins. MacKay [1992] has written a persuasive introduction to Bayesian probabilistic modelling; a more elementary introduction to some of the attractive ideas behind Bayesian methods is Jefferys & Berger [1992].

2

Pairwise alignment

2.1 Introduction

The most basic sequence analysis task is to ask if two sequences are related. This is usually done by first aligning the sequences (or parts of them) and then deciding whether that alignment is more likely to have occurred because the sequences are related, or just by chance. The key issues are: (1) what sorts of alignment should be considered; (2) the scoring system used to rank alignments; (3) the algorithm used to find optimal (or good) scoring alignments; and (4) the statistical methods used to evaluate the significance of an alignment score.

Figure 2.1 shows an example of three pairwise alignments, all to the same region of the human alpha globin protein sequence (SWISS-PROT database identifier HBA_HUMAN). The central line in each alignment indicates identical positions with letters, and 'similar' positions with a plus sign. ('Similar' pairs of residues are those which have a positive score in the substitution matrix used to score the alignment; we will discuss substitution matrices shortly.) In the first

```
(a)
HBA_HUMAN    GSAQVKGHGKKVADALTNAVAHVDDMPNALSALSDLHAHKL
             G+ +VK+HGKKV  A+++++AH+D++ +++++LS+LH   KL
HBB_HUMAN    GNPKVKAHGKKVLGAFSDGLAHLDNLKGTFATLSELHCDKL

(b)
HBA_HUMAN    GSAQVKGHGKKVADALTNAVAHV---D--DMPNALSALSDLHAHKL
             ++ ++++H+ KV   + +A  ++            +L+ L+++H+ K
LGB2_LUPLU   NNPELQAHAGKVFKLVYEAAIQLQVTGVVVTDATLKNLGSVHVSKG

(c)
HBA_HUMAN    GSAQVKGHGKKVADALTNAVAHVDDMPNALSALSD----LHAHKL
             GS+ + G +   +D L  ++ H+ D+  A +AL D    ++AH+
F11G11.2     GSGYLVGDSLTFVDLL--VAQHTADLLAANAALLDEFPQFKAHQE
```

Figure 2.1 *Three sequence alignments to a fragment of human alpha globin. (a) Clear similarity to human beta globin. (b) A structurally plausible alignment to leghaemoglobin from yellow lupin. (c) A spurious high-scoring alignment to a nematode glutathione S-transferase homologue named F11G11.2.*

alignment there are many positions at which the two corresponding residues are identical; many others are functionally conservative, such as the pair D–E towards the end, representing an alignment of an aspartic acid residue with a glutamic acid residue, both negatively charged amino acids. Figure 2.1b also shows a biologically meaningful alignment, in that we know that these two sequences are evolutionarily related, have the same three-dimensional structure, and function in oxygen binding. However, in this case there are many fewer identities, and in a couple of places gaps have been inserted into the alpha globin sequence to maintain the alignment across regions where the leghaemoglobin has extra residues. Figure 2.1c shows an alignment with a similar number of identities or conservative changes. However, in this case we are looking at a spurious alignment to a protein that has a completely different structure and function.

How are we to distinguish cases like Figure 2.1b from those like Figure 2.1c? This is the challenge for pairwise alignment methods. We must give careful thought to the scoring system we use to evaluate alignments. The next section introduces the issues in how to score alignments, and then there is a series of sections on methods to find the best alignments according to the scoring scheme. The chapter finishes with a discussion of the statistical significance of matches, and more detail on parameterising the scoring scheme. Even so, it will not always be possible to distinguish true alignments from spurious alignments. For example, it is in fact extremely difficult to find significant similarity between the lupin leghaemoglobin and human alpha globin in Figure 2.1b using pairwise alignment methods.

2.2 The scoring model

When we compare sequences, we are looking for evidence that they have diverged from a common ancestor by a process of mutation and selection. The basic mutational processes that are considered are *substitutions*, which change residues in a sequence, and *insertions* and *deletions*, which add or remove residues. Insertions and deletions are together referred to as *gaps*. Natural selection has an effect on this process by screening the mutations, so that some sorts of change may be seen more than others.

The total score we assign to an alignment will be a sum of terms for each aligned pair of residues, plus terms for each gap. In our probabilistic interpretation, this will correspond to the logarithm of the relative likelihood that the sequences are related, compared to being unrelated. Informally, we expect identities and conservative substitutions to be more likely in alignments than we expect by chance, and so to contribute positive score terms; and non-conservative changes are expected to be observed less frequently in real alignments than we expect by chance, and so these contribute negative score terms.

Using an additive scoring scheme corresponds to an assumption that we can consider mutations at different sites in a sequence to have occurred independently (treating a gap of arbitrary length as a single mutation). All the algorithms in this chapter for finding optimal alignments depend on such a scoring scheme. The assumption of independence appears to be a reasonable approximation for DNA and protein sequences, although we know that interactions between residues play a very critical role in determining protein structure. However, it is seriously inaccurate for structural RNAs, where base pairing introduces very important long-range dependencies. It is possible to take these dependencies into account, but doing so gives rise to significant computational complexities; we will delay the subject of RNA alignment until the end of the book (Chapter 10).

Substitution matrices

We need score terms for each aligned residue pair. A biologist with a good intuition for proteins could invent a set of 210 scoring terms for all possible pairs of amino acids, but it is extremely useful to have a guiding theory for what the scores mean. We will derive substitution scores from a probabilistic model.

First, let us establish some notation. We will be considering a pair of sequences, x and y, of lengths n and m, respectively. Let x_i be the ith symbol in x and y_j be the jth symbol of y. These symbols will come from some alphabet $\mathcal{A}$; in the case of DNA this will be the four bases $\{A, G, C, T\}$, and in the case of proteins the twenty amino acids. We denote symbols from this alphabet by lower-case letters like a, b. For now we will only consider ungapped global pairwise alignments: that is, two completely aligned equal-length sequences as in Figure 2.1a.

Given a pair of aligned sequences, we want to assign a score to the alignment that gives a measure of the relative likelihood that the sequences are related as opposed to being unrelated. We do this by having models that assign a probability to the alignment in each of the two cases; we then consider the ratio of the two probabilities.

The unrelated or *random* model R is simplest. It assumes that letter a occurs independently with some frequency q_a, and hence the probability of the two sequences is just the product of the probabilities of each amino acid:

$$P(x, y|R) = \prod_i q_{x_i} \prod_j q_{y_j}. \tag{2.1}$$

In the alternative *match* model M, aligned pairs of residues occur with a joint probability p_{ab}. This value p_{ab} can be thought of as the probability that the residues a and b have each independently been derived from some unknown original residue c in their common ancestor (c might be the same as a and/or b). This gives a probability for the whole alignment of

$$P(x,y|M) = \prod_i p_{x_i y_i}.$$

The ratio of these two likelihoods is known as the *odds ratio*:

$$\frac{P(x,y|M)}{P(x,y|R)} = \frac{\prod_i p_{x_i y_i}}{\prod_i q_{x_i} \prod_i q_{y_i}} = \prod_i \frac{p_{x_i y_i}}{q_{x_i} q_{y_i}}.$$

In order to arrive at an additive scoring system, we take the logarithm of this ratio, known as the *log-odds ratio*:

$$S = \sum_i s(x_i, y_i), \tag{2.2}$$

where

$$s(a,b) = \log\left(\frac{p_{ab}}{q_a q_b}\right) \tag{2.3}$$

is the log likelihood ratio of the residue pair (a,b) occurring as an aligned pair, as opposed to an unaligned pair.

As we wanted, equation (2.2) is a sum of individual scores $s(a,b)$ for each aligned pair of residues. The $s(a,b)$ scores can be arranged in a matrix. For proteins, for instance, they form a 20×20 matrix, with $s(a_i, a_j)$ in position i, j in the matrix, where a_i, a_j are the ith and jth amino acids (in some numbering). This is known as a *score matrix* or a *substitution matrix*. An example of a substitution matrix derived essentially as above is the BLOSUM50 matrix, shown in Figure 2.2. We can use these values to score Figure 2.1a and get a score of 130. Another commonly used set of substitution matrices are called the PAM matrices. A detailed description of the way that the BLOSUM and PAM matrices are derived is given at the end of the chapter.

An important result is that even if an intuitive biologist were to write down an *ad hoc* substitution matrix, the substitution matrix implies 'target frequencies' p_{ab} according to the above theory [Altschul 1991]. Any substitution matrix is making a statement about the probability of observing ab pairs in real alignments.

Exercise

2.1   Amino acids D, E and K are all charged; V, I and L are all hydrophobic. What is the average BLOSUM50 score within the charged group of three? Within the hydrophobic group? Between the two groups? Suggest reasons for the pattern observed.

	A	R	N	D	C	Q	E	G	H	I	L	K	M	F	P	S	T	W	Y	V
A	**5**	-2	-1	-2	-1	-1	-1	0	-2	-1	-2	-1	-1	-3	-1	1	0	-3	-2	0
R	-2	**7**	-1	-2	-4	1	0	-3	0	-4	-3	3	-2	-3	-3	-1	-1	-3	-1	-3
N	-1	-1	**7**	2	-2	0	0	0	1	-3	-4	0	-2	-4	-2	1	0	-4	-2	-3
D	-2	-2	2	**8**	-4	0	2	-1	-1	-4	-4	-1	-4	-5	-1	0	-1	-5	-3	-4
C	-1	-4	-2	-4	**13**	-3	-3	-3	-3	-2	-2	-3	-2	-2	-4	-1	-1	-5	-3	-1
Q	-1	1	0	0	-3	**7**	2	-2	1	-3	-2	2	0	-4	-1	0	-1	-1	-1	-3
E	-1	0	0	2	-3	2	**6**	-3	0	-4	-3	1	-2	-3	-1	-1	-1	-3	-2	-3
G	0	-3	0	-1	-3	-2	-3	**8**	-2	-4	-4	-2	-3	-4	-2	0	-2	-3	-3	-4
H	-2	0	1	-1	-3	1	0	-2	**10**	-4	-3	0	-1	-1	-2	-1	-2	-3	2	-4
I	-1	-4	-3	-4	-2	-3	-4	-4	-4	**5**	2	-3	2	0	-3	-3	-1	-3	-1	4
L	-2	-3	-4	-4	-2	-2	-3	-4	-3	2	**5**	-3	3	1	-4	-3	-1	-2	-1	1
K	-1	3	0	-1	-3	2	1	-2	0	-3	-3	**6**	-2	-4	-1	0	-1	-3	-2	-3
M	-1	-2	-2	-4	-2	0	-2	-3	-1	2	3	-2	**7**	0	-3	-2	-1	-1	0	1
F	-3	-3	-4	-5	-2	-4	-3	-4	-1	0	1	-4	0	**8**	-4	-3	-2	1	4	-1
P	-1	-3	-2	-1	-4	-1	-1	-2	-2	-3	-4	-1	-3	-4	**10**	-1	-1	-4	-3	-3
S	1	-1	1	0	-1	0	-1	0	-1	-3	-3	0	-2	-3	-1	**5**	2	-4	-2	-2
T	0	-1	0	-1	-1	-1	-1	-2	-2	-1	-1	-1	-1	-2	-1	2	**5**	-3	-2	0
W	-3	-3	-4	-5	-5	-1	-3	-3	-3	-3	-2	-3	-1	1	-4	-4	-3	**15**	2	-3
Y	-2	-1	-2	-3	-3	-1	-2	-3	2	-1	-1	-2	0	4	-3	-2	-2	2	**8**	-1
V	0	-3	-3	-4	-1	-3	-3	-4	-4	4	1	-3	1	-1	-3	-2	0	-3	-1	**5**

Figure 2.2 *The* BLOSUM50 *substitution matrix. The log-odds values have been scaled and rounded to the nearest integer for purposes of computational efficiency. Entries on the main diagonal for identical residue pairs are highlighted in bold.*

Gap penalties

We expect to penalise gaps. The standard cost associated with a gap of length g is given either by a linear score

$$\gamma(g) = -gd \tag{2.4}$$

or an affine score

$$\gamma(g) = -d - (g-1)e \tag{2.5}$$

where d is called the *gap-open* penalty and e is called the *gap-extension* penalty. The gap-extension penalty e is usually set to something less than the gap-open penalty d, allowing long insertions and deletions to be penalised less than they would be by the linear gap cost. This is desirable when gaps of a few residues are expected almost as frequently as gaps of a single residue.

Gap penalties also correspond to a probabilistic model of alignment, although this is less widely recognised than the probabilistic basis of substitution matrices. We assume that the probability of a gap occurring at a particular site in a given sequence is the product of a function $f(g)$ of the length of the gap, and the combined probability of the set of inserted residues,

$$P(\text{gap}) = f(g) \prod_{i \text{ in gap}} q_{x_i}. \qquad (2.6)$$

The form of (2.6) as a product of $f(g)$ with the q_{x_i} terms corresponds to an assumption that the length of a gap is not correlated to the residues it contains.

The natural values for the q_a probabilities here are the same as those used in the random model, because they both correspond to unmatched independent residues. In this case, when we divide by the probability of this region according to the random model to form the odds ratio, the q_{x_i} terms cancel out, so we are left only with a term dependent on length $\gamma(g) = \log(f(g))$; gap penalties correspond to the log probability of a gap of that length.

On the other hand, if there is evidence for a different distribution of residues in gap regions then there should be residue-specific scores for the unaligned residues in gap regions, equal to the logs of the ratio of their frequencies in gapped versus aligned regions. This might happen if, for example, it is expected that polar amino acids are more likely to occur in gaps in protein alignments than indicated by their average frequency in protein sequences, because the gaps are more likely to be in loops on the surface of the protein structure than in the buried core.

Exercises

2.2   Show that the probability distributions $f(g)$ that correspond to the linear and affine gap schemes given in equations (2.4) and (2.5) are both geometric distributions, of the form $f(g) = ke^{-\lambda g}$.

2.3   Typical gap penalties used in practice are $d = 8$ for the linear case, or $d = 12, e = 2$ for the affine case, both expressed in half bits. A *bit* is the unit obtained when one takes log base 2 of a probability, so in natural log units these correspond to $d = (8\log 2)/2$ and $d = (12\log 2)/2$, $e = (2\log 2)/2$ respectively. What are the corresponding probabilities of a gap (of any length) starting at some position, and the distributions of gap length given that there is a gap?

2.4   Using the BLOSUM50 matrix in Figure 2.2 and an affine gap penalty of $d = 12$, $e = 2$, calculate the scores of the alignments in Figure 2.1b and Figure 2.1c.

2.3 Alignment algorithms

Given a scoring system, we need to have an algorithm for finding an optimal alignment for a pair of sequences. Where both sequences have the same length n, there is only one possible global alignment of the complete sequences, but things become more complicated once gaps are allowed (or once we start looking for local alignments between subsequences of two sequences). There are

$$\binom{2n}{n} = \frac{(2n)!}{(n!)^2} \simeq \frac{2^{2n}}{\sqrt{2\pi n}} \tag{2.7}$$

possible global alignments between two sequences of length n. It is clearly not computationally feasible to enumerate all these, even for moderate values of n.

The algorithm for finding optimal alignments given an additive alignment score of the type we have described is called *dynamic programming*. Dynamic programming algorithms are central to computational sequence analysis. All the remaining chapters in this book except the last, which covers mathematical methods, make use of dynamic programming algorithms. The simplest dynamic programming alignment algorithms to understand are pairwise sequence alignment algorithms. The reader should be sure to understand this section, because it lays an important foundation for the book as a whole. Dynamic programming algorithms are guaranteed to find the optimal scoring alignment or set of alignments. In most cases heuristic methods have also been developed to perform the same type of search. These can be very fast, but they make additional assumptions and will miss the best match for some sequence pairs. We will briefly discuss a few approaches to heuristic searching later in the chapter.

Because we introduced the scoring scheme as a log-odds ratio, better alignments will have higher scores, and so we want to maximise the score to find the optimal alignment. Sometimes scores are assigned by other means and interpreted as *costs* or *edit distances*, in which case we would seek to minimise the cost of an alignment. Both approaches have been used in the biological sequence comparison literature. Dynamic programming algorithms apply to either case; the differences are trivial exchanges of 'min' for 'max'.

We introduce four basic types of alignment. The type of alignment that we want to look for depends on the source of the sequences that we want to align. For each alignment type there is a slightly different dynamic programming algorithm. In this section, we will only describe pairwise alignment for linear gap scores, with cost d per gap residue. However, the algorithms we introduce here easily extend to more complex gap models, as we will see later in the chapter.

We will use two short amino acid sequences to illustrate the various alignment methods, HEAGAWGHEE and PAWHEAE. To score the alignments, we use the BLOSUM50 score matrix, and a gap cost per unaligned residue of $d = -8$. Figure 2.3 shows a matrix s_{ij} of the local score $s(x_i, y_j)$ of aligning each residue pair from the two example sequences. Identical or conserved residue pairs are highlighted in bold. Informally, the goal of an alignment algorithm is to incorporate as many of these positively scoring pairs as possible into the alignment, while minimising the cost from unconserved residue pairs, gaps, and other constraints.

	H	E	A	G	A	W	G	H	E	E
P	−2	−1	−1	−2	−1	−4	−2	−2	−1	−1
A	−2	−1	**5**	0	**5**	−3	0	−2	−1	−1
W	−3	−3	−3	−3	−3	**15**	−3	−3	−3	−3
H	**10**	0	−2	−2	−2	−3	−2	**10**	0	0
E	0	**6**	−1	−3	−1	−3	−3	0	**6**	**6**
A	−2	−1	**5**	0	**5**	−3	0	−2	−1	−1
E	0	**6**	−1	−3	−1	−3	−3	0	**6**	**6**

Figure 2.3 *The two example sequences we will use for illustrating dynamic programming alignment algorithms, arranged to show a matrix of corresponding* BLOSUM50 *values per aligned residue pair. Positive scores are in bold.*

Exercises

2.5   Show that the number of ways of intercalating two sequences of lengths n and m to give a single sequence of length $n + m$, while preserving the order of the symbols in each, is $\binom{n+m}{m}$.

2.6   By taking alternating symbols from the upper and lower sequences in an alignment, then discarding the gap characters, show that there is a one-to-one correspondence between gapped alignments of the two sequences and intercalated sequences of the type described in the previous exercise. Hence derive the first part of equation (2.7).

2.7   Use Stirling's formula ($x! \simeq \sqrt{2\pi}x^{x+\frac{1}{2}}e^{-x}$) to prove the second part of equation (2.7).

Global alignment: Needleman–Wunsch algorithm

The first problem we consider is that of obtaining the optimal global alignment between two sequences, allowing gaps. The dynamic programming algorithm for solving this problem is known in biological sequence analysis as the Needleman–Wunsch algorithm [Needleman & Wunsch 1970], but the more efficient version that we describe was introduced by Gotoh [1982].

The idea is to build up an optimal alignment using previous solutions for optimal alignments of smaller subsequences. We construct a matrix F indexed by i and j, one index for each sequence, where the value $F(i,j)$ is the score of the best alignment between the initial segment $x_{1...i}$ of x up to x_i and the initial segment $y_{1...j}$ of y up to y_j. We can build $F(i,j)$ recursively. We begin by initialising $F(0,0) = 0$. We then proceed to fill the matrix from top left to bottom right. If $F(i-1,j-1)$, $F(i-1,j)$ and $F(i,j-1)$ are known, it is possible to calculate $F(i,j)$. There are three possible ways that the best score

```
I G A x_i          A I G A x_i          G A x_i − −
L G V y_j          G V y_j − −          S L G V y_j
```

Figure 2.4 *The three ways an alignment can be extended up to (i,j): x_i aligned to y_j, x_i aligned to a gap, and y_j aligned to a gap.*

$F(i,j)$ of an alignment up to x_i, y_j could be obtained: x_i could be aligned to y_j, in which case $F(i,j) = F(i-1,j-1) + s(x_i, y_j)$; or x_i is aligned to a gap, in which case $F(i,j) = F(i-1,j) - d$; or y_j is aligned to a gap, in which case $F(i,j) = F(i,j-1) - d$ (see Figure 2.4). The best score up to (i,j) will be the largest of these three options.

Therefore, we have

$$F(i,j) = \max \begin{cases} F(i-1,j-1) + s(x_i, y_j), \\ F(i-1,j) - d, \\ F(i,j-1) - d. \end{cases} \tag{2.8}$$

This equation is applied repeatedly to fill in the matrix of $F(i,j)$ values, calculating the value in the bottom right-hand corner of each square of four cells from one of the other three values (above-left, left, or above) as in the following figure.

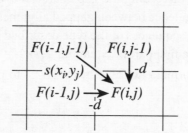

As we fill in the $F(i,j)$ values, we also keep a pointer in each cell back to the cell from which its $F(i,j)$ was derived, as shown in the example of the full dynamic programming matrix in Figure 2.5.

To complete our specification of the algorithm, we must deal with some boundary conditions. Along the top row, where $j = 0$, the values $F(i,j-1)$ and $F(i-1,j-1)$ are not defined so the values $F(i,0)$ must be handled specially. The values $F(i,0)$ represent alignments of a prefix of x to all gaps in y, so we can define $F(i,0) = -id$. Likewise down the left column $F(0,j) = -jd$.

The value in the final cell of the matrix, $F(n,m)$, is by definition the best score for an alignment of $x_{1...n}$ to $y_{1...m}$, which is what we want: the score of the best global alignment of x to y. To find the alignment itself, we must find the path of choices from (2.8) that led to this final value. The procedure for doing this is known as a *traceback*. It works by building the alignment in reverse, starting from the final cell, and following the pointers that we stored when building the

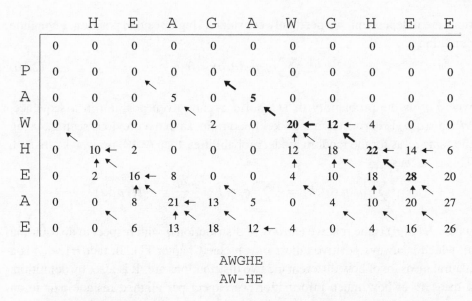

	H	E	A	G	A	W	G	H	E	E
	0	0	0	0	0	0	0	0	0	0
P	0	0	0	0	0	0	0	0	0	0
A	0	0	0	5	0	5	0	0	0	0
W	0	0	0	0	2	0	20	12	4	0	0
H	0	10	2	0	0	0	12	18	22	14	6
E	0	2	16	8	0	0	4	10	18	28	20
A	0	0	8	21	13	5	0	4	10	20	27
E	0	0	6	13	18	12	4	0	4	16	26

```
AWGHE
AW-HE
```

Figure 2.6 *Above, the local dynamic programming matrix for the example sequences. Below, the optimal local alignment, with score* 28.

Taking the option 0 corresponds to starting a new alignment. If the best alignment up to some point has a negative score, it is better to start a new one, rather than extend the old one. Note that a consequence of the 0 is that the top row and left column will now be filled with 0s, not $-id$ and $-jd$ as for global alignment.

The second change is that now an alignment can end anywhere in the matrix, so instead of taking the value in the bottom right corner, $F(n,m)$, for the best score, we look for the highest value of $F(i,j)$ over the whole matrix, and start the traceback from there. The traceback ends when we meet a cell with value 0, which corresponds to the start of the alignment. An example is given in Figure 2.6, which shows the best local alignment of the same two sequences whose best global alignment was found in Figure 2.5. In this case the local alignment is a subset of the global alignment, but that is not always the case.

For the local alignment algorithm to work, the expected score for a random match must be negative. If that is not true, then long matches between entirely unrelated sequences will have high scores, just based on their length. As a consequence, although the algorithm is local, the maximal scoring alignments would be global or nearly global. A true subsequence alignment would be likely to be masked by a longer but incorrect alignment, just because of its length. Similarly, there must be some $s(a,b)$ greater than 0, otherwise the algorithm won't find any alignment at all (it finds the best score or 0, whichever is higher).

What is the precise meaning of the requirement that the expected score of a random match be negative? In the ungapped case, the relevant quantity to consider is the expected value of a fixed length alignment. Because successive positions are independent, we need only consider a single residue position, giving the condition

$$\sum_{a,b} q_a q_b s(a,b) < 0, \tag{2.10}$$

where q_a is the probability of symbol a at any given position in a sequence. When $s(a,b)$ is derived as a log likelihood ratio, as in the previous section, using the same q_a as for the random model probabilities, then (2.10) is always satisfied. This is because

$$\sum_{a,b} q_a q_b s(a,b) = -\sum_{a,b} q_a q_b \log \frac{q_a q_b}{p_{ab}} = -H(p||q)$$

where $H(p||q)$ is the relative entropy of distribution q with respect to distribution p, which is always positive unless $q = p$ (see Chapter 11). In fact $H(p||q)$ is a natural measure of how different the two distributions are. It is also, by definition, a measure of how much information we expect per aligned residue pair in an alignment.

Unfortunately we cannot give an equivalent analysis for optimal gapped alignments. There is no analytical method for predicting what gap scores will result in local versus global alignment behaviour. However, this is a question of practical importance when setting parameter values in the scoring system (the match and gap scores $s(a,b)$ and $\gamma(g)$), and tables have been generated for standard scoring schemes showing local/global behaviour, along with other statistical properties [Altschul & Gish 1996]. We will return to this subject later, when considering the statistical significance of scores.

The local version of the dynamic programming sequence alignment algorithm was developed in the early 1980s. It is frequently known as the *Smith–Waterman* algorithm, after Smith & Waterman [1981]. Gotoh [1982] formulated the efficient affine gap cost version that is normally used (affine gap alignment algorithms are discussed on page 29).

Repeated matches

The procedure in the previous section gave the best single local match between two sequences. If one or both of the sequences are long, it is quite possible that there are many different local alignments with a significant score, and in most cases we would be interested in all of these. An example would be where there are many copies of a repeated domain or motif in a protein. We give here a method for finding such matches. This method is asymmetric: it finds one or more non-overlapping copies of sections of one sequence (e.g. the domain or motif) in the other. There is another widely used approach for finding multiple matches due to Waterman & Eggert [1987], which will be described in Chapter 4.

		H	E	A	G	A	W	G	H	E	E	
	0	0	0	0	1	1	1	1	1	3	**9**	← 9
P	0	0	0	0	1	1	1	1	1	3	9
A	0	0	0	5	1	6	1	1	1	3	9
W	0	0	0	0	2	1	21	13	5	3	9
H	0	10	2	0	1	1	13	19	23	15	9
E	0	2	16	8	1	1	5	11	19	29	21
A	0	0	8	21	13	6	1	5	11	21	28
E	0	0	6	13	18	12	4	1	5	17	27

```
HEAGAWGHEE
HEA.AW-HE.
```

Figure 2.7 *Above, the repeat dynamic programming matrix for the example sequences, for T = 20 . Below, the optimal alignment, with total score 9 = 29 − 20. There are two separate match regions, with scores 1 and 8. Dots are used to indicate unmatched regions of x.*

Let us assume that we are only interested in matches scoring higher than some threshold T. This will be true in general, because there are always short local alignments with small positive scores even between entirely unrelated sequences. Let y be the sequence containing the domain or motif, and x be the sequence in which we are looking for multiple matches.

An example of the repeat algorithm is given in Figure 2.7. We again use the matrix F, but the recurrence is now different, as is the meaning of $F(i,j)$. In the final alignment, x will be partitioned into regions that match parts of y in gapped alignments, and regions that are unmatched. We will talk about the score of a completed match region as being its standard gapped alignment score minus the threshold T. All these match scores will be positive. $F(i,j)$ for $j \geq 1$ is now the best sum of match scores to $x_{1...i}$, assuming that x_i is in a matched region, and the corresponding match ends in x_i and y_j (they may not actually be aligned, if this is a gapped section of the match). $F(i,0)$ is the best sum of completed match scores to the subsequence $x_{1...i}$, i.e. assuming that x_i is in an unmatched region.

To achieve the desired goal, we start by initialising $F(0,0) = 0$ as usual, and then fill the matrix using the following recurrence relations:

$$F(i,0) \quad = \quad \max \begin{cases} F(i-1,0), \\ F(i-1,j)-T, \qquad j=1,\ldots,m; \end{cases} \qquad (2.11)$$

$$F(i,j) = \max \begin{cases} F(i,0), \\ F(i-1,j-1)+s(x_i,y_j), \\ F(i-1,j)-d, \\ F(i,j-1)-d. \end{cases} \tag{2.12}$$

Equation (2.11) handles unmatched regions and ends of matches, only allowing matches to end when they have score at least T. Equation (2.12) handles starts of matches and extensions. The total score of all the matches is obtained by adding an extra cell to the matrix, $F(n+1,0)$, using (2.11). This score will have T subtracted for each match; if there were no matches of score greater than T it will be 0, obtained by repeated application of the first option in (2.11).

The individual match alignments can be obtained by tracing back from cell $(n+1,0)$ to $(0,0)$, at each point going back to the cell that was the source of the score in the current cell in the max() operation. This traceback procedure is a global procedure, showing what each residue in x will be aligned to. The resulting global alignment will contain sections of more conventional gapped local alignments of subsequences of x to subsequences of y.

Note that the algorithm obtains all the local matches in one pass. It finds the maximal scoring set of matches, in the sense of maximising the combined total of the excess of each match score above the threshold T. Changing the value of T changes what the algorithm finds. Increasing T may exclude matches. Decreasing it may split them, as well as finding new weaker ones. A locally optimal match in the sense of the preceding section will be split into pieces if it contains internal subalignments scoring less than $-T$. However, this may be what is wanted: given two similar high scoring sections significant in their own right, separated by a non-matching section with a strongly negative score, it is not clear whether it is preferable to report one match or two.

Overlap matches

Another type of search is appropriate when we expect that one sequence contains the other, or that they overlap. This often occurs when comparing fragments of genomic DNA sequence to each other, or to larger chromosomal sequences. Several different types of configuration can occur, as shown here:

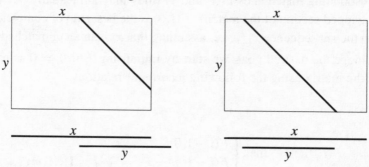

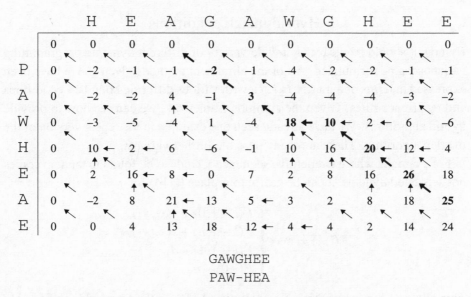

GAWGHEE

PAW-HEA

Figure 2.8 *Above, the overlap dynamic programming matrix for the example sequences. Below, the optimal overlap alignment, with score 25.*

What we want is really a type of global alignment, but one that does not penalise overhanging ends. This gives a clue to what sort of algorithm to use: we want a match to start on the top or left border of the matrix, and finish on the right or bottom border. The initialisation equations are therefore that $F(i,0) = 0$ for $i = 1,\ldots,n$ and $F(0,j) = 0$ for $j = 1,\ldots,m$, and the recurrence relations within the matrix are simply those for a global alignment (2.8). We set $F_{\max}$ to be the maximum value on the right border $(i,m), i = 1,\ldots,n$, and the bottom border $(n,j), j = 1,\ldots,m$. The traceback starts from the maximum point and continues until the top or left edge is reached.

There is a repeat version of this overlap match algorithm, in which the analogues of (2.11) and (2.12) are

$$F(i,0) = \max \begin{cases} F(i-1,0), \\ F(i-1,m)-T; \end{cases} \tag{2.13}$$

$$F(i,j) = \max \begin{cases} F(i-1,j-1)+s(x_i,y_j), \\ F(i-1,j)-d, \\ F(i,j-1)-d. \end{cases} \tag{2.14}$$

Note that the line (2.13) in the recursion for $F(i,0)$ is now just looking at complete matches to $y_{1\ldots m}$, rather than all possible subsequences of y as in (2.11) in the previous section. However, (2.11) is still used in its original form for obtaining $F(n+1,0)$, so that matches of initial subsequences of y to the end of x can be obtained.

Hybrid match conditions

By now it should be clear that a wide variety of different dynamic programming variants can be formulated. All of the alignment methods given above have been expressed in terms of a matrix $F(i, j)$, with various differing boundary conditions and recurrence rules. Given the common framework, we can see how to provide hybrid algorithms. We have already seen one example in the repeat version of the overlap algorithm. There are many possible further variants.

For example, where a repetitive sequence y tends to be found in tandem copies not separated by gaps, it can be useful to replace (2.14) for $j = 1$ with

$$F(i, 1) = \max \begin{cases} F(i-1,0) + s(x_i, y_1), \\ F(i-1,n) + s(x_i, y_1), \\ F(i-1,1) - d, \\ F(i,0) - d. \end{cases}$$

This allows a bypass of the $-T$ penalty in (2.11), so the threshold applies only once to each tandem cluster of repeats, not once to each repeat.

Another example might be if we are looking for a match that starts at the beginning of both sequences but can end at any point. This would be implemented by setting only $F(0,0) = 0$, using (2.8) in the recurrence, but allowing the match to end at the largest value in the whole matrix.

In fact, it is even possible to consider mixed boundary conditions where, for example, there is thought to be a significant prior probability that an entire copy of a sequence will be found in a larger sequence, but also some probability that only a fragment will be present. In this case we would set penalties on the boundaries or for starting internal matches, calculating the penalty costs as the logarithms of the respective probabilities. Such a model would be appropriate when looking for members of a repeat family in genomic DNA, since normally these are whole copies of the repeat, but sometimes only fragments are seen.

When performing a sequence similarity search we should ideally always consider what types of match we are looking for, and use the most appropriate algorithm for that case. In practice, there are often only good implementations available of a few of the standard cases, and it is often more convenient to use those, and postprocess the resulting matches afterwards.

2.4 Dynamic programming with more complex models

So far we have only considered the simplest gap model, in which the gap score $\gamma(g)$ is a simple multiple of the length. This type of scoring scheme is not ideal for biological sequences: it penalises additional gap steps as much as the first, whereas, when gaps do occur, they are often longer than one residue. If we are given a general function for $\gamma(g)$ then we can still use all the dynamic programming versions described in Section 2.3, with adjustments to the recurrence relations as typified by the following:

$$F(i,j) = \max \begin{cases} F(i-1,j-1)+s(x_i,y_j), & \\ F(k,j)+\gamma(i-k), & k=0,\ldots,i-1, \\ F(i,k)+\gamma(j-k), & k=0,\ldots,j-1. \end{cases} \quad (2.15)$$

which gives a replacement for the basic global dynamic relation. However, this procedure now requires $O(n^3)$ operations to align two sequences of length n, rather than $O(n^2)$ for the linear gap cost version, because in each cell (i,j) we have to look at $i+j+1$ potential precursors, not just three as previously. This is a prohibitively costly increase in computational time in many cases. Under some conditions on the properties of $\gamma()$ the search in k can be bounded, returning the expected computational time to $O(n^2)$, although the constant of proportionality is higher in these cases [Miller & Myers 1988].

Alignment with affine gap scores

The standard alternative to using (2.15) is to assume an affine gap cost structure as in (2.5): $\gamma(g) = -d - (g-1)e$. For this form of gap cost there is once again an $O(n^2)$ implementation of dynamic programming. However, we now have to keep track of multiple values for each pair of residue coefficients (i,j) in place of the single value $F(i,j)$. We will initially explain the process in terms of three variables corresponding to the three separate situations shown in Figure 2.4, which we show again here for convenience.

```
 I G A x_i          A I G A x_i          G A x_i  -  -
 L G V y_j          G V y_j  -  -        S L G V y_j
```

Let $M(i,j)$ be the best score up to (i,j) given that x_i is aligned to y_j (left case), $I_x(i,j)$ be the best score given that x_i is aligned to a gap (in an insertion with respect to y, central case), and finally $I_y(i,j)$ be the best score given that y_j is in an insertion with respect to x (right case).

The recurrence relations corresponding to (2.15) now become

$$M(i,j) = \max \begin{cases} M(i-1,j-1)+s(x_i,y_j), & \\ I_x(i-1,j-1)+s(x_i,y_j), & \\ I_y(i-1,j-1)+s(x_i,y_j); \end{cases} \quad (2.16)$$

$$I_x(i,j) = \max \begin{cases} M(i-1,j)-d, \\ I_x(i-1,j)-e; \end{cases}$$

$$I_y(i,j) = \max \begin{cases} M(i,j-1)-d, \\ I_y(i,j-1)-e. \end{cases}$$

In these equations, we assume that a deletion will not be followed directly by an

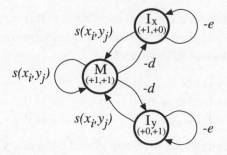

Figure 2.9 *A diagram of the relationships between the three states used for affine gap alignment.*

insertion. This will be true for the optimal path if $-d - e$ is less than the lowest mismatch score. As previously, we can find the alignment itself using a traceback procedure.

The system defined by equations (2.16) can be described very elegantly by the diagram in Figure 2.9. This shows a state for each of the three matrix values, with transition arrows between states. The transitions each carry a score increment, and the states each specify a $\Delta(i, j)$ pair, which is used to determine the change in indices i and j when that state is entered. The recurrence relation for updating each matrix value can be read directly from the diagram (compare Figure 2.9 with equations (2.16)). The new value for a state variable at (i, j) is the maximum of the scores corresponding to the transitions coming into the state. Each transition score is given by the value of the source state at the offsets specified by the $\Delta(i, j)$ pair of the target state, plus the specified score increment. This type of description corresponds to a *finite state automaton* (FSA) in computer science. An alignment corresponds to a path through the states, with symbols from the underlying pair of sequences being transferred to the alignment according to the $\Delta(i, j)$ values in the states. An example of a short alignment and corresponding state path through the affine gap model is shown in Figure 2.10.

It is in fact frequent practice to implement an affine gap cost algorithm using

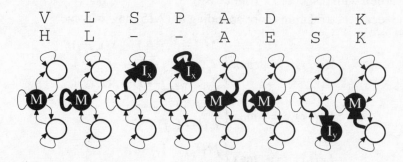

Figure 2.10 *An example of the state assignments for an alignment using the affine gap model.*

only two states, M and I, where I represents the possibility of being in a gapped region. Technically, this is only guaranteed to provide the correct result if the lowest mismatch score is greater than or equal to $-2e$. However, even if there are mismatch scores below $-2e$, the chances of a different alignment are very small. Furthermore, if one does occur it would not matter much, because the alignment differences would be in a very poorly matching gapped region. The recurrence relations for this version are

$$M(i,j) = \max \begin{cases} M(i-1,j-1)+s(x_i,y_j), \\ I(i-1,j-1)+s(x_i,y_j); \end{cases}$$

$$I(i,j) = \max \begin{cases} M(i,j-1)-d, \\ I(i,j-1)-e, \\ M(i-1,j)-d, \\ I(i-1,j)-e. \end{cases}$$

These equations do not correspond to an FSA diagram as described above, because the I state may be used for $\Delta(1,0)$ or $\Delta(0,1)$ steps. There is, however, an alternative FSA formulation in which the $\Delta(i,j)$ values are associated with the transitions, rather than the states. This type of automaton can account for the two-state affine gap algorithm, using extra transitions for the deletion and insertion alternatives. In fact, the standard one-state algorithm for linear gap costs can be expressed as a single-state transition emitting FSA with three transitions corresponding to different $\Delta(i,j)$ values ($\Delta(1,1)$, $\Delta(1,0)$ and $\Delta(0,1)$). For those interested in pursuing the subject, the simpler state-based automata are called *Moore machines* in the computer science literature, and the transition-emitting systems are called *Mealy machines* (see Chapter 9).

More complex FSA models

One advantage of the FSA description of dynamic programming algorithms is that it is easy to see how to generate new types of algorithm. An example is given in Figure 2.11, which shows a four-state FSA with two match states. The idea here is that there may be high fidelity regions of alignment without gaps, corresponding to match state A, separated by lower fidelity regions with gaps, corresponding to match state B and gap states I_x and I_y. The substitution scores $s(a,b)$ and $t(a,b)$ can be chosen to reflect the expected degrees of similarity in the different regions. Similarly, FSA algorithms can be built for alignments of transmembrane proteins with separate match states for intracellular, extracellular or transmembrane regions, or for other more complex scenarios [Birney & Durbin 1997]. Searls & Murphy [1995] give a more abstract definition of such FSAs and have developed interactive tools for building them.

One feature of these more complex algorithms is that, given an alignment path, there is also an implicit attachment of labels to the symbols in the original se-

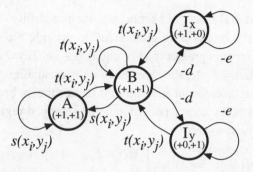

Figure 2.11 *The four-state finite state automaton with separate match states A and B for high and low fidelity regions. Note that this FSA emits on transitions with costs* $s(x_i, y_j)$ *and* $t(x_i, y_j)$, *rather than emitting on states, a distinction discussed earlier in the text.*

quences, indicating which state was used to match them. For example, with the transmembrane protein matching model, the alignment will assign sections of each protein to be transmembrane, intracellular or extracellular at the same time as finding the optimal alignment. In many cases this labelling of the sequence may be as important as the alignment information itself.

We will return to state models for pairwise alignment in Chapter 4.

Exercise

2.10   Calculate the score of the example alignment in Figure 2.10, with $d = 12, e = 2$.

2.5   Heuristic alignment algorithms

So far, all the alignment algorithms we have considered have been 'correct', in the sense that they are guaranteed to find the optimal score according to the specified scoring scheme. In particular, the affine gap versions described in the last section are generally regarded as providing the most sensitive sequence matching methods available. However, they are not the fastest available sequence alignment methods, and in many cases speed is an issue. The dynamic programming algorithms described so far have time complexity of the order of $O(nm)$, the product of the sequence lengths. The current protein database contains of the order of 100 million residues, so for a sequence of length one thousand, approximately 10^{11} matrix cells must be evaluated to search the complete database. At ten million matrix cells a second, which is reasonable for a single workstation at the time this is being written, this would take 10^4 seconds, or around three hours. If we want to search with many different sequences, time rapidly becomes an important issue.

For this reason, there have been many attempts to produce faster algorithms than straight dynamic programming. The goal of these methods is to search as small a fraction as possible of the cells in the dynamic programming matrix, while still looking at all the high scoring alignments. In cases where sequences are very similar, there are a number of methods based on extending computer science exact match string searching algorithms to non-exact cases, that provably find the optimal match [Chang & Lawler 1990; Wu & Manber 1992; Myers 1994]. However, for the scoring matrices used to find distant matches, these exact methods become intractable, and we must use heuristic approaches that sacrifice some sensitivity, in that there are cases where they can miss the best scoring alignment. A number of heuristic techniques are available. We give here brief descriptions of two of the best-known algorithms, BLAST and FASTA, to illustrate the types of approaches and trade offs that can be made. However, a detailed analysis of heuristic algorithms is beyond the scope of this book.

BLAST

The BLAST package [Altschul *et al.* 1990] provides programs for finding high scoring local alignments between a query sequence and a target database, both of which can be either DNA or protein. The idea behind the BLAST algorithm is that true match alignments are very likely to contain somewhere within them a short stretch of identities, or very high scoring matches. We can therefore look initially for such short stretches and use them as 'seeds', from which to extend out in search of a good longer alignment. By keeping the seed segments short, it is possible to pre-process the query sequence to make a table of all possible seeds with their corresponding start points.

BLAST makes a list of all 'neighbourhood words' of a fixed length (by default 3 for protein sequences, and 11 for nucleic acids), that would match the query sequence somewhere with score higher than some threshold, typically around 2 bits per residue. It then scans through the database, and whenever it finds a word in this set, it starts a 'hit extension' process to extend the possible match as an ungapped alignment in both directions, stopping at the maximum scoring extension (in fact, because of the way this is done, there is a small chance that it will stop short of the true maximal extension).

The most widely used implementation of BLAST finds ungapped alignments only. Perhaps surprisingly, restricting to ungapped alignments misses only a small proportion of significant matches, in part because the expected best score of unrelated sequences drops, so partial ungapped scores can still be significant, and also because BLAST can find and report more than one high scoring match per sequence pair and can give significance values for combined scores [Karlin & Altschul 1993]. Nonetheless, new versions of BLAST have recently become available that give gapped alignments [Altschul & Gish 1996; Altschul *et al.* 1997].

FASTA

Another widely used heuristic sequence searching package is FASTA [Pearson & Lipman 1988]. It uses a multistep approach to finding local high scoring alignments, starting from exact short word matches, through maximal scoring ungapped extensions, to finally identify gapped alignments.

The first step uses a lookup table to locate all identically matching words of length *ktup* between the two sequences. For proteins, *ktup* is typically 1 or 2, for DNA it may be 4 or 6. It then looks for diagonals with many mutually supporting word matches. This is a very fast operation, which for example can be done by sorting the matches on the difference of indices $(i - j)$.

The best diagonals are pursued further in step (2), which is analogous to the hit extension step of the BLAST algorithm, extending the exact word matches to find maximal scoring ungapped regions (and in the process possibly joining together several seed matches).

Step (3) then checks to see if any of these ungapped regions can be joined by a gapped region, allowing for gap costs. In the final step, the highest scoring candidate matches in a database search are realigned using the full dynamic programming algorithm, but restricted to a subregion of the dynamic programming matrix forming a band around the candidate heuristic match.

Because the last stage of FASTA uses standard dynamic programming, the scores it produces can be handled exactly like those from the full algorithms described earlier in the chapter. There is a tradeoff between speed and sensitivity in the choice of the parameter *ktup*: higher values of *ktup* are faster, but more likely to miss true significant matches. To achieve sensitivities close to those of full local dynamic programming for protein sequences it is necessary to set *ktup* = 1.

2.6 Linear space alignments

Aside from time, another computational resource that can limit dynamic programming alignment is memory usage. All the algorithms described so far calculate score matrices such as $F(i, j)$, which have overall size nm, the product of the sequence lengths. For two protein sequences, of typical length a few hundred residues, this is well within the capacity of modern desktop computers; but if one or both of the sequences is a DNA sequence tens or hundreds of thousands of bases long, the required memory for the full matrix can exceed a machine's physical capacity. Fortunately, we are in a better situation with memory than speed: there are techniques that give the optimal alignment in limited memory, of order $n + m$ rather than nm, with no more than a doubling in time. These are commonly referred to as *linear space* methods. Underlying them is an important basic technique in pairwise sequence dynamic programming.

In fact, if only the maximal score is needed, the problem is simple. Since the recurrence relation for $F(i, j)$ is local, depending only on entries one row back, we can throw away rows of the matrix that are further than one back from the current point. If looking for a local alignment we need to find the maximum score in the whole matrix, but it is easy to keep track of the maximum value as the matrix is being built. However, while this will get us the score, it will not find the alignment; if we throw away rows to avoid $O(nm)$ storage, then we also lose the traceback pointers. A new approach must be used to obtain the alignment.

Let us assume for now that we are looking for the optimal global alignment, using linear gap scoring. The method will extend easily to the other types of alignment. We use the principle of *divide and conquer*.

Let $u = \lfloor \frac{n}{2} \rfloor$, the integer part of $\frac{n}{2}$. Let us suppose for now that we can identify a v such that cell (u, v) is on the optimal alignment, i.e. v is the row where the alignment crosses the $i = u$ column of the matrix. Then we can split the dynamic programming problem into two parts, from top left $(0, 0)$ to (u, v), and from (u, v) to (n, m). The optimal alignment for the whole matrix will be the concatenation of the optimal alignments for these two separate submatrices. (For this to work precisely, define the alignment not to include the origin.) Once we have split the alignment once, we can fill in the whole alignment recursively, by successively halving each region, at every step pinning down one more aligned pair of residues. This can either continue down until sequences of zero length are being aligned, which is trivial and means that the region is completely specified, or alternatively, when the sequences are short enough, the standard $O(n^2)$ alignment and traceback method can be used.

So how do we find v? For $i > u$ let us define $c(i, j)$ such that $(u, c(i, j))$ is on the optimal path from $(1, 1)$ to (i, j). We can update $c(i, j)$ as we calculate $F(i, j)$. If (i', j') is the preceding cell to (i, j) from which $F(i, j)$ is derived, then set $c(i, j) = j'$ if $i = u$, else $c(i, j) = c(i', j')$. Clearly this is a local operation, for which we only need to maintain the previous row of $c()$, just as we only maintain the previous row of $F()$. We can now read out from the final cell of the matrix the value we desire: $v = c(n, m)$.

As far as we are aware, this procedure for finding v has not been published by any of the people who use it. A more widely known procedure first appeared in the computer science literature [Hirschberg 1975] and was introduced into computational biology by Myers & Miller [1988], and thus is usually called the *Myers–Miller* algorithm in the sequence analysis field. The Myers–Miller algorithm does not propagate the traceback pointer $c(i, j)$, but instead finds the alignment midpoint (u, v) by combining the results of forward and backward dynamic programming passes at row u (see their paper for details). Myers–Miller is an elegant recursive algorithm, but it is a little more difficult to explain in detail. Waterman [1995, p. 211] gives a third linear space approach. Chao, Hardison & Miller [1994] give a review of linear space algorithms in pairwise alignment.

Exercises

2.11   Fill in the correct values of $c(i, j)$ for the global alignment of the example
        pair of sequences in Figure 2.5 for the first pass of the algorithm ($u = 5$).
2.12   Show that the time required by the linear space algorithm is only about
        twice that of the standard $O(nm)$ algorithm.

2.7   Significance of scores

Now that we know how to find an optimal alignment, how can we assess the significance of its score? That is, how do we decide if it is a biologically meaningful
alignment giving evidence for a homology, or just the best alignment between
two entirely unrelated sequences? There are two possible approaches. One is
Bayesian in flavour, based on the comparison of different models. The other is
based on the traditional statistical approach of calculating the chance of a match
score greater than the observed value, assuming a null model, which in this case
is that the underlying sequences were unrelated.

The Bayesian approach: model comparison

We gave the log-odds ratio on p. 15 as the relevant score without much motivation. We might argue that what is really wanted is the probability that the
sequences are related as opposed to being unrelated, which would be $P(M|x, y)$,
rather than the likelihood calculated above, $P(x, y|M)$. $P(M|x, y)$ can be calcu-
lated using Bayes' rule, once we state some more assumptions. First we must
specify the *a priori* probabilities of the two models. These reflect our expectation
that the sequences are related before we actually see them. We will write these
as $P(M)$, the prior probability that the sequences are related, and hence that the
match model is correct, and $P(R) = 1 - P(M)$, the prior probability that the random model is correct. Then once we have seen the data the posterior probability
that the match model is correct, and hence that the sequences are related, is

$$P(M|x, y) = \frac{P(x, y|M)P(M)}{P(x, y)}$$

$$= \frac{P(x, y|M)P(M)}{P(x, y|M)P(M) + P(x, y|R)P(R)}$$

$$= \frac{P(x, y|M)P(M)/P(x, y|R)P(R)}{1 + P(x, y|M)P(M)/P(x, y|R)P(R)}.$$

Let

$$S' = S + \log\left(\frac{P(M)}{P(R)}\right) \tag{2.17}$$

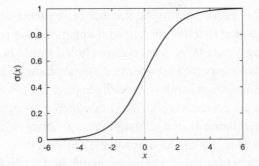

Figure 2.12 *The logistic function.*

where

$$S = \log \left(\frac{P(x,y|M)}{P(x,y|R)} \right)$$

is the log-odds score of the alignment. Then

$$P(M|x,y) = \sigma(S')$$

where

$$\sigma(x) = \frac{e^x}{1+e^x}.$$

$\sigma(x)$ is known as the *logistic* function. It is a sigmoid function, tending to 1 as x tends to infinity, to 0 as x tends to minus infinity, and with value $\frac{1}{2}$ at $x = 0$ (see Figure 2.12). The logistic function is widely used in neural network analysis to convert scores built from sums into probabilities – not entirely a coincidence.

From (2.17) we can see that we should add the prior log-odds ratio, $\log \left(\frac{P(M)}{P(R)} \right)$, to the standard score of the alignment. This corresponds to multiplying the likelihood ratio by the prior odds ratio, which makes intuitive sense. Once this has been done we can in principle compare the resulting value with 0 to indicate whether the sequences are related. For this to work, we have to be very careful that all the expressions we use really are probabilities, and in particular that when we sum them over all possible pairs of sequences that might have been given they sum to 1. When a scoring scheme is constructed in an *ad hoc* fashion this is unlikely to be true.

A particular example of where the prior odds ratio becomes important is when we are looking at a large number of different alignments for a possible significant match. This is the typical situation when searching a database. It is clear that if we have a fixed prior odds ratio, then even if all the database sequences are unrelated, as the number of sequences we try to match increases, the probability of one of the matches looking significant by chance will also increase. In fact, given a fixed prior odds ratio, the expected number of (falsely) significant observations will increase linearly. If we want it to stay fixed, then we must set the prior

odds ratio in inverse proportion to the number of sequences in the database N.
The effect of this is that to maintain a fixed number of false positives we should
compare S with $\log N$, not 0. A conservative choice would be to choose a score
that corresponds to an expected number of false positives of say 0.1 or 0.01. Of
course, this type of approach is not necessarily appropriate. For example, we may
believe that 1% of all proteins are kinases, in which case the prior odds should
be $\frac{1}{100}$, and the expectation is that although false positives will increase as more
sequences are looked at, so will true positives. On the other hand, if we believe
that we will be looking for cases where one match in the whole database will be
significant, then the $\log N$ comparison is more reasonable.

At this point we can turn to consider the statistical significance of a score obtained from the local match algorithm. In this case we have to correct for the fact
that we are looking at the best of many possible different local matches between
subsequences of the two sequences. A simple estimate of the number of start
points of local matches is the product of the lengths of the sequences, nm. If all
matches were constant length and all start points gave independent matches, this
would result in a requirement to compare the best score S with $\log(nm)$. However, these assumptions are both clearly wrong (for instance, match segments at
consecutive points along a diagonal are not independent), with the consequence
that a further small correction factor should be added to S, dependent only on
the scoring function s, but not on n and m. There is no analytical theory for this
effect, but for scoring systems typically used when comparing protein sequences
it seems that a multiplicative factor of around 0.1 is appropriate. Since what we
care about is an additive term of the logarithm of this factor, the effect is comparatively small.

The classical approach: the extreme value distribution

There is an alternative way to consider significance in such situations, using a
more classical statistical framework. We can look at the distribution of the maximum of N match scores to independent random sequences. If the probability
of this maximum being greater than the observed best score is small, then the
observation is considered significant.

In the simple case of a fixed ungapped alignment (2.2), the score of a match
to a random sequence is the sum of many similar random variables, and so will
be very well approximated by a normal distribution. The asymptotic distribution
of the maximum M_N of a series of N independent normal random variables is
known, and has the form

$$P(M_N \leq x) \simeq \exp(-KNe^{\lambda(x-\mu)}) \tag{2.18}$$

for some constants K, λ. This form of limiting distribution is called the *extreme
value distribution* or EVD (Chapter 11). We can use equation (2.18) to calculate

the probability that the best match from a search of a large number N of unrelated sequences has score greater than our observed maximal score, S. If this is less than some small value, such as 0.05 or 0.01, then we can conclude that it is unlikely that the sequence giving rise to the observed maximal score is unrelated, i.e. it is likely that it is related.

It turns out that, even when the individual scores are not normally distributed, the extreme value distribution is still the correct limiting distribution for the maximum of a large number of separate scores (see Chapter 11). Because of this, the same type of significance test can be used for any search method that looks for the best score from a large set of equivalent possibilities. Indeed, for best local match scores from the local alignment algorithm, the best score between two (significantly long) sequences will itself be distributed according to the extreme value distribution, because in this case we are effectively comparing the outcomes of $O(nm)$ distinct random starts within the single matrix.

For local ungapped alignments, Karlin & Altschul [1990] derived the appropriate EVD distribution analytically, using results given more fully in Dembo & Karlin [1991]. We give this here in two steps. First, the number of unrelated matches with score greater than S is approximately Poisson distributed, with mean

$$E(S) = Kmne^{-\lambda S}, \tag{2.19}$$

where λ is the positive root of

$$\sum_{a,b} q_a q_b e^{\lambda s(a,b)} = 1, \tag{2.20}$$

and K is a constant given by a geometrically convergent series also dependent only on the q_a and $s(a,b)$. This K corresponds directly to the multiplicative factor we described at the end of the previous section; it corrects for the non-independence of possible starting points for matches. The value λ is really a scale parameter, to convert the $s(a,b)$ into a natural scale. Note that if the $s(a,b)$ were initially derived as log likelihood quantities using equation (2.3) then $\lambda = 1$, because $e^{\lambda s(a,b)} = p_{ab}/q_a q_b$.

The probability that there is a match of score greater than S is then

$$P(x > S) = 1 - e^{-E(S)}. \tag{2.21}$$

It is easy to see that combining equations (2.19) and (2.21) gives a distribution of the same EVD form as (2.18), but without μ. In fact, it is common not to bother with calculating a probability, but just to use a requirement that $E(S)$ is significantly less than 1. This converts into a requirement that

$$S > T + \frac{\log mn}{\lambda} \tag{2.22}$$

for some fixed constant T. This corresponds to the Bayesian analysis in the

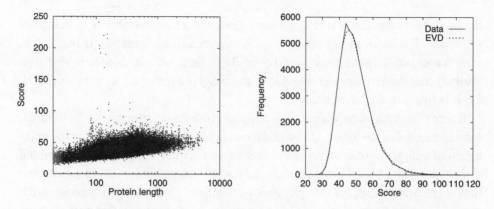

Figure 2.13 *Left, a scatter plot of the distribution of local match scores obtained from comparing human cytochrome C (SWISS-PROT accession code P000001) against the SWISS-PROT34 protein database with the Smith–Waterman implementation SSEARCH [Pearson 1996]. Right, the corresponding length-normalised distribution of scores, showing the fit to an EVD distribution.*

previous section suggesting that we should compare S with $\log mn$, but in this case we can assign a precise meaning to the value of T that we use.

Although no corresponding analytical theory has yet been derived for gapped alignments, Mott [1992] suggested that gapped alignment scores for random sequences follow the same form of extreme value distribution as ungapped scores, and there is now considerable empirical evidence to support this. Altschul & Gish [1996] have fit λ and K values for (2.19) for a range of standard protein alignment scoring schemes, using a large amount of randomly generated sample data.

Correcting for length

When searching a database of mixed length sequences, the best local matches to longer database sequences tend to have higher scores than the best local matches to shorter sequences, even when all the sequences are unrelated. An example is shown in Figure 2.13. This is not surprising: if our search sequence has length n and the database sequences have length m_i, then there are more possible start points in the nm_i matrix for larger m_i. However, if our prior expectation is that a match to any database entry should be equally likely, then we want random match scores to be comparable independent of length.

A theoretically justifiable correction for length dependence is that we should adjust the best score for each database entry by subtracting $\log(m_i)$. This follows from the expression for S' in the previous section. An alternative, which appears to perform slightly better in practice and is easily carried out when there are large numbers of sequences being searched, is to bin all the database entries by length, and then fit a linear function of the log sequence length [Pearson 1995] (the separation of 'background' from signal makes this a little tricky to implement).

Why use the alignment score as the test statistic?

So far in this section we have always assumed that we will use the same alignment score as a test statistic for the alignment's significance as was used to find the best match during the search phase. It might seem attractive to search for a match with one criterion, then evaluate it with another, uncorrelated one. This would seem to prevent the problem that the search phase increases the background level when testing. However, we need both the search and significance test to have as much discriminative power as possible. It is important to use the best available statistic for both. If we miss a genuinely related alignment in the search phase, then we obviously can't consider it when testing for significance.

A consequence of using the test statistic for searching is that the best match in unrelated sequences will tend to look qualitatively like a real match. As a striking example of this, Karlin & Altschul [1990] showed that when optimal local ungapped alignments are found between random sequences, the frequency of observing residue a aligned to residue b in these alignments will be $q_a q_b e^{\lambda s(a,b)}$, i.e. exactly the frequency p_{ab} with which we expect to observe a being aligned to b in our true, evolutionarily matched model. The only property we can use to discriminate true from false matches is the magnitude of the score, the expectation of which is proportional to the length of the match.

Of course, it may be that there are complex calculations involved in the most sensitive scoring scheme, which could not practically be implemented during the search stage. In this case, it may be necessary to search with a simpler score, but keep several alternative high scoring alignments, rather than simply the best one. We give methods for obtaining such suboptimal alignments in Chapter 4.

2.8 Deriving score parameters from alignment data

We finish this chapter by returning to the subject of the first section: how to determine the components of the scoring model, the substitution and gap scores. There we described how to derive scores for pairwise alignment algorithms from probabilities. However, this left open the issue of how to estimate the probabilities. It should be clear that the performance of our whole alignment system will depend on the values of these parameters, so considerable care has gone into their estimation.

A simple and obvious approach would be to count the frequencies of aligned residue pairs and of gaps in confirmed alignments, and to set the probabilities

p_{ab}, q_a and $f(g)$ to the normalised frequencies. (This corresponds to obtaining maximum likelihood estimates for the probabilities; see Chapter 11.)

There are two difficulties with this simple approach. The first is that of obtaining a good random sample of confirmed alignments. Alignments tend not to be independent from each other because protein sequences come in families. The second is more subtle. In truth, different pairs of sequences have diverged by different amounts. When two sequences have diverged from a common ancestor very recently, we expect many of their residues to be identical. The probability p_{ab} for $a \neq b$ should be small, and hence $s(a,b)$ should be strongly negative unless $a = b$. At the other extreme, when a long time has passed since two sequences diverged, we expect p_{ab} to tend to the background frequency $q_a q_b$, so $s(a,b)$ should be close to zero for all a,b. This suggests that we should use scores that are matched to the expected divergence of the sequences we wish to compare.

Dayhoff PAM matrices

Dayhoff, Schwartz & Orcutt [1978] took both these difficulties into consideration when defining their PAM matrices, which have been very widely used for practical protein sequence alignment. The basis of their approach is to obtain substitution data from alignments between very similar proteins, allowing for the evolutionary relationships of the proteins in families, and then extrapolate this information to longer evolutionary distances.

They started by constructing hypothetical phylogenetic trees relating the sequences in 71 families, where each pair of sequences differed by no more than 15% of their residues. To build the trees they used the parsimony method (Chapter 7), which provides a list of the residues that are most likely to have occurred at each position in each ancestral sequence. From this they could accumulate an array A_{ab} containing the frequencies of all pairings of residues a and b between sequences and their immediate ancestors on the tree. The evolutionary direction of this pairing was ignored, both A_{ab} and A_{ba} being incremented each time either an a in the ancestral sequence was replaced by a b in the descendant, or vice versa. Basing the counts on the tree avoided overcounting substitutions because of evolutionary relatedness.

Because they wanted to extrapolate to longer times, the primary value that they needed to estimate was not the joint probability p_{ab} of seeing a aligned to b, but instead the conditional probability $P(b|a,t)$ that residue a is substituted by b in time t. $P(b|a,t) = p_{ab}(t)/q_a$. We can calculate conditional probabilities for a long time interval by multiplying those for a short interval, as shown below. These conditional probabilities are known as *substitution probabilities*; they play an important part in phylogenetic tree building (see Chapter 8). The short time interval estimates for $P(b|a)$ can be derived from the A_{ab} matrix by setting $P(b|a) = B_{a,b} = A_{ab}/\sum_c A_{ac}$.

These values must next be adjusted to correct for divergence time t. The expected number of substitutions in a 'typical' protein, where the residue a occurs at the frequency q_a, is $\sum_{a,b} q_a q_b B_{ab}$. Dayhoff *et al.* defined a substitution matrix to be a 1 PAM matrix (an acronym for 'point accepted mutation') if the expected number of substitutions was 1%, i.e. if $\sum_{a,b} q_a q_b B_{ab} = 0.01$. To turn their B matrix into a 1 PAM matrix of substitution probabilities, they scaled the off-diagonal terms by a factor σ and adjusted the diagonal terms to keep the sum of a row equal to 1. More precisely, they defined $C_{ab} = \sigma B_{ab}$ for $a \neq b$, and $C_{aa} = \sigma B_{aa} + (1 - \sigma)$, with σ chosen to make C into a 1 PAM matrix; we will denote this 1 PAM C by $S(1)$. Its entries can be regarded as the probability of substituting a with b in unit time, $P(b|a, t = 1)$.

To generate substitution matrices appropriate to longer times, $S(1)$ is raised to a power n (multiplying the matrix by itself n times), giving $S(n) = S(1)^n$. For instance, $S(2)$, the matrix product of $S(1)$ with itself, has entries $P(a|b, t = 2) = \sum_c P(a|c, t = 1) P(c|b, t = 1)$, which are the probabilities of the substitution of b by a occurring via some intermediate, c. For small n, the off-diagonal entries increase approximately linearly with n. Another way to view this is that the matrix $S(n)$ represents the result of n steps of a Markov chain with 20 states, corresponding to the 20 amino acids, each step having transition probabilities given by $S(1)$ (Markov chains will be introduced fully in Chapter 3).

Finally, a matrix of scores is obtained from $S(t)$. Since $P(b|a) = p_{ab}/q_a$, the entries of the score matrix for time t are given by

$$s(a, b|t) = \log \frac{P(b|a, t)}{q_b}.$$

These values are scaled and rounded to the nearest integer for computational convenience. The most widely used matrix is PAM250, which is scaled by $3/\log 2$ to give scores in third-bits.

BLOSUM matrices

The Dayhoff matrices have been one of the mainstays of sequence comparison techniques, but they do have their limitations. The entries in $S(1)$ arise mostly from short time interval substitutions, and raising $S(1)$ to a higher power, to give for instance a PAM250 matrix, does not capture the true difference between short time substitutions and long term ones [Gonnet, Cohen & Benner 1992]. The former are dominated by amino acid substitutions that arise from single base changes in codon triplets, for example L $\leftrightarrow$ I, L $\leftrightarrow$ V or Y $\leftrightarrow$ F, whereas the latter show all types of codon changes.

Since the PAM matrices were made, databases have been formed containing multiple alignments of more distantly related proteins, and these can be used to derive score matrices more directly. One such set of score matrices that is widely used is the BLOSUM matrix set [Henikoff & Henikoff 1992]. In detail, they were derived from a set of aligned, ungapped regions from protein families called the BLOCKS database [Henikoff & Henikoff 1991]. The sequences from each block were clustered, putting two sequences into the same cluster whenever their percentage of identical residues exceeded some level $L\%$. Henikoff & Henikoff then calculated the frequencies A_{ab} of observing residue a in one cluster aligned against residue b in another cluster, correcting for the sizes of the clusters by weighting each occurrence by $1/(n_1 n_2)$, where n_1 and n_2 are the respective cluster sizes.

From the A_{ab}, they estimated q_a and p_{ab} by $q_a = \sum_b A_{ab} / \sum_{cd} A_{cd}$, i.e. the fraction of pairings that include an a, and $p_{ab} = A_{ab} / \sum_{cd} A_{cd}$, i.e. the fraction of pairings between a and b out of all observed pairings. From these they derived the score matrix entries using the standard equation $s(a,b) = \log p_{ab} / q_a q_b$ (2.3). Again, the resulting log-odds score matrices were scaled and rounded to the nearest integer value. The matrices for $L = 62$ and $L = 50$ in particular are widely used for pairwise alignment and database searching, BLOSUM62 being standard for ungapped matching, and BLOSUM50 being perhaps better for alignment with gaps [Pearson 1996]. BLOSUM62 is scaled so that its values are in half-bits, i.e. the log-odds values were multiplied by $2/\log 2$, and BLOSUM50 is given in third-bits. Note that lower L values correspond to longer evolutionary time, and are applicable for more distant searches.

Estimating gap penalties

There is no similar standard set of time-dependent gap models. If there were a time-dependent gap score model, one reasonable assumption might be that the expected number of gaps would increase linearly with time, but their length distribution would stay constant. In an affine gap model, this corresponds to making the gap-open score d linear in $\log t$, while the gap-extend score e would remain constant. Gonnet, Cohen & Benner [1992] derive a similar distribution from empirical data. In fact, they suggest that a better fit is obtained by the form $\gamma(g) = A + B \log t + C \log g$, although there is some circularity in their approach because the data come from a complete comparison of the protein database against itself using sequence alignment algorithms.

In practice, people choose gap costs empirically once they have chosen their substitution scores. This is possible because there are only two affine gap parameters, whereas there are 210 substitution score parameters for proteins. A careful discussion of the factors involved in choosing gap penalties can be found in Vingron & Waterman [1994].

There is a final twist to be added once we have a combined substitution and gap model. Now that there is a possibility of a gap occurring in a sequence at a given position, it is no longer inevitable that there will be a match. It can be argued that we should include in our substitution score a term for the probability that a gap has not opened. The probability that there is a gap in a particular position in sequence x is $\sum_{i \geq 1} f(i)$, and likewise there is the same probability that there is a gap in y at that position. From this we can derive the probability that there is a no gap, i.e. that there is a match:

$$P(\text{no gap}) = 1 - 2\sum_{i \geq 1} f(i). \tag{2.23}$$

As a consequence, the substitution score, which corresponds to a match, should not be $s(a,b)$ but instead $s'(a,b) = s(a,b) + \log P(\text{no gap})$. The effect of this would be to reduce the pairwise scores as gaps become more likely, i.e. as gap penalties decrease. This correction is, however, small, and is not normally made when deriving a scoring system from alignment frequencies.

2.9 Further reading

Good reviews of dynamic programming methods for biological sequence comparison include Pearson [1996] and Pearson & Miller [1992]. The sensitivity of dynamic programming methods has been evaluated and compared to the fast heuristic methods BLAST and FASTA by Pearson [1995] and Shpaer et al. [1996].

Bucher & Hofmann [1996] have described a probabilistic version of the Smith–Waterman algorithm, which is related to the methods we will discuss in Chapter 4.

Interesting areas in pairwise dynamic programming alignment that we have not covered include fast 'banded' dynamic programming algorithms [Chao, Pearson & Miller 1992], the problem of aligning protein query sequences to DNA target sequences [Huang & Zhang 1996], and the problem of recovering not only the optimal alignment but also 'suboptimal' or 'near-optimal' alignments [Zuker 1991; Vingron 1996].

3

Markov chains and hidden Markov models

Having introduced some methods for pairwise alignment in Chapter 2, the emphasis will switch in this chapter to questions about a single sequence. The main aim of the chapter is to develop the theory for a very general form of probabilistic model for sequences of symbols, called a hidden Markov model (abbreviated HMM). The types of question we can use HMMs and their simpler cousins, Markov models, to consider are: 'Does this sequence belong to a particular family?' or 'Assuming the sequence does come from some family, what can we say about its internal structure?' An example of the second type of problem would be to try to identify alpha helix or beta sheet regions in a protein sequence.

As well as giving examples from the biological sequence world, we also give the mathematics and algorithms for many of the operations on HMMs in a more general form. These methods, or close analogues of them, are applied in many other sections of the book. This chapter therefore contains a fairly large amount of mathematically technical material. We have tried to organise it so that the first half, approximately, leads the reader through the essential algorithms using a single biological example. In the later sections we introduce a variety of other examples to illustrate more complex extensions of the basic approaches.

In the next chapter, we will see how HMMs can also be applied to the types of alignment problem discussed in Chapter 2, in Chapter 5 they are applied to searching databases for protein families, and in Chapter 6 to alignment of several sequences simultaneously. In fact, the search and alignment applications constitute probably the best-known use of HMMs for biological sequence analysis. However, we present HMM theory here in a less specialised context in order to emphasise its much broader applicability, which goes far beyond that of sequence alignment.

The overwhelming majority of papers on HMMs belong to the speech recognition literature, where they were applied first in the early 1970s. One of the best general introductions to the subject is the review by Rabiner [1989], which also covers the history of the topic. Although there will be quite a bit of overlap between that and the present chapter, there will be important differences in focus.

Before going on to introduce HMMs for biological sequence analysis, it is perhaps interesting to look briefly at how they are used for speech recognition [Rabiner & Juang 1993]. After recording, a speech signal is divided into pieces (called frames) of 10–20 milliseconds. After some preprocessing each frame is assigned to one out of a large number of predefined categories by a process known as vector quantisation. Typically there are 256 such categories. The speech signal is then represented as a long sequence of category labels and from that the speech recogniser has to find out what sequence of phonemes (or words) was spoken. The problems are that there are variations in the actual sound uttered, and there are also variations in the time taken to say the various parts of the word.

Many problems in biological sequence analysis have the same structure: based on a sequence of symbols from some alphabet, find out what the sequence represents. For proteins the sequences consist of symbols from the alphabet of 20 amino acids, and we typically want to know what protein family a given sequence belongs to. Here the primary sequence of amino acids is analogous to the speech signal and the protein family to the spoken word it represents. The time-variation of the speech signal corresponds to having insertions and deletions in the protein sequences.

Let us turn to a simpler example, which we will use to introduce first standard Markov models, of the non-hidden variety, then a simple hidden Markov model.

Example: CpG islands

In the human genome wherever the dinucleotide CG occurs (frequently written CpG to distinguish it from the C-G base pair across the two strands) the C nucleotide (cytosine) is typically chemically modified by methylation. There is a relatively high chance of this methyl-C mutating into a T, with the consequence that in general CpG dinucleotides are rarer in the genome than would be expected from the independent probabilities of C and G. For biologically important reasons the methylation process is suppressed in short stretches of the genome, such as around the promoters or 'start' regions of many genes. In these regions we see many more CpG dinucleotides than elsewhere, and in fact more C and G nucleotides in general. Such regions are called CpG islands [Bird 1987]. They are typically a few hundred to a few thousand bases long.

We will consider two questions: Given a short stretch of genomic sequence, how would we decide if it comes from a CpG island or not? Second, given a long piece of sequence, how would we find the CpG islands in it, if there are any? Let us start with the first question. □

3.1 Markov chains

What sort of probabilistic model might we use for CpG island regions? We know
that dinucleotides are important. We therefore want a model that generates sequences in which the probability of a symbol depends on the previous symbol.
The simplest such model is a classical Markov chain. We like to show a Markov
chain graphically as a collection of 'states', each of which corresponds to a particular residue, with arrows between the states. A Markov chain for DNA can be
drawn like this:

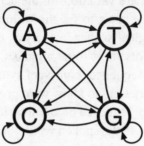

where we see a state for each of the four letters A, C, G, and T in the DNA alphabet. A probability parameter is associated with each arrow in the figure, which
determines the probability of a certain residue following another residue, or one
state following another state. These probability parameters are called the *transition probabilities*, which we will write a_{st}:

$$a_{st} = P(x_i = t | x_{i-1} = s). \tag{3.1}$$

For any probabilistic model of sequences we can write the probability of the
sequence as

$$
\begin{aligned}
P(x) &= P(x_L, x_{L-1}, \ldots, x_1) \\
&= P(x_L | x_{L-1}, \ldots, x_1) P(x_{L-1} | x_{L-2}, \ldots, x_1) \cdots P(x_1)
\end{aligned}
$$

by applying $P(X, Y) = P(X|Y)P(Y)$ many times. The key property of a Markov
chain is that the probability of each symbol x_i depends only on the value of the
preceding symbol x_{i-1}, not on the entire previous sequence, i.e. $P(x_i | x_{i-1}, \ldots, x_1)$
$= P(x_i | x_{i-1}) = a_{x_{i-1} x_i}$. The previous equation therefore becomes

$$
\begin{aligned}
P(x) &= P(x_L | x_{L-1}) P(x_{L-1} | x_{L-2}) \cdots P(x_2 | x_1) P(x_1) \\
&= P(x_1) \prod_{i=2}^{L} a_{x_{i-1} x_i}. \tag{3.2}
\end{aligned}
$$

Although we have derived this equation in the context of CpG islands in DNA
sequences, it is in fact the general equation for the probability of a specific sequence from any Markov chain. There is a large literature on Markov chains, see
for example Cox & Miller [1965].

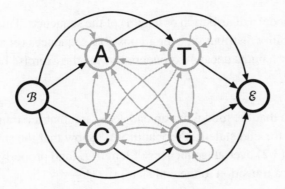

Figure 3.1 *Begin and end states can be added to a Markov chain (grey model) for modelling both ends of a sequence.*

Exercise

3.1   The sum of the probabilities of all possible sequences of length L can be written (using (3.2))

$$\sum_{\{x\}} P(x) = \sum_{x_1} \sum_{x_2} \cdots \sum_{x_L} P(x_1) \prod_{i=2}^{L} a_{x_{i-1}x_i}.$$

Show that this is equal to 1.

Modelling the beginning and end of sequences

Notice that as well as specifying the transition probabilities we must also give the probability $P(x_1)$ of starting in a particular state. To avoid the inhomogeneity of (3.2) introduced by the starting probabilities, it is possible to add an extra *begin state* to the model. At the same time we add a letter to the alphabet, which we will call $\mathcal{B}$. By defining $x_0 = \mathcal{B}$ the beginning of a sequence is also included in (3.2), so for instance the probability of the first letter in the sequence is

$$P(x_1 = s) = a_{\mathcal{B}s}.$$

Similarly we can add a symbol $\mathcal{E}$ to the end of a sequence to ensure the end is modelled. Then the probability of ending with residue t is

$$P(\mathcal{E}|x_L = t) = a_{t\mathcal{E}}.$$

To match the new symbols, we add begin and end states to the DNA model (see Figure 3.1). In fact, we need not explicitly add any letters to the alphabet, but instead can treat the two new states as 'silent' states that just serve as start and end points.

Traditionally the end of a sequence is not modelled in Markov chains; it is assumed that the sequence can end anywhere. The effect of adding an explicit end state is to model a distribution of lengths of the sequence. This way the model defines a probability distribution over all possible sequences (of any length). The distribution over lengths decays exponentially; see the exercise below.

Exercises

3.2   Assume that the model has an end state, and that the transition from any state to the end state has probability τ. Show that the sum of the probabilities (3.2) over all sequences of length L (and properly terminating by making a transition to the end state) is $\tau(1-\tau)^{L-1}$.

3.3   Show that the sum of the probability over all possible sequences of any length is 1. This proves that the Markov chain really describes a proper probability distribution over the whole space of sequences. (Hint: Use the result that, for $0 < x < 1$, $\sum_{i=0}^{\infty} x^i = 1/(1-x)$.)

Using Markov chains for discrimination

A primary use for equation (3.2) is to calculate the values for a likelihood ratio test. We illustrate this here using real data for the CpG island example. From a set of human DNA sequences we extracted a total of 48 putative CpG islands and derived two Markov chain models, one for the regions labelled as CpG islands (the '+' model) and the other from the remainder of the sequence (the '−' model). The transition probabilities for each model were set using the equation

$$a_{st}^{+} = \frac{c_{st}^{+}}{\sum_{t'} c_{st'}^{+}}, \tag{3.3}$$

and its analogue for a_{st}^{-}, where c_{st}^{+} is the number of times letter t followed letter s in the labelled regions. These are the maximum likelihood (ML) estimators for the transition probabilities, as described in Chapter 1.

(In this case there were almost 60 000 nucleotides, and ML estimators are adequate. If the number of counts of each type had been small, then a Bayesian estimation process would have been more appropriate, as discussed in Chapter 11 and below for HMMs.) The resulting tables are

+	A	C	G	T		−	A	C	G	T
A	0.180	0.274	0.426	0.120		A	0.300	0.205	0.285	0.210
C	0.171	0.368	0.274	0.188		C	0.322	0.298	0.078	0.302
G	0.161	0.339	0.375	0.125		G	0.248	0.246	0.298	0.208
T	0.079	0.355	0.384	0.182		T	0.177	0.239	0.292	0.292

where the first row in each case contains the frequencies with which an A is followed by each of the four bases, and so on for the other rows, so each row sums to one. These numbers are not the same; for example, G following A is much more common than T following A. Notice also that the tables are asymmetric. In both tables the probability for G following C is lower than that for C following G, although the effect is stronger in the '−' table, as expected.

To use these models for discrimination, we calculate the log-odds ratio

$$S(x) = \log \frac{P(x|\text{model} +)}{P(x|\text{model} -)} = \sum_{i=1}^{L} \log \frac{a^+_{x_{i-1}x_i}}{a^-_{x_{i-1}x_i}}$$

$$= \sum_{i=1}^{L} \beta_{x_{i-1}x_i}$$

where x is the sequence and $\beta_{x_{i-1}x_i}$ are the log likelihood ratios of corresponding transition probabilities. A table for β is given below:

β	A	C	G	T
A	−0.740	0.419	0.580	−0.803
C	−0.913	0.302	1.812	−0.685
G	−0.624	0.461	0.331	−0.730
T	−1.169	0.573	0.393	−0.679

Figure 3.2 shows the distribution of scores, $S(x)$, normalised by dividing by their length, i.e. as an average number of bits per molecule.[1] If we had not normalised by length, the distribution would have been much more spread out.

We see a reasonable discrimination between regions labelled CpG island and other regions. The discrimination is not very much influenced by the length normalisation. If we wanted to pursue this further and investigate the cases of misclassification, it is worth remembering that the error could either be due to an inadequate or incorrectly parameterised model, or to mislabelling of the training data.

3.2 Hidden Markov models

There are a number of extensions to classical Markov chains, which we will come back to later in the chapter. Here, however, we will proceed immediately to hidden Markov models. We will motivate this by turning to the second of the two questions posed initially for CpG islands: How do we find them in a long unannotated sequence? The Markov chain models that we have just built could be used for this purpose, by calculating the log-odds score for a window of, say, 100 nucleotides around every nucleotide in the sequence and plotting it. We would

---

[1] Base 2 logarithms were used, in which case the unit is called a bit. See Chapter 11.

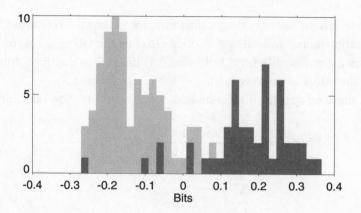

Figure 3.2 *The histogram of the length-normalised scores for all the sequences. CpG islands are shown with dark grey and non-CpG with light grey.*

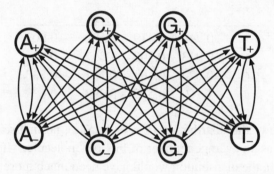

Figure 3.3 *An HMM for CpG islands. In addition to the transitions shown, there is also a complete set of transitions within each set, as in the earlier simple Markov chains.*

then expect CpG islands to stand out with positive values. However, this is somewhat unsatisfactory if we believe that in fact CpG islands have sharp boundaries, and are of variable length. Why use a window size of 100? A more satisfactory approach is to build a single model for the entire sequence that incorporates both Markov chains.

To simulate in one model the 'islands' in a 'sea' of non-island genomic sequence, we want to have both the Markov chains of the last section present in the same model, with a small probability of switching from one chain to the other at each transition point. However, this introduces the complication that we now have two states corresponding to each nucleotide symbol. We resolve this by re-labelling the states. We now have A_+, C_+, G_+ and T_+ which emit A, C, G and T respectively in CpG island regions, and A_-, C_-, G_- and T_- correspondingly in non-island regions; see Figure 3.3.

The transition probabilities in this model are set so that within each group they are close to the transition probabilities of the original component model, but there is also a small but finite chance of switching into the other component. Overall there is more chance of switching from '+' to '−' than vice versa, so if left to run free, the model will spend more of its time in the '−' non-island states than in the island states.

The relabelling is the critical step. The essential difference between a Markov chain and a hidden Markov model is that for a hidden Markov model there is not a one-to-one correspondence between the states and the symbols. It is no longer possible to tell what state the model was in when x_i was generated just by looking at x_i. In our example there is no way to tell by looking at a single symbol C in isolation whether it was emitted by state C_+ or state C_-

Formal definition of an HMM

Let us formalise the notation for hidden Markov models, and derive the probability of a particular sequence of states and symbols. We now need to distinguish the sequence of states from the sequence of symbols. Let us call the state sequence the *path*, π. The path itself follows a simple Markov chain, so the probability of a state depends only on the previous state. The ith state in the path is called π_i. The chain is characterised by parameters

$$a_{kl} = P(\pi_i = l | \pi_{i-1} = k). \tag{3.4}$$

To model the beginning of the process we introduce a begin state, as was introduced earlier to model the beginning of sequences in Markov chains (Figure 3.1). The transition probability a_{0k} from this begin state to state k can be thought of as the probability of starting in state k. It is also possible to model ends as before by always ending a state sequence with a transition into an end state. For convenience we label both begin and end states as 0 (there is no conflict here because you can only transit out of the begin state, and only into the end state, so variables are not used more than once).

Because we have decoupled the symbols b from the states k, we must introduce a new set of parameters for the model, $e_k(b)$. For our CpG model each state is associated with a single symbol, but this is not a requirement; in general a state can produce a symbol from a distribution over all possible symbols. We therefore define

$$e_k(b) = P(x_i = b | \pi_i = k), \tag{3.5}$$

the probability that symbol b is seen when in state k. These are known as the *emission* probabilities.

For our CpG island model the emission probabilities are all 0 or 1. To illustrate emission probabilities we reintroduce here the casino example from Chapter 1.

Example: The occasionally dishonest casino, part 1

Let us consider an example from Chapter 1. In a casino they use a fair die most of the time, but occasionally they switch to a loaded die. The loaded die has probability 0.5 of a six and probability 0.1 for the numbers one to five. Assume that the casino switches from a fair to a loaded die with probability 0.05 before each roll, and that the probability of switching back is 0.1. Then the switch between dice is a Markov process. In each state of the Markov process the outcomes of a roll have different probabilities, and thus the whole processs is an example of a hidden Markov model. We can draw it like this:

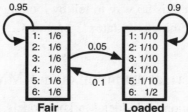

where the emission probabilities $e()$ are shown in the state boxes.   □

What is hidden in the above model? If you can just see a sequence of rolls (the sequence of observations) you do not know which rolls used a loaded die and which used a fair one, because that is kept secret by the casino; that is, the *state sequence is hidden*. In a Markov chain you always know exactly in which state a given observation belongs. Obviously the casino wouldn't tell you that they use loaded dice and what the various probabilities are. Yet for this more complicated situation, which we will return to later, it is possible to estimate the probabilities in the above HMM (once you have a suspicion that they use two different dice).

The reason for the name *emission* probabilities is that it is often convenient to think of HMMs as generative models, that generate or emit sequences. For instance we can generate random sequences of rolls from the model of the fair/-loaded dice above by simulating the successive choices of die, then rolls of the chosen die. More generally a sequence can be generated from an HMM as follows: First a state π_1 is chosen according to the probabilities a_{0i}. In that state an observation is emitted according to the distribution e_{π_1} for that state. Then a new state π_2 is chosen according to the transition probabilities $a_{\pi_1 i}$ and so forth. This way a sequence of random, artificial observations are generated. Therefore, we will sometimes say things like $P(x)$ is the probability that x was *generated* by the model.

It is now easy to write down the joint probability of an observed sequence x and a state sequence π:

$$P(x,\pi) = a_{0\pi_1} \prod_{i=1}^{L} e_{\pi_i}(x_i) a_{\pi_i \pi_{i+1}}, \qquad (3.6)$$

where we require $\pi_{L+1} = 0$. For example, the probability of sequence CGCG being emitted by the state sequence (C_+, G_-, C_-, G_+) in our model is

$$a_{0,C_+} \times 1 \times a_{C_+,G_-} \times 1 \times a_{G_-,C_-} \times 1 \times a_{C_+,G_+} \times 1 \times a_{C_-,0}.$$

Equation (3.6) is the HMM analogue of equation (3.2). However, it is not so useful in practice because in general we do not know the path. In the following sections we describe how to estimate the path, either by finding the most likely one, or alternatively by using an *a posteriori* distribution over states. Then we go on to show how to estimate the parameters for an HMM.

Most probable state path: the Viterbi algorithm

Although it is no longer possible to tell what state the system is in by looking at the corresponding symbol, it is often the sequence of underlying states that we are interested in. To find out what the observation sequence 'means' by considering the underlying states is called *decoding* in the jargon of speech recognition. There are several approaches to decoding. Here we will describe the most common one, called the Viterbi algorithm. It is a dynamic programming algorithm closely related to the ones covered in Chapter 2.

In general there may now be many state sequences that could give rise to any particular sequence of symbols. For example, in our CpG model the state sequences (C_+, G_+, C_+, G_+), (C_-, G_-, C_-, G_-) and (C_+, G_-, C_+, G_-) would all generate the symbol sequence CGCG. However, they do so with very different probabilities. The third is the product of multiple small probabilities of switching back and forth between the components, and hence is much smaller than the first two. The second is itself significantly smaller than the first because it contains two C to G transitions which are significantly less probable in the '$-$' component than in the '$+$' component. Of these three choices, therefore, it is most likely that the sequence CGCG came from a set of '$+$' states.

A predicted path through the HMM will tell us which part of the sequence is predicted as a CpG island, because we assumed above that each state was assigned to model either CpG islands or other regions. If we are to choose just one path for our prediction, perhaps the one with the highest probability should be chosen,

$$\pi^* = \underset{\pi}{\mathrm{argmax}} \, P(x, \pi). \tag{3.7}$$

The most probable path π^* can be found recursively. Suppose the probability $v_k(i)$ of the most probable path ending in state k with observation i is known for all the states k. Then these probabilities can be calculated for observation x_{i+1} as

$$v_l(i+1) = e_l(x_{i+1}) \max_k (v_k(i) a_{kl}). \tag{3.8}$$

All sequences have to start in state 0 (the begin state), so the initial condition is

v		C	G	C	G
$\mathcal{B}$	1	0	0	0	0
A_+	0	0	0	0	0
C_+	0	**0.13**	0	0.012	0
G_+	0	0	**0.034**	0	0.0032
T_+	0	0	0	0	0
A_-	0	0	0	0	0
C_-	0	0.13	0	0.0026	0
G_-	0	0	0.010	0	0.00021
T_-	0	0	0	0	0

Figure 3.4 *For the model of CpG islands shown in Figure 3.3 and the sequence* CGCG, *this is the resulting table of v. The most probable path is shown with bold face.*

that $v_0(0) = 1$. By keeping pointers backwards, the actual state sequence can be found by backtracking. The full algorithm is:

Algorithm: Viterbi

Initialisation ($i = 0$): $v_0(0) = 1$, $v_k(0) = 0$ for $k > 0$.

Recursion ($i = 1 \ldots L$): $v_l(i) = e_l(x_i) \max_k (v_k(i-1) a_{kl})$;
$$\text{ptr}_i(l) = \text{argmax}_k (v_k(i-1) a_{kl}).$$

Termination: $P(x, \pi^*) = \max_k (v_k(L) a_{k0})$;
$$\pi_L^* = \text{argmax}_k (v_k(L) a_{k0}).$$

Traceback ($i = L \ldots 1$): $\pi_{i-1}^* = \text{ptr}_i(\pi_i^*)$.   ◁

Note that an end state is assumed, which is the reason for a_{k0} in the termination step. If ends are not modelled, this a will disappear.

There are some implementational issues both for the Viterbi algorithm and the algorithms described later. The most severe practical problem is that multiplying many probabilities always yields very small numbers that will give underflow errors on any computer. For this reason the Viterbi algorithm should always be done in log space, i.e. calculating $\log(v_l(i))$, which will make the products become sums and the numbers stay reasonable. This is discussed in Section 3.6.

Figure 3.4 shows the full table of values of v for the sequence CGCG and the CpG island model. When we apply the same algorithm to a longer sequence the derived optimal path π^* will switch between the '+' and the '−' components of the model, and thereby give the precise boundaries of the predicted CpG island regions.

Example: The occasionally dishonest casino, part 2

For a sequence of dice rolls we can now find the most probable path through the model shown on p. 54. A total of 300 random rolls were generated from

```
Rolls    315116246446644245311321631164152133625144543631656626566666
Die      FFFFFFFFFFFFFFFFFFFFFFFFFFFFFFFFFFFFFFFFFFFFFFLLLLLLLLLLLLLLLL
Viterbi  FFFFFFFFFFFFFFFFFFFFFFFFFFFFFFFFFFFFFFFFFFFFFFFFLLLLLLLLLLLLLL

Rolls    651166453132651245636666463163666316232264552362666666625151631
Die      LLLLLLFFFFFFFFFFFFFFLLLLLLLLLLLLLLLLFFFLLLLLLLLLLLLLLLLFFFFFFFFF
Viterbi  LLLLLLFFFFFFFFFFFFFFLLLLLLLLLLLLLLLLLLLLLLLLLLLLLLLLLLFFFFFFFFF

Rolls    222555441666566563564324364131513465146353411126414626253356
Die      FFFFFFFFLLLLLLLLLLLLLLFFFFFFFFFFFFFFFFFFFFFFFFFFFFFFFFFFFFFFLL
Viterbi  FFFFFFFFFFFFFFFFFFFFFFFFFFFFFFFFFFFFFFFFFFFFFFFFFFFFFFFFFFFFFL

Rolls    366163666466232534413661661163252562462255265252266435353336
Die      LLLLLLLLFFFFFFFFFFFFFFFFFFFFFFFFFFFFFFFFFFFFFFFFFFFFFFFFFFFFFF
Viterbi  LLLLLLLLLLLLFFFFFFFFFFFFFFFFFFFFFFFFFFFFFFFFFFFFFFFFFFFFFFFFFF

Rolls    233121625364414432335163243633665562466662632666612355245242
Die      FFFFFFFFFFFFFFFFFFFFFFFFFFFFFFFFLLLLLLLLLLLLLLLLLLLLLFFFFFFFFFF
Viterbi  FFFFFFFFFFFFFFFFFFFFFFFFFFFFFFFFFLLLLLLLLLLLLLLLLLLLLLFFFFFFFFFF
```

Figure 3.5 *The numbers show 300 rolls of a die as described in the example. Below is shown which die was actually used for that roll (F for fair and L for loaded). Under that the prediction by the Viterbi algorithm is shown.*

the model as described earlier. Each roll was generated either with the fair die (F) or the loaded one (L), as shown below the outcome of the roll in Figure 3.5. The Viterbi algorithm was used to predict the state sequence, i.e. which die was used for each of the rolls. Generally, as you can see, the Viterbi algorithm has recovered the state sequence fairly well. □

Exercise

3.4   Show that $\pi^* = \underset{\pi}{\mathrm{argmax}}\, P(\pi|x)$ is equivalent to (3.7).

The forward algorithm

For Markov chains we calculated the probability of a sequence, $P(x)$, with equation (3.2). The resulting values were used to distinguish between CpG islands and other DNA for instance. We want to be able to calculate this probability for an HMM as well. Because many different state paths can give rise to the same sequence x, we must add the probabilities for all possible paths to obtain the full probability of x,

$$P(x) = \sum_{\pi} P(x, \pi). \tag{3.9}$$

The number of possible paths π increases exponentially with the length of the sequence, so brute force evaluation of (3.9) by enumerating all paths is not practical. One approach is to use equation (3.6) evaluated at the most probable path π^* obtained in the last section as an approximation to $P(x)$. This implicitly assumes that the only path with significant probability is π^*, a somewhat startling assumption which however in many cases is surprisingly good. In fact the approximation is unnecessary, because the full probability can itself be calculated by a similar dynamic programming procedure to the Viterbi algorithm, replacing the maximisation steps with sums. This is called the *forward* algorithm.

The quantity corresponding to the Viterbi variable $v_k(i)$ in the forward algorithm is

$$f_k(i) = P(x_1 \ldots x_i, \pi_i = k), \tag{3.10}$$

which is the probability of the observed sequence up to and including x_i, requiring that $\pi_i = k$. The recursion equation is

$$f_l(i+1) = e_l(x_{i+1}) \sum_k f_k(i) a_{kl}. \tag{3.11}$$

The full algorithm is:

Algorithm: Forward algorithm

Initialisation $(i = 0)$: $f_0(0) = 1$, $f_k(0) = 0$ for $k > 0$.

Recursion $(i = 1 \ldots L)$: $f_l(i) = e_l(x_i) \sum_k f_k(i-1) a_{kl}.$

Termination: $P(x) = \sum_k f_k(L) a_{k0}.$ ◁

Like the Viterbi algorithm, the forward algorithm (and the backward algorithm in the next section) can give underflow errors when implemented on a computer. Again this can be solved by working in log space, although not as elegantly as for Viterbi. Alternatively a scaling method can be used. Both approaches are described in Section 3.6.

As well as their use in the forward algorithm, the quantities $f_k(i)$ have a number of other uses, including those described in the next two sections.

The backward algorithm and posterior state probabilities

The Viterbi algorithm finds the most probable path through the model, but as we remarked at the time, this may not always be the most appropriate basis for further inference about the sequence. We might for instance want to know what the most probable state is for an observation x_i. More generally, we may want the probability that observation x_i came from state k given the observed sequence, i.e. $P(\pi_i = k|x)$. This is the posterior probability of state k at time i when the emitted sequence is known.

Our approach to the posterior probability is a little indirect. We first calculate the probability of producing the entire observed sequence with the ith symbol being produced by state k:

$$
\begin{aligned}
P(x, \pi_i = k) &= P(x_1 \ldots x_i, \pi_i = k) P(x_{i+1} \ldots x_L | x_1 \ldots x_i, \pi_i = k) \\
&= P(x_1 \ldots x_i, \pi_i = k) P(x_{i+1} \ldots x_L | \pi_i = k), \quad (3.12)
\end{aligned}
$$

the second row following because everything after k only depends on the state at k. The first term in this is recognised as $f_k(i)$ from (3.10) that was calculated by the forward algorithm of the previous section. The second term is called $b_k(i)$,

$$
b_k(i) = P(x_{i+1} \ldots x_L | \pi_i = k). \quad (3.13)
$$

It is analogous to the forward variable, but instead obtained by a backward recursion starting at the end of the sequence:

Algorithm: Backward algorithm

Initialisation $(i = L)$: $b_k(L) = a_{k0}$ for all k.

Recursion $(i = L - 1, \ldots, 1)$: $b_k(i) = \sum_l a_{kl} e_l(x_{i+1}) b_l(i + 1)$.

Termination: $P(x) = \sum_l a_{0l} e_l(x_1) b_l(1)$. ◁

The termination step is rarely needed, because $P(x)$ is usually found by the forward algorithm, and it is just shown for completeness.

Equation (3.12) can now be written as $P(x, \pi_i = k) = f_k(i) b_k(i)$, and from it we obtain the required posterior probabilities by straightforward conditioning,

$$
P(\pi_i = k | x) = \frac{f_k(i) b_k(i)}{P(x)}, \quad (3.14)
$$

where $P(x)$ is the result of the forward (or backward) calculation.

Example: The occasionally dishonest casino, part 3

In Figure 3.6 the posterior probability for the die being fair is shown for the sequence of rolls shown in Figure 3.5. Notice that the posterior probability does not reflect which die was actually used in some places. This is to be expected, simply because a misleading sequence of rolls can occur at random. □

Posterior decoding

A major use of the $P(\pi_i = k | x)$ is for two alternative forms of decoding in addition to the Viterbi decoding we introduced in the previous section. These are particularly useful when many different paths have almost the same probability as the most probable one, because then it is not well justified to consider only the most probable path.

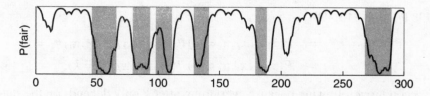

Figure 3.6 *The posterior probability of being in the state corresponding to the fair die in the casino example. The x axis shows the number of the roll. The shaded areas show when the roll was generated by the loaded die.*

The first approach is to define a state sequence $\hat{\pi}_i$ that can be used in place of π_i^*,

$$\hat{\pi}_i = \underset{k}{\text{argmax}}\, P(\pi_i = k | x). \qquad (3.15)$$

As suggested by its definition, this state sequence may be more appropriate when we are interested in the state assignment at a particular point i, rather than the complete path. In fact, the state sequence defined by $\hat{\pi}_i$ may not be particularly likely as a path through the entire model; it may even not be a legitimate path at all if some transitions are not permitted, which is normally the case.

The second, and perhaps more important, new decoding approach arises when it is not the state sequence itself which is of interest, but some other property derived from it. Assume we have a function $g(k)$ defined on the states. The natural value to look at then is

$$G(i|x) = \sum_k P(\pi_i = k | x) g(k). \qquad (3.16)$$

An important special case of this is where $g(k)$ takes the value 1 for a subset of the states and 0 for the rest. In this case, $G(i|x)$ is the posterior probability of the symbol i coming from a state in the specified set. For example, with our CpG island model, what really concerns us is whether a base is part of an island or not. For this purpose we want to define $g(k) = 1$ for $k \in \{A_+, C_+, G_+, T_+\}$ and $g(k) = 0$ for $k \in \{A_-, C_-, G_-, T_-\}$. Then $G(i|x)$ is precisely the posterior probability according to the model that base i is in a CpG island.

In the case where we have a labelling of the states defining a partition of them (as we in fact have with the CpG island model, labelling them as '+' or '−') it is possible to use (3.16) to find the most probable label at each position of the sequence. This is not quite the most probable global labelling of a given sequence. That, however, is not entirely straightforward. See Schwartz & Chow [1990] and Krogh [1997b] for further discussion of this.

Example: Prediction of CpG islands

Now CpG islands can be predicted from our model. By the Viterbi algorithm we can find the most probable path through the model. When this path goes through the + states, a CpG island is predicted. For the set of 41 sequences, each with a putative CpG island, all the islands are found except for two (false negatives), and 121 new ones are predicted (false positives). The real CpG islands are quite long (of the order of 1000 bases), whereas the predicted ones are short, and a CpG island is usually predicted as several short ones. By applying the two simple post-processing steps (1) concatenate predictions less than 500 bases apart (2) discard predictions shorter than 500, the number of false positives are reduced to 67.

Using posterior decoding, the same two CpG islands are missed and 236 false positives are predicted. Using the same post-processing as above this number is reduced to 83. For this problem, there is not a big difference between the two methods, except that the posterior decoding predicts even more very short islands. It is possible that some of the false positives are real CpG islands. The two false negatives are perhaps wrongly labelled, but it is also possible that a more sophisticated model is needed for capturing all the features of these signals.

□

Example: The occasionally dishonest casino, part 4

The model for the casino is changed, so there is only a probability of 0.01 for switching from fair to loaded. Obviously the probability of staying with the fair die must then be 0.99, but all other probabilities are unchanged. From this model 1000 random rolls are generated. From these rolls the most probable path found by the Viterbi algorithm never visits the loaded die state. In Figure 3.7 the posterior probability for the dice being fair is shown for these rolls. Although not perfect, posterior decoding would predict something reasonably close to the truth.

□

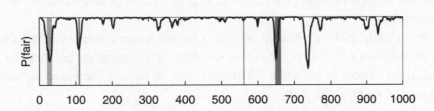

Figure 3.7 *The posterior probability of the die being fair, but using probability 0.01 for switching to the loaded die (cf. Figure 3.6).*

3.3 Parameter estimation for HMMs

Probably the most difficult problem faced when using HMMs is that of specifying the model in the first place. There are two parts to this: the design of the structure, i.e. what states there are and how they are connected, and the assignment of parameter values, the transition and emission probabilities a_{kl} and $e_k(b)$.

In this section we will discuss the parameter estimation problem, for which there is a well-developed theory. In the next section we will consider model structure design, which is more of an art.

The framework in which we will be working is to assume that we have a set of example sequences of the type that we want the model to fit well, known as *training* sequences. Let these be $x^1, \ldots, x^n$. We assume that they are independent, and thus that the joint probability of all the sequences given a particular assignment of parameters is the product of the probabilities of the individual sequences. In fact, we work in log space, and so with the log probability of the sequences,

$$l(x^1, \ldots, x^n | \theta) = \log P(x^1, \ldots, x^n | \theta) = \sum_{j=1}^{n} \log P(x^j | \theta), \qquad (3.17)$$

where θ represents the entire current set of values of the parameters in the model (all the as and es). This is equal to the log likelihood of the model; see Chapter 11.

Estimation when the state sequence is known

Just as it was easier to write down the probability of a sequence when the path was known, so it is easier to estimate the probability parameters when the paths are known for all the examples. Frequently this is the case. An example would be if we were given a set of genomic sequences in which the CpG islands were already labelled, based on experimental data. Other examples would be for an HMM that predicted secondary structure, with training sequences obtained from the set of proteins with known structures, or for an HMM predicting genes from genomic sequences, where the transcript structure has been determined by cDNA sequencing.

When all the paths are known, we can count the number of times each particular transition or emission is used in the set of training sequences. Let these be A_{kl} and $E_k(b)$. Then, as shown in Chapter 11, the maximum likelihood estimators for a_{kl} and $e_k(b)$ are given by

$$a_{kl} = \frac{A_{kl}}{\sum_{l'} A_{kl'}} \quad \text{and} \quad e_k(b) = \frac{E_k(b)}{\sum_{b'} E_k(b')}. \qquad (3.18)$$

The estimation equation for a_{kl} is exactly the same as for a simple Markov chain.

As always, maximum likelihood estimators are vulnerable to overfitting if there are insufficient data. Indeed if there is a state k that is never used in the set of example sequences, then the estimation equations are undefined for that state, because both the numerator and denominator will have value zero. To avoid such problems it is preferable to add predetermined pseudocounts to the A_{kl} and $E_k(b)$ before using (3.18).

$$A_{kl} \quad = \quad \text{number of transitions } k \text{ to } l \text{ in training data} + r_{kl},$$

$$E_k(b) \quad = \quad \text{number of emissions of } b \text{ from } k \text{ in training data} + r_k(b).$$

The pseudocounts r_{kl} and $r_k(b)$ should reflect our prior biases about the probability values. In fact they have a natural probabilistic interpretation as the parameters of Bayesian Dirichlet prior distributions on the probabilities for each state (see Chapter 11). They must be positive, but do not need to be integers. Small total values $\sum_{l'} r_{kl'}$ or $\sum_{b'} r_k(b')$ indicate weak prior knowledge, whereas larger total values indicate more definite prior knowledge, which requires more data to modify it.

Estimation when paths are unknown: Baum–Welch and Viterbi training

When the paths are unknown for the training sequences, there is no longer a direct closed-form equation for the estimated parameter values, and some form of iterative procedure must be used. All the standard algorithms for optimisation of continuous functions can be used; see for example Press *et al.* [1992]. However, there is a particular iteration method that is standardly used, known as the Baum–Welch algorithm [Baum 1972]. This has a natural probabilistic interpretation. Informally, it first estimates the A_{kl} and $E_k(b)$ by considering probable paths for the training sequences using the current values of a_{kl} and $e_k(b)$. Then (3.18) is used to derive new values of the as and es. This process is iterated until some stopping criterion is reached.

It is possible to show that the overall log likelihood of the model is increased by the iteration, and hence that the process will converge to a local maximum. Unfortunately, there are usually many local maxima, and which one you end up with depends strongly on the starting values of the parameters. The problem of local maxima is particularly severe when estimating large HMMs, and later we will discuss various ways to help deal with it.

More formally, the Baum–Welch algorithm calculates A_{kl} and $E_k(b)$ as the *expected* number of times each transition or emission is used, given the training sequences. To do this it uses the same forward and backward values as the posterior probability decoding method. The probability that a_{kl} is used at position i in sequence x is (see Exercise 3.5)

$$P(\pi_i = k, \pi_{i+1} = l | x, \theta) = \frac{f_k(i)a_{kl}e_l(x_{i+1})b_l(i+1)}{P(x)}. \tag{3.19}$$

From this we can derive the expected number of times that a_{kl} is used by summing over all positions and over all training sequences,

$$A_{kl} = \sum_j \frac{1}{P(x^j)} \sum_i f_k^j(i)a_{kl}e_l(x_{i+1}^j)b_l^j(i+1), \tag{3.20}$$

where $f_k^j(i)$ is the forward variable $f_k(i)$ defined in (3.10) calculated for sequence

j, and $b_i^j(i)$ is the corresponding backward variable. Similarly, we can find the expected number of times that letter b appears in state k,

$$E_k(b) = \sum_j \frac{1}{P(x^j)} \sum_{\{i|x_i^j=b\}} f_k^j(i)b_k^j(i), \tag{3.21}$$

where the inner sum is only over those positions i for which the symbol emitted is b.

Having calculated these expectations, the new model parameters are calculated just as before using (3.18). We can iterate using the new values of the parameters to obtain new values of the As and Es as before, but in this case we are converging in a continuous-valued space, and so will never in fact reach the maximum. It is therefore necessary to set a convergence criterion, typically stopping when the change in total log likelihood is sufficiently small. Other stop criteria than the log likelihood change can be used for the iteration. For instance the log likelihood can be normalised by the number of sequences n and maybe also by the sequence lengths, so that you consider the change in the average log likelihood per residue. We can summarise the Baum–Welch algorithm like this:

Algorithm: Baum–Welch

Initialisation: Pick arbitrary model parameters.
Recurrence:
    Set all the A and E variables to their pseudocount values r (or to zero).
    For each sequence $j = 1 \ldots n$:
        Calculate $f_k(i)$ for sequence j using the forward algorithm (p. 58).
        Calculate $b_k(i)$ for sequence j using the backward algorithm (p. 59).
        Add the contribution of sequence j to A (3.20) and E (3.21).
    Calculate the new model parameters using (3.18).
    Calculate the new log likelihood of the model.
Termination:
    Stop if the change in log likelihood is less than some predefined threshold
    or the maximum number of iterations is exceeded.   ◁

As indicated here, it is normal to add pseudocounts to the A and E values just as in the case where the state paths are known. This works well, but the normal Bayesian interpretation in terms of Dirichlet priors does not carry through rigorously in this case; see Chapter 11.

The Baum–Welch algorithm is a special case of a very powerful general approach to probabilistic parameter estimation called the EM algorithm. This algorithm and the derivation of Baum–Welch is given in Section 11.6 of Chapter 11.

An alternative to the Baum–Welch algorithm is frequently used, which we will call Viterbi training. In this approach, the most probable paths for the training sequences are derived using the Viterbi algorithm given above, and these are used in the re-estimation process given in the previous section. Again, the process is iterated when the new parameter values are obtained. In this case the algorithm converges precisely, because the assignment of paths is a discrete process, and we can continue until none of the paths change. At this point the parameter estimates will not change either, because they are determined completely by the paths. Unlike Baum–Welch, this procedure does not maximise the true likelihood, i.e. $P(x^1, \ldots, x^n | \theta)$ regarded as a function of of the model parameters θ. Instead, it finds the value of θ that maximises the contribution to the likelihood $P(x^1, \ldots, x^n | \theta, \pi^*(x^1), \ldots, \pi^*(x^n))$ from the most probable paths for all the sequences. Probably for this reason, Viterbi training performs less well in general than Baum–Welch. However, it is widely used, and it can be argued that when the primary use of the HMM is to produce decodings via Viterbi alignments, then it is good to train using them.

Example: The occasionally dishonest casino, part 5

We are suspicious that a casino is operated as described in the example on p. 54, but we do not know for certain. Night after night we collect data by simply observing rolls. When we have enough, we want to estimate a model. Assume the data we collected were the 300 rolls shown in Figure 3.5. From this sequence of observations a model was estimated by the Baum–Welch algorithm. Initially all the probabilities were set to random numbers. Here are diagrams of the model that generated the data (identical to the one in the example on p. 54) and the estimated model.

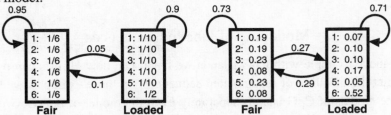

You can see they are fairly similar, although the estimated transition probabilities are quite different from the real ones. This is partly a problem of local minima, and by trying more times it is actually possible to obtain a model closer to the correct one. However, from a limited amount of data it is never possible to estimate the parameters exactly.

To illustrate the last point, 30 000 random rolls were generated (data are not shown!), and a model was estimated. This came very close to the correct one:

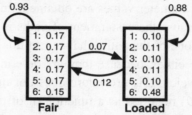

To see how good these models are compared to just assuming a fair die all the time, the log-odds per roll was calculated using the 300 observations for the three models:

The correct model	0.101 bits
Model estimated from 300 rolls	0.097 bits
Model estimated from 30 000 rolls	0.100 bits

The worst model estimated from 300 rolls has almost the same log-odds as the two other models. That is because it is being tested on the same data as it was estimated from. Testing it on an independent set of rolls yields significantly lower log-odds than the other two models.   □

Exercises

3.5   Derive the result (3.19). Use the fact that

$$P(\pi_i = k, \pi_{i+1} = l | x, \theta) = \frac{1}{P(x|\theta)} P(x, \pi_i = k, \pi_{i+1} = l | \theta),$$

and that this again can be written in terms of $P(x_1, \ldots, x_i, \pi_i = k | \theta)$ and

$$P(x_{i+1}, \ldots, x_L, \pi_{i+1} = l | x_1, \ldots, x_i, \theta, \pi_i = k)$$
$$= P(x_{i+1}, \ldots, x_L, \pi_{i+1} = l | \theta, \pi_i = k).$$

3.6   Derive (3.21).

Modelling of labelled sequences

In the above example with CpG islands we have seen how HMMs can be used to predict the labelling of unannotated sequences. In these examples we had to train the models of CpG islands separately from the model of non-CpG islands and then combine them into a larger HMM afterwards. This separate estimation can be quite tedious, especially if there are more than two different classes involved. Also, if the transitions between the submodels are ambiguous, so for instance a given sequence can use more than one transition from the CpG submodel to the other submodel, then the estimation of the transitions is not a simple counting problem. There is, however, a more straightforward method to estimate everything at once, which we will describe now.

Sequence	x_1	x_2	x_3	x_4	x_5	x_6	x_7	x_8	x_9	x_{10} ...
Labels	−	−	−	+	+	+	+	−	−	− ...

State					
1 −	f			f	
2 −	calculated		$f=0$	calculated	
3 −	as usual			as usual	
4 −					
5 +		f			
6 +	$f=0$	calculated		$f=0$	
7 +		as usual			
8 +					

Figure 3.8 *The forward table for a model with four states labelled + and four labelled −. Each column corresponds to an observation and each row to a state of the model. The first ten residues shown, $x_1,\ldots,x_{10}$, are assumed to be labelled − − − + + + + − − −.*

The starting point is the combined model of all the classes, where we have assigned a class label to each state. To model CpG islands the natural labels are '+' for the island states and '−' for the non-island states. We also have labels on the observations $x = x_1,\ldots,x_L$, which we we call $y = y_1,\ldots,y_L$. The y_i is '+' if x_i is part of a CpG island and '−' otherwise. In the Baum–Welch algorithm (or the Viterbi alternative) we now only allow *valid* paths through the model when calculating the fs and bs. A valid path is one where the state labels and sequence labels are the same, i.e., π_i has label y_i. During the forward and backward algorithms this corresponds to setting $f_l(i) = 0$ and $b_l(i) = 0$ for all the states l with a label different from y_i (see Figure 3.8).

Discriminative estimation

Unless there are ambiguous transitions between submodels, the above estimation procedure gives the same result as if the submodels were estimated separately by the Baum–Welch algorithm and then combined with appropriate transitions afterwards. This actually corresponds to maximising the likelihood

$$\theta^{ML} = \underset{\theta}{\operatorname{argmax}}\, P(x,y|\theta).$$

Usually our primary interest is in obtaining good predictions of y, so it is preferable to maximise $P(y|x,\theta)$ instead. This is called *conditional maximum likelihood* (CML),

$$\theta^{CML} = \underset{\theta}{\operatorname{argmax}}\, P(y|x,\theta); \tag{3.22}$$

see for example Juang & Rabiner [1991] and Krogh [1994]. A related criterion is called *maximum mutual information* or MMI [Bahl *et al.* 1986].

The likelihood $P(y|x,\theta)$ can be rewritten as

$$P(y|x,\theta) = \frac{P(x,y|\theta)}{P(x|\theta)},$$

where $P(x, y|\theta)$ is the probability calculated by the forward algorithm for labelled sequences described above, and $P(x|\theta)$ is the probability calculated by the standard forward algorithm disregarding all the labels. There is no EM algorithm for optimising this likelihood, and the estimation becomes more complex; see for example Normandin & Morgera [1991] and the references above.

3.4 HMM model structure

Choice of model topology

So far we have assumed that transitions are possible from any state to any other state. Although it is tempting to start with a fully connected model, i.e. one in which all transitions are allowed, and 'let the model find out for itself' which transitions to use, it almost never works in practice. For problems of any realistic size it will usually lead to very bad models, even with plenty of training data. Here the problem is not over fitting, but local maxima. The less constrained the model is, the more severe the local maximum problem becomes. There are methods that attempt to adapt the model topology based on the data by adding and removing transitions and states [Stolcke & Omohundro 1993; Fujiwara, Asogawa & Konagaya 1994]. However, in practice successful HMMs are constructed by carefully deciding which transitions are to be allowed in the model, based on knowledge about the problem under investigation.

To disable the transition from state k to state l corresponds to setting $a_{kl} = 0$. If we use Baum–Welch estimation (or the Viterbi approximation) then a_{kl} will still be zero after the re-estimation process, because when the probability is zero the expected number of transitions from k to l will also be zero. Therefore all the mathematics is unchanged even if not all transitions are possible.

We should choose a model which has an interpretation in terms of our knowledge of the problem. For instance, to model CpG islands it was important that the model was capable of giving a different probability to a CG dinucleotide in the island states from in the non-island states, because that was expected to be the main determinator for CpG islands.

Duration modelling

When modelling a phenomenon where for instance the nucleotide distribution does not change for a certain length of DNA, the simplest model design is to make a state with a transition to itself with probability p. We did this with both our CpG island and our dishonest casino example. After entering the state there is a probability $1 - p$ of leaving it, so the probability of staying in the state for l

residues is

$$P(l \text{ residues}) = (1-p)p^{l-1}. \tag{3.23}$$

(The emission probabilities are disregarded.) This exponentially decaying distribution on lengths (called a geometric distribution) can be inappropriate in some applications, where the distribution of lengths is important and significantly different from exponential. More complex length distributions can be modelled by introducing several states with the same distribution over residues and transitions between each other. For instance a (sub-) model like this:

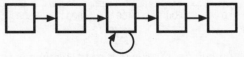

will give sequences of a minimum length of 5 residues and an exponentially decaying distribution over longer sequences. Similarly, a model like this:

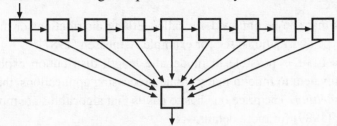

can model *any* distribution of lengths between 2 and 10.

A more subtle way of obtaining a non-geometric length distribution is to use an array of n states, each with a transition to itself of probability p and a transition to the next of probability $1 - p$:

Obviously the smallest sequence length such a model can capture is n. For any given path of length l through the model, the probability of all its transitions is $p^{l-n}(1-p)^n$ (we are disregarding emission probabilities for now, as above). The number of possible paths through the states is $\binom{l-1}{n-1}$, so the total probability summed over all possible paths is

$$P(l) = \binom{l-1}{n-1}p^{l-n}(1-p)^n. \tag{3.24}$$

This distribution is called a negative binomial and it is shown in Figure 3.9 for $p = 0.99$ and $n \leq 5$. For small lengths the number of paths through the model grows faster than the geometrical distribution decays, and therefore the distribution becomes bell-shaped. The number of paths depends on the model topology, and it is possible to make more general models where the number of paths has a different dependence on n and l. For continuous Markov processes the types of

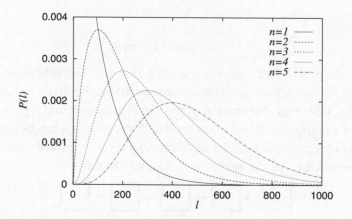

Figure 3.9 *The probability distribution over lengths for models with p =
0.99 and n identical states, with n ranging from* 1 *to* 5.

distributions that can be obtained are called Erlang distributions or more generally phase-type distributions, see for example Asmussen [1987].

Alternatively, it is possible to model the length distribution explicitly. As
length is equivalent to time in many signal processing applications, this is called
duration modelling. The price one has to pay is that algorithms are much slower.
See Rabiner [1989] for more details.

Silent states

We have already seen examples of states that do not emit symbols in an HMM,
the begin and end states. Such states are called *silent states* or *null states*, and
they can also be useful in other places in an HMM. In Chapter 5 we will see an
example where all states in a chain of states need to be connected to all states later
in the chain. The length of such a chain is often 200 states or more, and connecting them appropriately with transitions would require roughly 20 000 transition
probabilities (assuming 200 states). This number is too large to be reliably estimated from realistic datasets. Instead, by using silent states, we can get away
with around 600 transitions.

The situation is as follows: to allow for arbitrary deletions a chain of states
needs to be completely 'forward connected'.

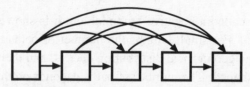

Instead we can connect all the states to a parallel chain of silent states, represented
here by circles.

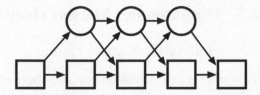

Because the silent states do not emit any letters, it is possible to get from any 'real' state to any later 'real' state without emitting any letters.

A price is paid for the reduction in the number of parameters. The fully connected model can have for instance high probability transitions from state 1 to state 5 and from state 2 to state 4, but low probability ones for transitions 1 to 4 and 2 to 5. This would not be possible with the model using silent states.

So long as there are no loops consisting entirely of silent states, it is easy to extend all the HMM algorithms to incorporate them. The condition that there are no loops mean that the states can be numbered so that any transition between silent states goes from a lower to a higher numbered state. For the forward algorithm, the change is as follows:

(i) For all 'real' states l, calculate $f_l(i+1)$ as before from $f_k(i)$ for states k.
(ii) For any silent state l, add $\sum_k f_k(i+1)a_{kl}$ to $f_l(i+1)$ for 'real' states k.
(iii) Starting from the lowest numbered silent state l add $\sum_k f_k(i+1)a_{kl}$ to $f_l(i+1)$ for all silent states $k < l$.

The change to the Viterbi algorithm is exactly the same (sums replaced by maximisation of course), and for the backward algorithm the change is essentially the same except in the third step the silent states are updated in reverse order.

If there are loops consisting entirely of silent states, the situation gets a little more complicated. It is possible to eliminate the silent states from the calculation by calculating (exactly) the effective transition probabilities between real states in the model, which involves inverting the transition matrix for the Markov model of silent states [Cox & Miller 1965]. Often, however, these effective transitions correspond to a fully connected model, and this leads to a substantial increase in the complexity of the model. Usually it is best to simply make sure such loops do not exist.

Exercises

3.7 Calculate the total number of transitions needed in a forward connected model as the one shown above with a length of L. Calculate the same number for a model with silent states (as above).

3.8 Show that the number of paths through an array of n states is indeed $\binom{l-1}{n-1}$ for length l as in (3.24).

3.9 What is the probability distribution over lengths for a model with an array of n states with self-loops if using the Viterbi algorithm?

3.5 More complex Markov chains

High order Markov chains

An nth order Markov process is a stochastic process where each event depends on the previous n events, so

$$P(x_i|x_{i-1}, x_{i-2}, \ldots, x_1) = P(x_i|x_{i-1}, \ldots, x_{i-n}). \tag{3.25}$$

The Markov chains we have discussed so far are of order 1.

An nth order Markov chain over some alphabet $\mathcal{A}$ is equivalent to a first order Markov chain over the alphabet $\mathcal{A}^n$ of n-tuples. This follows from the simple fact that $P(x_k|x_{k-1} \ldots x_{k-n}) = P(x_k, x_{k-1} \ldots x_{k-n+1}|x_{k-1} \ldots x_{k-n})$ (the probability of A and B given B is the probability of A given B). That is, the probability of x_k given the n-tuple ending in x_{k-1} is equal to the probability of the n-tuple ending in x_k given the n-tuple ending in x_{k-1}.

Consider the simple example of a second order Markov chain for sequences of only two different characters A and B. A sequence is translated to a sequence of pairs, so for instance the sequence ABBAB becomes AB-BB-BA-AB. The equivalent four-state first order Markov chain will look like this:

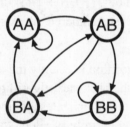

In this equivalent model not all transitions are allowed (or alternatively, some of the transition probabilities are zero). This is because only two different pairs can follow a given letter; the state AB for instance can only be followed by the states BA and BB. No sequence exists that can go from state AB to state AA. Similarly, a second order model for DNA is equivalent to a first order model over an alphabet of the 16 dinucleotides. A sequence of five bases, CGTCA, corresponds to a chain of four states, CG-GT-TC-CA, in a dinucleotide model.

Despite the theoretical equivalence between an nth order model and a first order model, the framework of high order models (meaning models of order greater than 1) is sometimes more convenient. Theoretically the high order models are treated in a way completely equivalent to first order models.

Finding prokaryotic genes

An example is given by a model for identifying prokaryotic genes. Genes of prokaryotes (bacteria) have a very simple one-dimensional structure. A gene coding for a protein starts with a start codon, then has a number of codons coding

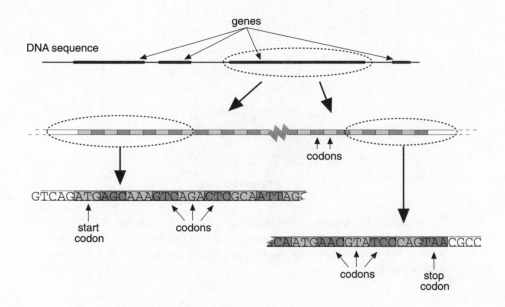

Figure 3.10 *The organisation of genes in prokaryotes.*

for amino acids, and ends with a stop codon; see Figure 3.10. Codons are DNA nucleotide triplets of which 61 code for amino acids and three are stop codons. In order to focus on the modelling, many complications such as frame shifts and non-protein genes are ignored here.

It is very easy to find good gene candidates by simply looking for stretches of DNA with the correct structure, i.e. starting with one of the three possible start codons, continuing with a number of non-stop codons and ending with one of the three stop codons. Such a gene candidate is called an *open reading frame* or just an ORF. Usually there are many overlapping ORFs that have the same stop codon, but different start codons. (The term ORF is often used for the maximal open reading frame between two stop codons, but we shall use it for all possible gene candidates.) There are many more ORFs than real genes, and here we will sketch possible ways of distinguishing between a non-coding ORF and a real gene.

In this example DNA from the bacterium *E. coli* is used (the dataset is described in detail in Krogh, Mian & Haussler [1994]). We consider only genes more than 100 nucleotides long. In the dataset there are 1100 such genes. This set is arbitrarily divided into a training set of 900 for training our models, and a test set containing the remaining 200 genes.

We estimate a first order model just as we did for the CpG islands early in this chapter and test how well it discriminates genes from other ORFs. In the test set we found roughly 6500 ORFs with a length of more than 100 bases. ORFs that share the stop codon with a known real gene were not included, because they

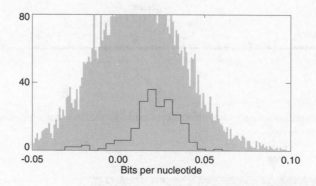

Figure 3.11 *Histograms of the log-odds per nucleotide for all NORFs (grey) and genes (black line) according to a first order Markov chain. Because of the large number of NORFs, the histogram bin size is five times smaller for the NORFs.*

would generally score very well and make our subsequent analysis more difficult. The remaining ORFs that are not labelled as coding will be called NORFs (for non-coding ORFs).

In Figure 3.11 a histogram is shown of the log-odds per nucleotide. As the null model for calculating log-odds we used the simplest possible, with the probability for each nucleotide equal to the frequency by which it occurs in all the data. The average log-odds per nucleotide for all the genes is 0.018, whereas it is half as much (0.009) for the NORFs, but the variance makes it almost useless for discrimination. You could fool yourself into thinking that the model had a decent discriminative power if you plotted the histogram of log-odds without dividing by the sequence length, because the genes are longer on average than NORFS, and therefore also the total log-odds is larger for the NORFs. Almost all the apparent information about genes would come from the length distribution and not from the model.

It is worth noticing that the average of the histogram is *not* at 0 bits, and that the averages of the two distributions (genes and NORFs) are quite close. This indicates that the Markov chain has indeed found a non-random correlation between nucleotide pairs, but it is essentially the same in coding and non-coding regions. In a second order chain, the probability of a nucleotide depends on the two previous ones, so it spans the length of a codon. Therefore we also tried a second order model, but the result is almost identical to the one for the first order model, so we do not show the histogram. It would probably not help much to switch to a Markov chain of even higher order, because these models do not separate the three reading frames, i.e. the three different nucleotide positions in the codon.

It is possible to make a high order inhomogeneous Markov chain (discussed in the next section) for modelling the bases in three different reading frames, but since our goal is to score ORFs, we will do it differently. The sequences are transformed to sequences of codons. An arbitrary symbol is assigned to each of the 64 codons, and all genes and NORFs are translated to this alphabet (yielding sequences of one-third the length of the nucleotide sequences). Notice that this transformation is slightly different from the one above for transforming an nth order model into a first order one, because the triplets are non-overlapping.

A 64-state first order Markov chain was estimated from the translated sequences and tested on the genes in the test set and the NORFs in exactly the same way as the models above. The result is shown in Figure 3.12. Although the separation is not perfect, we see that it is much better than for the other model. Notice that the distribution we compare to in the log-odds score now is a uniform distribution over codons. The grey peak is centred around 0, indicating that the Markov chain has found a signal that is special to coding regions, and that codon usage is essentially random in the average NORF, and that a significant fraction of the NORFs scoring highly represent real genes that are not labelled as such in our data. It is likely that most of the ORFs scoring above 0.3–0.35 bits in this plot are overlapping with real genes. The NORF histogram uses a smaller bin size (as in Figure 3.11), and if the same bin size was used, the NORF histogram would be about five times higher.

If the log-odds is not normalised by sequence length the discrimination improves significantly, because real genes tend to be longer than NORFs, see Figure 3.12.

Exercises

3.10   Calculate the number of parameters in the above codon model. The dataset contains on the order of 300 000 codons. Would it be feasible to estimate a second order Markov chain from this dataset?

3.11   How can the above gene model be improved?

Inhomogeneous Markov chains

As we saw above, a successful Markov model of genes needs to model the codon statistics. This can also be done without translating to another alphabet. It is well known that in genes the three codon positions have quite different statistics, and therefore it is natural to use three different Markov chains to model coding regions. The three models are numbered 1 to 3 according to the position in the codon. Assuming that x_1 is in codon position 3, the probability of $x_2, x_3, \ldots$ would then be

$$a^1_{x_1 x_2} a^2_{x_2 x_3} a^3_{x_3 x_4} a^1_{x_4 x_5} a^2_{x_5 x_6} \cdots$$

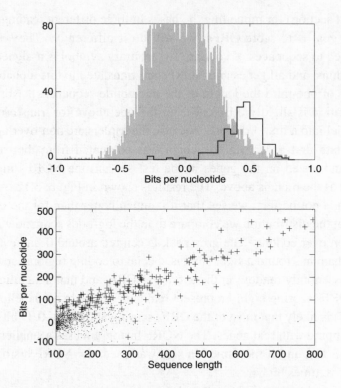

Figure 3.12 *The top plot shows the histograms of NORFs and genes for the Markov chain of codons (cf. Figure 3.11). Below, the log-odds is shown as a function of length for genes (+) and NORFs (·).*

where the parameters for model k are called a^k. This is called an *inhomogeneous Markov chain*. Here we assumed the chain was first order, but it is of course possible to extend it to order n. The estimation of the parameters is a straightforward extension of the estimation of the homogeneous models described in Section 3.1: for a second order inhomogeneous Markov chain as above the parameters of model 1 are estimated by counting the triplets with the last base in codon position 1, and similarly for model 2 and 3.

Inhomogeneous Markov chains are used extensively in the GENEMARK genefinding program [Borodovsky & McIninch 1993], which is currently the most widely used method for prokaryotic genefinding. Inhomogeneous models of order up to five of coding regions have been combined with homogeneous models of the non-coding regions to localise genes in a number of different bacterial genomes.

The first order model described above can also be constructed as an HMM, with the number of states equal to three times the length of the alphabet (a total of 12 for DNA). Higher order models can be made by adding many additional states to the HMM. However, it is also possible to have nth order Markov emission probabilities in the states of an HMM, in which the emission probabilities are conditioned on the n previous characters, so the emission probabilities (3.5) become

$$e_k(b|b_1,\ldots,b_n) = P(x_i|\pi_i = k, x_{i-1} = b_1,\ldots,x_{i-n} = b_n).$$

All the algorithms derived for standard HMMs can be used with only obvious alterations for models with these emissions. Such models are also being used for genefinding [Krogh 1998].

Exercise

3.12   Draw the HMM that corresponds to the first order inhomogeneous Markov chain given above.

3.6 Numerical stability of HMM algorithms

Even on modern floating point processors we will run into numerical problems when multiplying many probabilities in the Viterbi, forward, or backward algorithms. For DNA for instance, we might want to model genomic sequences of 100 000 bases or more. Assuming that the product of one emission and one transition probability is typically 0.1, the probability of the Viterbi path would then be of the order of $10^{-100000}$. Most computers would behave badly with such numbers: either an underflow error would occur and the program would crash; or, worse, the program would keep running and produce arbitrary wrong numbers. There are two different ways of dealing with this problem.

The log transformation

For the Viterbi algorithm we should always use the logarithm of all probabilities. Since the log of a product is the sum of the logs, all the products are turned into sums. Assuming the logarithm base 10, the log of the above probability of $10^{-100000}$ is just -100000. Thus, the underflow problem is essentially solved. Additionally, the sum operation is faster on some computers than the product, so on these computers the algorithm will also run faster.

We will put a tilde on all the model parameters after taking the log, so for example $\tilde{a}_{kl} = \log a_{kl}$. Then the recursion relation for the Viterbi algorithm (3.8) becomes

$$V_l(i+1) = \tilde{e}_l(x_{i+1}) + \max_k(V_k(i) + \tilde{a}_{kl}),$$

where we use V for the logarithm of v. The base of the logarithm is not important as long as it is larger than 1 (such as 2, e, and 10).

It is more efficient to take the log of all the model parameters before running the Viterbi algorithm, to avoid calling the logarithm function repeatedly during the dynamic programming iteration.

For the forward and backward algorithms there is a problem with the log transformation: the logarithm of a sum of probabilities cannot be calculated from the logs of the probabilities without using exponentiation and log functions, which are computationally expensive. However, the situation is not in practice so bad. Assume you want to calculate $\tilde{r} = \log(p + q)$ from the log of the probabilities, $\tilde{p} = \log p$ and $\tilde{q} = \log q$. The direct way is to do $\tilde{r} = \log(\exp(\tilde{p}) + \exp(\tilde{q}))$. By pulling out $\tilde{p}$, one can write this as

$$\tilde{r} = \tilde{p} + \log(1 + \exp(\tilde{q} - \tilde{p})).$$

It is possible to approximate the function $\log(1 + \exp(x))$ by interpolation from a table. For a reasonable level of accuracy, the table can actually be quite small, assuming we always pull out the largest of $\tilde{p}$ and $\tilde{q}$, because $\exp(\tilde{q} - \tilde{p})$ rapidly approaches zero for large $(\tilde{p} - \tilde{q})$.

Scaling of probabilities

An alternative to using the log transformation is to rescale the f and b variables, so they stay within a manageable numerical interval [Rabiner 1989]. For each i define a scaling variable s_i, and define new f variables

$$\tilde{f}_l(i) = \frac{f_l(i)}{\prod_{j=1}^{i} s_j}. \tag{3.26}$$

From this it is easy to see that

$$\tilde{f}_l(i+1) = \frac{1}{s_{i+1}} e_l(x_{i+1}) \sum_k \tilde{f}_k(i) a_{kl},$$

so the forward recursion (3.11) is only changed slightly. This will work however we define s_i, but a convenient choice is one that makes $\sum_l \tilde{f}_l(i) = 1$, which means that

$$s_{i+1} = \sum_l e_l(x_{i+1}) \sum_k \tilde{f}_k(i) a_{kl}.$$

The b variables have to be scaled with the same numbers, so the recursion step in (3.3) becomes

$$\tilde{b}_k(i) = \frac{1}{s_i} \sum_l a_{kl} \tilde{b}_l(i+1) e_l(x_{i+1})$$

This scaling method normally works well, but in models with many silent states, such as the one we describe in Chapter 5, underflow errors can still occur.

Exercises

3.13   Use (3.26) to prove that $P(x) = \prod_{j=1}^{L} s_j$ with the above choice of s_i. It is
        of course wiser to calculate $\log P(x) = \sum_j \log s_j$.

3.14   Use the result of the previous exercise to show that the equation (3.20)
        actually simplifies when using the scaled f and b variables. Also, derive
        the result (3.21) for the scaled variables.

3.7   Further reading

More basic introductions to HMMs include Rabiner & Juang [1986] and Krogh
[1998].

Some early applications of HMM-like models to sequence analysis was done
by Borodovsky *et al.* [1986a; 1986b; 1986c] who used inhomogeneous Markov
chains as described on p. 75. This later led to the GENEMARK genefinder program
[Borodovsky & McIninch 1993]. Cardon & Stormo [1992] introduced an expectation maximisation (EM) method, which has many similarities with an HMM,
for modelling protein binding motifs. Later applications of HMMs to genefinding
include Krogh, Mian & Haussler [1994], Henderson, Salzberg & Fasman [1997],
and Krogh [1997a,1997b,1998] as well as systems combining neural networks
and HMMs [Stormo & Haussler 1996; Kulp *et al.* 1996; Reese *et al.* 1997;
Burge & Karlin 1997]. Such hybrid systems are also becoming quite popular
for other applications; see for instance Bengio *et al.* [1992], Frasconi & Bengio
[1994], Renals *et al.* [1994], Baldi & Chauvin [1995], and Riis & Krogh [1997].

Churchill [1989] used HMMs for modelling compositional differences between DNA from mitochondria and from the human X chromosome and bacteriophage lambda, and later for studying the compositional structure of genomes
[Churchill 1992]. Other applications include a three-state HMM for prediction
of protein secondary structure [Asai, Hayamizu & Handa 1993], a HMM with
ten states in a ring for modelling an oscillatory pattern in nucleosomes [Baldi *et
al.* 1996], detection of short protein coding regions and analysis of translation
initiation sites in cyanobacteria [Yada & Hirosawa 1996; Yada, Sazuka & Hirosawa 1997], characterization of prokaryotic and eukaryotic promoters [Pedersen
et al. 1996], and recognition of branch points [Tolstrup, Rouzé & Brunak 1997].
Several other applications of HMMs will be discussed in the context of profile
HMMs in Chapters 5 and 6.

4

Pairwise alignment using HMMs

Now that we have acquired new technical machinery from hidden Markov model theory, we return for a brief chapter to pairwise sequence alignment. In Chapter 2 we introduced finite state automata with multiple states as a convenient description of more complex dynamic programming algorithms for pairwise alignment. It is also possible to consider them as a basis for a probabilistic interpretation of the gapped alignment process, by converting them into HMMs. One advantage of this approach is that we will be able to use the resulting probabilistic model to explore questions about the reliability of the alignment obtained by dynamic programming, and to explore alternative (suboptimal) alignments. Indeed, by weighting all alternatives probabilistically, we will be able to score the similarity of two sequences independent of any specific alignment. We can also build more specialised probabilistic models out of simple pieces, to model more complex versions of sequence alignment, as discussed previously for FSAs.

Let us first review briefly the finite state automaton that we introduced for pairwise alignment with affine gap penalties. We required three states, M corresponding to a match, and two states corresponding to inserts, which we name here X and Y as shown in Figure 4.1. The recurrence relations for updating the

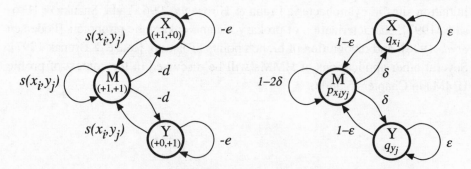

Figure 4.1 *A finite state machine diagram for affine gap alignment on the left, and the corresponding probabilistic model on the right.*

values of these states in the dynamic programming matrix are

$$V^{M}(i,j) = s(x_i, y_j) + \max \begin{cases} V^{M}(i-1, j-1), \\ V^{X}(i-1, j-1), \\ V^{Y}(i-1, j-1); \end{cases} \quad (4.1)$$

$$V^{X}(i,j) = \max \begin{cases} V^{M}(i-1, j) - d, \\ V^{X}(i-1, j) - e; \end{cases}$$

$$V^{Y}(i,j) = \max \begin{cases} V^{M}(i, j-1) - d, \\ V^{Y}(i, j-1) - e. \end{cases}$$

These equations are appropriate for global alignment. As previously, we will generally give detailed equations for global alignment, while indicating what changes need to be made for local alignment.

4.1 Pair HMMs

We need to make two sets of changes to an FSA as shown on the left side of Figure 4.1 to turn it into an HMM. First, as shown on the right of Figure 4.1, we must give probabilities both for emissions of symbols from the states, and for transitions between states. For example, state M has emission probability distribution p_{ab} for emitting an aligned pair $a{:}b$, and states X and Y will have distributions q_a for emitting symbol a against a gap. Because state X emits symbols x_i from sequence x, we write q_{x_i} inside the circle representing state X. We also specify transition probabilities between the states, which must satisfy the requirement that the probabilities for all the transitions leaving each state sum to one. Allowing for symmetry, there are two free parameters for the transition probabilities between the three main states. We denote the transition from M to an insert state (X or Y) by δ, and the probability of staying in an insert state by ε.

However, the resulting model shown on the right side of Figure 4.1 does not generate a full model that will provide a probability distribution over all possible sequences. To do that, we need to define a Begin and an End state, as shown in Figure 4.2. In effect these formalise the initialisation and termination conditions that we needed for the dynamic programming algorithms in Chapter 2. We will see below that more complex arrangements of Begin and End states can correspond to local and other types of alignments. Adding an explicit End state introduces the need for another parameter, the probability of a transition into the End state, which we assume for now to be the same from each of M, X and Y; we call it τ. This will in effect determine the average length of an alignment from the model. For now, we will set the transitions from the Begin state to be the same as from the M state (we could have just said that we will start in M, but we wanted to make clear that initialisation can be given independent consideration as well as termination).

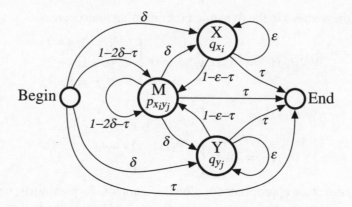

Figure 4.2 *The full probabilistic version of Figure 4.1.*

This gives us a probabilistic model that is very similar to a hidden Markov model as we defined it in Chapter 3. The difference is that instead of emitting a single sequence it emits a pairwise alignment. We will call this type of model a *pair HMM* to distinguish it from the more standard types of HMMs that emit single sequences. All the algorithms from Chapter 3 carry across to pair HMMs, although they need an extra dimension of search space because of the extra emit-ted sequence. For example, instead of writing $v_k(i)$ for the Viterbi probabilities, we write $v^k(i, j)$. We will give below the explicit sets of equations for the key algorithms, applied to the basic pair HMM shown in Figure 4.2.

Just as a standard HMM can generate a sequence, our pair HMM can generate an aligned pair of sequences. This is done by starting in the Begin state, and cycling over the following two steps: (1) pick the next state according to the distribution of transition probabilities leaving the current state; (2) pick a symbol pair to be added to the alignment according to the emission distribution in the new state. The process stops when a transition is made into the End state. Because we have probabilities for each step, we can also keep track of the total probability of generating a particular alignment that we have made. This is just the product of the probabilities of each individual step.

The most probable path is the optimal FSA alignment

The Viterbi algorithm from Chapter 3 will allow us to find the most probable path through a pair HMM given sequences x and y. The correct form for the global pair HMM of Figure 4.2 is as follows. To make the equations simpler, we define the Begin state to be M. As in the previous chapter, we use lower-case symbols $v^\bullet(i, j)$ for probability values, and upper-case $V^\bullet(i, j)$ for log-odds scores. We give the Viterbi algorithm first in terms of probabilities:

Algorithm: Viterbi algorithm for pair HMMs

Initialisation:

$v^M(0,0) = 1$. All other $v^{\bullet}(i,0), v^{\bullet}(0,j)$ are set to 0.

Recurrence: $i = 1, \ldots, n, j = 1, \ldots, m$;

$$v^M(i,j) = p_{x_i y_j} \max \begin{cases} (1-2\delta-\tau)v^M(i-1,j-1), \\ (1-\varepsilon-\tau)v^X(i-1,j-1), \\ (1-\varepsilon-\tau)v^Y(i-1,j-1); \end{cases}$$

$$v^X(i,j) = q_{x_i} \max \begin{cases} \delta v^M(i-1,j), \\ \varepsilon v^X(i-1,j); \end{cases}$$

$$v^Y(i,j) = q_{y_j}, \max \begin{cases} \delta v^M(i,j-1) \\ \varepsilon v^Y(i,j-1). \end{cases}$$

Termination:

$$v^E = \tau \max(v^M(n,m), v^X(n,m), v^Y(n,m)). \qquad \triangleleft$$

To find the best alignment, we keep pointers and trace back as usual. Of course, to get the alignment itself we keep track of which residues are emitted at each step in the path during the traceback, as in Chapter 2, as well as (or even in place of) the sequence of states as for the type of HMM described in Chapter 3.

Although it is clear that the recurrence equations of the pair HMM Viterbi algorithm have the same sort of form as those for the state machine version of pairwise alignment (4.1), it is instructive to see the exact form of the correspondence.

First, we have to transform into log-odds ratios with respect to the random model. In fact, now we have a full probabilistic model for our alignment, we should also have one for our random model, with a proper termination condition. Previously we have ignored the fact that our random model could not produce sequences of varying length in a proper probabilistic fashion. Here is a new random model, which is also a pair HMM.

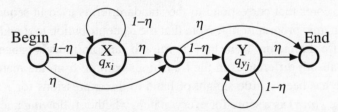

The main states are X and Y, which emit the two sequences in turn, independently of each other. Each has a loop back onto itself with probability $(1-\eta)$. As well as Begin and End states, there is also a silent state in between X and Y, indicated by a smaller circle. This does not emit any symbols, but is used to gather inputs from both the X and Begin states (see the section on silent states on p. 70 for further information on how these are used). When defined this way the model allows zero-length sequences x or y, just as the pair HMM model in Figure 4.2 does, and generates a simple form for the random model distribution over sequences. The probability of a pair of sequences x and y according to this model is

$$P(x, y|R) \;=\; \eta(1-\eta)^n \prod_{i=1}^{n} q_{x_i} \, \eta(1-\eta)^m \prod_{j=1}^{m} q_{x_j}$$

$$\;=\; \eta^2 (1-\eta)^{n+m} \prod_{i=1}^{n} q_{x_i} \prod_{j=1}^{m} q_{x_j}. \tag{4.2}$$

We now want to allocate the terms in this expression to those that make up the probability of the Viterbi alignment, so that the odds ratio for the whole alignment can be expressed as a product of odds ratios of individual terms (and, correspondingly, so that the log-odds ratio of the alignment is a sum of log-odds terms). We do this by allocating one factor of $(1-\eta)$ and the corresponding q_a factor to each residue that is emitted in a Viterbi step. So the match transitions will be allocated $(1-\eta)^2 q_a q_b$ where a and b are the two residues matched, and the insert states $(1-\eta)q_a$ where a is the residue inserted. Because the Viterbi path must account for all the residues, exactly $(n+m)$ terms will be used, and all of (4.2) except the initial factor of η^2 is accounted for.

In log-odds terms, we can now compute in terms of an additive model with log-odds emission scores and log-odds transition scores. In practice this is normally the most practical way to implement pair HMMs. From this, it is possible to merge the emission scores with the transitions as shown here:

$$s(a,b) \;=\; \log \frac{p_{ab}}{q_a q_b} + \log \frac{(1 - 2\delta - \tau)}{(1-\eta)^2},$$

$$d \;=\; -\log \frac{\delta(1 - \varepsilon - \tau)}{(1-\eta)(1 - 2\delta - \tau)},$$

$$e \;=\; -\log \frac{\varepsilon}{1-\tau},$$

to produce scores that correspond to the standard terms used in sequence alignment by dynamic programming. Note that the q_a contribution to d and e has vanished because the factors from the Viterbi and random models cancelled. Also in order to absorb differences in the transitions coming from the match and gap states, there has been a little sleight of hand in the expressions for s and d. We intend to use $s(a,b)$ as a score for every match, whether following another match or an insertion. In order to make this work correctly, we have built into d an adjustment to correct for the difference in match score when returning back from an insertion. This means that the dynamic programming matrix terms for the insertions no longer correspond exactly to the log-odds ratios of being in those states, although the final result will be correct.

We can now give the log-odds version of the Viterbi alignment algorithm in a form that looks like standard pairwise dynamic programming.

Algorithm: Optimal log-odds alignment

Initialisation:

$$V^M(0,0) = 2\log\eta, V^X(0,0) = V^Y(0,0) = -\infty.$$

All $V^{\bullet}(i,-1), V^{\bullet}(-1,j)$ are set to $-\infty$.

Recursion: $i = 0,\ldots,n, j = 0,\ldots,m$ except $(0,0)$;

$$V^M(i,j) = s(x_i,y_j) + \max \begin{cases} V^M(i-1,j-1), \\ V^X(i-1,j-1), \\ V^Y(i-1,j-1); \end{cases}$$

$$V^X(i,j) = \max \begin{cases} V^M(i-1,j)-d, \\ V^X(i-1,j)-e; \end{cases}$$

$$V^Y(i,j) = \max \begin{cases} V^M(i,j-1)-d, \\ V^Y(i,j-1)-e. \end{cases}$$

Termination:

$$V = \max(V^M(n,m), V^X(n,m)+c, V^Y(n,m)+c). \qquad \triangleleft$$

These are identical to (4.1) except for the constant $2\log\eta$ in the initialisation, and the constant $c = \log(1 - 2\delta - \tau) - \log(1 - \varepsilon - \tau)$ in the termination, which is needed to correct back for our adjustment described above in d. In fact the latter correction is only a result of having used the same exit probability τ for match and insert states. If the exit transition probabilities from the gap states are set to $(1 - \varepsilon)\tau/(1 - \tau)$ then c will be zero, and hence the log-odds algorithm will have exactly the same form as our standard pairwise affine gap alignment algorithm, with a single additive constant coming from the initialisation conditions.

The procedure as we have described it shows how for any pair HMM of the type shown in Figure 4.2 we can derive an equivalent FSA for obtaining the most probable alignment. This allows us to see a rigorous probability-based interpretation for the terms used in sequence alignment. To do the reverse, i.e. to go from a dynamic programming algorithm expressed as an FSA to a pair HMM, is more complicated. There will in general be a need for a new parameter λ which will act as a global scaling factor for the scores, and for any given set of scores there may be constraints on the choice of η and τ.

A pair HMM for local alignment

The model shown in Figure 4.2 is appropriate to finding a global match between sequences. As described in Chapter 2, many of the most sensitive pairwise searches are local. When we introduced the local alignment algorithm, and other variants such as the repeat and overlap algorithms, we explained them in terms of changes in the update equations and boundary conditions. Both of these are made explicit in the pair HMM formalism by adding states and transitions. We can therefore draw a separate pair HMM model for each variant. In Figure 4.3 we

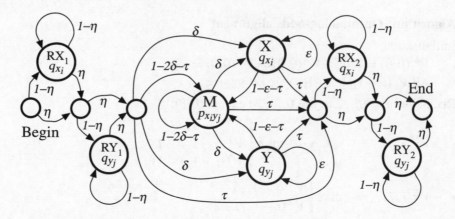

Figure 4.3 *A pair HMM for local alignment. This is composed of the global model (states* M, X *and* Y*) flanked by two copies of the random model (states* RX_1, RY_1 *and* RX_2, RY_2).*

show a model for local alignment. This looks more complicated than the global model in Figure 4.2, but it is made up of simpler pieces in a straightforward fashion.

A complete probabilistic model must account for all of the sequences x and y: not only the local alignment between x and y, but also the unaligned flanking sequences. We therefore add extra model sections before and after the three-state matching segment from Figure 4.2. Each flanking segment is a copy of the complete random background model, because the sequences in the flanking regions are unaligned. Most terms in the likelihood contributions of these sections will cancel out with equivalent terms in the random model when calculating the log-odds scores of a match in comparison to the random model, leaving only the local matching score from the central part of the model, and some extra one-off terms. Similar composite models can be built for overlap and repeat models, and the various hybrids discussed in Chapter 2.

Exercises

4.1   What is the probability that sequence x has length t under the full random model?

4.2   What is the expected length of sequences from the full random model? How should the parameter η be set?

4.2 The full probability of x and y, summing over all paths

Having a pair HMM allows us to do more than provide an alternative rationale for standard pairwise alignment by dynamic programming. One issue that we raised when discussing the significance of matches in Chapter 2 was that, when similarity is weak, it is hard to identify the correct alignment to score and test for significance. Now we can bypass this problem (and the approach taken throughout the whole of Chapter 2) by calculating the probability that a given pair of sequences are related according to the HMM by *any* alignment. We do this by summing over alignments,

$$P(x,y) = \sum_{\text{alignments } \pi} P(x,y,\pi).$$

How do we calculate this sum? Again, there is a standard HMM algorithm, described in Chapter 3 as the *forward* algorithm. The way this works out for pair HMMs is that we can again use the same dynamic programming idea that we used for finding the maximal scoring alignment, but add rather than take the maximum at each step. The probability version of the forward algorithms is given below, using $f^k(i,j)$ to represent the combined probability of all alignments up to (i,j) that end in state k. As before, we give this only for the global model of Figure 4.2; the extension to other types of pairwise alignment model such as the local model described above is straightforward.

Algorithm: Forward calculation for pair HMMs

Initialisation:

$f^M(0,0) = 1. \ f^X(0,0) = f^Y(0,0) = 0.$

All $f^{\bullet}(i,-1), f^{\bullet}(-1,j)$ are set to 0.

Recursion: $i = 0,\ldots,n, j = 0,\ldots,m$ except $(0,0)$;

$$\begin{aligned}
f^M(i,j) &= p_{x_i y_j}\big[(1-2\delta-\tau)f^M(i-1,j-1)+ \\
&\qquad (1-\varepsilon-\tau)(f^X(i-1,j-1)+f^Y(i-1,j-1))\big]; \\
f^X(i,j) &= q_{x_i}\big[\delta f^M(i-1,j)+\varepsilon f^X(i-1,j)\big]; \\
f^Y(i,j) &= q_{y_j}\big[\delta f^M(i,j-1)+\varepsilon f^Y(i,j-1)\big].
\end{aligned}$$

Termination:

$$f^E(n,m) = \tau\big[f^M(n,m)+f^X(n,m)+f^Y(n,m)\big]. \qquad \lhd$$

We can now consider the log-odds ratio of the resulting full probability $P(x,y) = f^E(n,m)$ to the null model probability given by (4.2). This is a measure of the likelihood that the two sequences are related to each other by some unspecified alignment, as opposed to being unrelated. In doing this we have not assumed any specific alignment. Of course, if there is an unambiguous best alignment, almost all the probability in the total sum will be contributed by the single

```
HBA_HUMAN    KVADALTNAVAHVD-----DMPNALSALSDLH
             KV   + +A  ++            +L+ L+++H
LGB2_LUPLU   KVFKLVYEAAIQLQVTGVVVTDATLKNLGSVH


HBA_HUMAN    KVADALTNAVAHVDDM-----PNALSALSDLH
             KV   + +A  ++            +L+ L+++H
LGB2_LUPLU   KVFKLVYEAAIQLQVTGVVVTDATLKNLGSVH


HBA_HUMAN    KVADALTNA-----VAHVDDMPNALSALSDLH
             KV   + +A      V V      +L+ L+++H
LGB2_LUPLU   KVFKLVYEAAIQLQVTGVVVTDATLKNLGSVH
```

Figure 4.4 *An example of uncertainty in positioning a gap: three significantly different gap placements in the globin alignment from Figure 2.1(b),
with very similar scores.*

path corresponding to this best alignment. However, the full score will always
be higher than that for the optimal alignment (using the same scoring scheme),
and it can be significantly different when there are many comparable alternative
alignments, or alignment variations.

An important use of the full probability is to define a posterior distribution
$P(\pi|x,y)$ over alignments π given a pair of sequences x, y. This is given by

$$P(\pi|x,y) = \frac{P(x,y,\pi)}{P(x,y)}. \tag{4.3}$$

If we set $\pi = \pi^{\star}$, the Viterbi path, in (4.3), then we obtain the posterior probability according to the model of the Viterbi path $v^{E}(n,m)/f^{E}(n,m)$, which we can
interpret as the probability that the optimal scoring alignment is 'correct'. Frequently this is vanishingly small! For example for the alignment of alpha globin
to leghaemoglobin in Figure 2.1(b) it is 4.6×10^{-6}. This observation, although
perhaps alarming if one was hoping that the standard alignment algorithms would
find the 'correct' alignment, is not surprising. There are many small variants of
the best alignment that have nearly the same score, or equivalently are nearly
equally likely. In particular, where there is a gap there is often a choice of where
the gap should be placed; moving it left or right by a residue or so frequently
leads to no change or a seemingly random fluctuation.

Figure 4.4 shows an example of this behaviour with corresponding sections of
the human alpha globin and lupin leghaemoglobin sequences. The first alignment
shown is close to the structurally verified alignment, and has score 3 (BLOSUM50,
gap-open -12, gap-extend minus -2). The next has the same score, although the
gap is offset by two positions. The third has score 6, although the gap is misplaced
by five residues. The difference in scores of 3 coresponds to an increase in relative
likelihood of a factor of two according to the alignment model, since BLOSUM50
scores are given in third-bits. It is clear that simple sequence alignment is not an

accurate way to determine the alignment in this case, which is admittedly highly diverged.

Exercise

4.3   The relative scores for gap position variants such as shown in Figure 4.4 depend only on the substitution scores, not the gap scores. Why is this, and what are the consequences for alignment accuracy using dynamic programming algorithms?

4.3   Suboptimal alignment

Given that there are frequently alternative alignments with nearly the same probability (or more generally nearly the same score) as the best alignment, it is naturally of interest to see what they are. Such alignments are known as *suboptimal* alignments. There are a number of different approaches to examining and characterising suboptimal alignments. First let us consider more carefully what we might expect to find.

One class of alignments with scores close to the optimal score will be those mentioned above that only differ in a few positions from the optimal alignment (e.g. those in Figure 4.4). Because minor variations at different places in the alignment can be combined independently, the number of these 'local' variants grows exponentially as the difference in score from the optimal score increases. It is therefore impractical to give all such variants. However, the flexibility in varying the alignment can vary substantially with position along the alignment. There are sampling methods that illustrate typical variants, and methods that show for each cell in the dynamic programming matrix how 'close' it is to being in the alignment. Examples of both of these are given below.

Another type of suboptimal alignment is one that differs substantially, or perhaps completely, from the optimal alignment. Methods for finding this type of suboptimal alignment can be used where one suspects that more than one correct alignment may be present, for instance where there are repeats in one or both of the sequences. In general, this is more relevant when searching for local alignments, which only align together a part of each sequence.

Probabilistic sampling of alignments

We first give a method for sampling alignments from the posterior distribution defined in (4.3). Recall that this gave a probability to each possible alignment of the two sequences, according to its likelihood of being correct under the model. An ensemble of such samples will give a picture of the type of alignment information that is reliably retrievable from a given sequence pair. Any particular property of direct interest can be estimated by averaging over the sample, as suggested in the section on posterior decoding of HMMs (p. 60). This is a powerful general strategy for using similarity information when the alignment is uncertain in detail; for example it is used later in the book in Chapter 8.

To generate a sample alignment, we trace back through the matrix of $f^k(i, j)$ values, but instead of taking the highest scoring choice at each step, we make a probabilistic choice based on the relative strengths of the three components. To illustrate how this is done, let us imagine we are part way through the traceback, in state M at position (i, j), which we call cell M(i, j). We know from the forward algorithm that

$$f^M(i, j) = p_{x_i y_j}\left[(1 - 2\delta - \tau)f^M(i-1, j-1) + \right.$$
$$\left. (1 - \varepsilon - \tau)(f^X(i-1, j-1) + f^Y(i-1, j-1))\right].$$

We choose the next step to be

M$(i - 1, j - 1)$ with prob. $\dfrac{p_{x_i y_j}(1 - 2\delta - \tau)f^M(i-1, j-1)}{f^M(i, j)}$,

X$(i - 1, j - 1)$ with prob. $\dfrac{p_{x_i y_j}(1 - \varepsilon - \tau)f^X(i-1, j-1)}{f^M(i, j)}$,

Y$(i - 1, j - 1)$ with prob. $\dfrac{p_{x_i y_j}(1 - \varepsilon - \tau)f^Y(i-1, j-1)}{f^M(i, j)}$.

The corresponding distribution if in cell X(i, j) would be to choose

M$(i - 1, j)$ with prob. $\dfrac{q_{x_i}\delta f^M(i-1, j)}{f^X(i, j)}$,

X$(i - 1, j)$ with prob. $\dfrac{q_{x_i}\varepsilon f^X(i-1, j)}{f^X(i, j)}$,

and similarly for cell Y(i, j).

A set of sample global alignments from our simple example data is given here:

```
HEAGAWGHEE        HEAGAWGHE-E        HEAGAWGHEE
-P-A-WHEAE        -PA--W-HEAE        -P-A-WHEAE


HEAGAWGHEE        HEAGAWGHEE         HEAGAWGHE-E
P---AWHEAE        -P--AWHEAE         -P--AW-HEAE


HEAGAWGHE-E       HEAGAWGHE-E        HEAGAWGHEE
-P--AW-HEAE       --P-AW-HEAE        --PA-WHEAE
```

You can see that alternatives are more likely where gaps are required and evidence for the alignment is weak, as at the beginning of the sequences. Pairings that contribute strongly to the score, such as the Ws, or that come in blocks, as at the end of the sequence, are more stable. The frequency of a pairing in such samples can be used as a natural indicator of its reliability in the alignment. Below we present a direct way of calculating the expected value of this frequency, i.e. the probability that any particular pair of residues should be aligned, according to the model.

The same type of sampling approach that we have used here will be used later in the book when building multiple alignments (Chapter 5).

Finding distinct suboptimal alignments

As mentioned above, a number of different methods have been given for finding alignments that are not simply minor variants of the optimal alignment. One approach is to use the 'repeat' algorithm in Chapter 2. This found the optimal set of high-scoring matches between one sequence and multiple non-overlapping segments of the other sequence. However, for the current purposes, this is unsatisfactory because it treats the two sequences differently. Also, the best single alignment may not even be present in the set.

The most widely used method for searching for distinct suboptimal alignments is due to Waterman & Eggert [1987], who give an algorithm to find the next best alignment that has no aligned residue pairs in common with any previously determined alignment. Once the top match has been obtained, the standard (Viterbi) dynamic programming matrix is recalculated, with the additional step during the recurrence that cells corresponding to residue pairs contained in the best match are set to zero, preventing them from contributing to the next alignment. The resulting matrix and score will therefore contain information about the second best alignment. This procedure can be repeated, zeroing all the cells for any match obtained so far each time, until the next score is below T (see Figure 4.5). In fact, if the matrix is stored in memory then it is not necessary to recalculate the complete matrix each iteration: a marking procedure can be used to indicate which cells need to be updated. For references to some of the other approaches to finding suboptimal alignments, see Section 4.6.

4.4 The posterior probability that x_i is aligned to y_j

If the probability of any single complete path being entirely correct is small, can we say anything about the local accuracy of an alignment? Often part of an alignment is fairly clear, and other regions are less certain. The degree of conservation varies depending on structural and functional contraints, so that core sequences may be well conserved, while loop regions are not reliably alignable.

	H	E	A	G	A	W	G	H	E	E
	0	0	0	0	0	0	0	0	0	0	0
P	0	0	0	0	0	0	0	0	0	0	0
A	0	0	0	5	0	5	0	0	0	0	0
W	0	0	0	0	2	0	20	12	4	0	0
H	0	10	2	0	0	0	12	18	22	14	6
E	0	2	16	8	0	0	4	10	18	28	20
A	0	0	8	21	13	5	0	4	10	20	27
E	0	0	6	13	18	12	4	0	4	16	26

	H	E	A	G	A	W	G	H	E	E
	0	0	0	0	0	0	0	0	0	0	0
P	0	0	0	0	0	0	0	0	0	0	0
A	0	0	0	5	0	0	0	0	0	0	0
W	0	0	0	0	2	0	0	0	0	0	0
H	0	10	2	0	0	0	0	0	0	0	0
E	0	2	16	8	0	0	0	0	0	0	6
A	0	0	8	21	13	5	0	0	0	0	0
E	0	0	6	13	18	12	4	0	0	6	6

Figure 4.5 *The Waterman–Eggert algorithm applied to our standard data example. Above, the standard local alignment matrix exactly as in Figure 2.6. Below, the best local match has been zeroed out so that the second best alignment can be obtained.*

Given this situation, it can be useful to be able to give a reliability measure for each part of an alignment.

The HMM formalism allows us to do this. The idea is that we calculate the combined probability of all the alignments that pass through a specified matched pair of residues (x_i, y_j). We then compare this value with the full probability of all alignments of the pair of sequences, calculated in the previous section. If the ratio is near to one, then we can say that that match is highly reliable; if near zero, then the match is unreliable. This method used to do this is very closely related to the algorithm given for posterior decoding in Chapter 3.

Let us introduce a new notation $x_i \diamond y_j$ to mean that x_i is aligned to y_j. Then from standard conditional probability theory we have

$$
\begin{aligned}
P(x, y, x_i \diamond y_j) &= P(x_{1...i}, y_{1...j}, x_i \diamond y_j) P(x_{i+1...n}, y_{j+1...m} | x_{1...i}, y_{1...j}, x_i \diamond y_j) \\
&= P(x_{1...i}, y_{1...j}, x_i \diamond y_j) P(x_{i+1...n}, y_{j+1...m} | x_i \diamond y_j)
\end{aligned}
$$

The first term is the forward probability $f^{M}(i,j)$ calculated above by the forward algorithm. The second is the corresponding backward probability $b^{M}(i,j)$ which is calculated by the corresponding backward algorithm.

Algorithm: Backward calculation for pair HMMs

Initialisation:

$b^{M}(n,m) = b^{X}(n,m) = b^{Y}(n,m) = \tau.$

All $b^{\bullet}(i,m+1), b^{\bullet}(n+1,j)$ are set to 0.

Recursion: $i = n, \ldots, 1, j = m, \ldots, 1$ except (n,m);

$$
\begin{aligned}
b^{M}(i,j) &= (1-2\delta-\tau)p_{x_{i+1}y_{j+1}}b^{M}(i+1,j+1) \\
&\quad + \delta\left[q_{x_{i+1}}b^{X}(i+1,j)+q_{y_{j+1}}b^{Y}(i,j+1)\right]; \\
b^{X}(i,j) &= (1-\varepsilon-\tau)p_{x_{i+1}y_{j+1}}b^{M}(i+1,j+1)+\varepsilon q_{x_{i+1}}b^{X}(i+1,j); \\
b^{Y}(i,j) &= (1-\varepsilon-\tau)p_{x_{i+1}y_{j+1}}b^{M}(i+1,j+1)+\varepsilon q_{y_{j+1}}b^{Y}(i,j+1). \quad \triangleleft
\end{aligned}
$$

There is no special termination step needed, because we only need the $b^{\bullet}(i,j)$ values for $i,j \geq 1$.

We can now use Bayes' rule to obtain

$$
P(x_i \diamond y_j|x,y) = \frac{P(x,y,x_i \diamond y_j)}{P(x,y)},
$$

and can also obtain similar values for the posterior probabilities of using specific insert states. Figure 4.6 shows the results of this procedure applied to the example sequences that we used in Chapter 2.

Miyazawa [1994] describes essentially the same approach, and goes on to define what he calls a 'probability alignment'. It might seem attractive to define an alignment of x to y by finding for each i the j that maximises $P(x_i \diamond y_j)$ (we drop explicit conditioning with respect to x and y from here on, since it will always be present). However, this is not guaranteed to produce a well-formed alignment; it may contain aligned pairs $(i_1,j_1),(i_2,j_2)$ which are inconsistent with the sequence orders, i.e. for which $i_2 > i_1$ and $j_1 < j_2$. Miyazawa pointed out that if we restrict ourselves to pairs (i,j) for which $P(x_i \diamond y_j) > 0.5$, then these will always be consistent, and will also only align each x_i to at most one y_j. In places where the alignment is clear, it will be covered by this condition. On the other hand, where it is not clear, for example in corresponding loop regions of distantly related proteins, there will be gaps in both sequences where no particular pairs of residues are strongly supported as being aligned.

The expected accuracy of an alignment

Miyazawa's approach typically gives rise to incomplete alignments, in that there may be significant sections where no $P(x_i \diamond y_j) > 0.5$. Although this may be

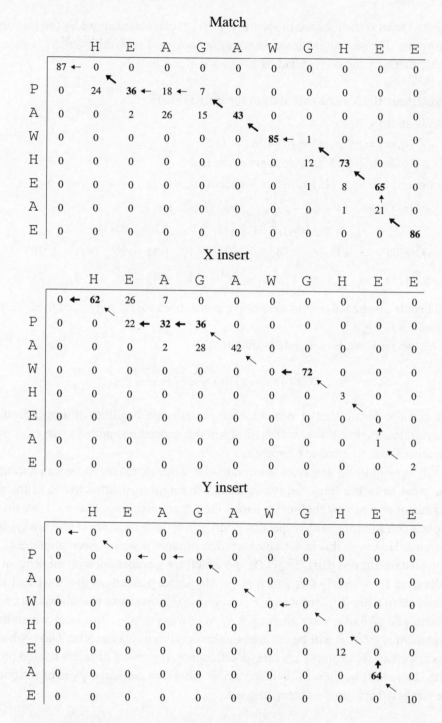

Figure 4.6 *Posterior probabilities for the example data used in Chapter 2. The three tables show the posterior probabilities of using the* M, X *or* Y *states respectively at each (i,j) position. Values are shown as percentages, i.e. 100 times the relevant probability rounded to the nearest integer. The path indicated is the optimal accuracy path in the sense of (4.4).*

what is wanted, it is also possible to use the posterior match probabilities to give a complete alignment with maximal overall accuracy, in the sense outlined below. We first note that we can calculate the expected overlap $\mathcal{A}(\pi)$ between a given alignment π and paths sampled from the posterior distribution. This is equivalently the expected number of correct matches in π, which is a natural measure of the overall accuracy of π.

$$\mathcal{A}(\pi) = \sum_{(i,j)\in\pi} P(x_i \diamond y_j),$$

where the sum is over all aligned pairs in π. For the alpha globin/leghaemoglobin alignment of Figure 2.1(b) $\mathcal{A}(\pi) = 16.48$, or on average 0.40 per aligned residue.

Given this new type of score for an alignment, can we find the alignment between two sequences with the highest accuracy? We might hope that this, while perhaps not providing the most discriminative score for use in detecting whether two sequences are related, would give a more accurate alignment if they are. The method required is surprisingly simple. We perform standard dynamic programming using score values given by the posterior probabilities of pair matches, without gap costs. The recursion equations are:

$$A(i,j) = \max \begin{cases} A(i-1,j-1) + P(x_i \diamond y_j), \\ A(i-1,j), \\ A(i,j-1), \end{cases} \tag{4.4}$$

and the standard traceback procedure will produce the best alignment. It is clear that this procedure will optimise the sum of the $P(x_i \diamond y_j)$ terms in a legitimate alignment. Interestingly the same algorithm works for any sort of gap score; what will change with different scores are the $P(x_i \diamond y_j)$ terms themselves, which are obtained from the standard, scoring scheme-specific dynamic programming procedures described above.

The optimal accuracy path for the short sequences used as examples in Chapter 2 is shown in Figure 4.6. Note that it is not the same as the most likely, or Viterbi path. The initial P in the shorter sequence is clearly preferably aligned to the E and not the A of the longer sequence, although the individual scores for aligning P to E and A are the same. Intuitively, the reason for this is that aligning to the E allows more options in where the subsequent gap can be placed.

4.5 Pair HMMs versus FSAs for searching

One of the strong points of probabilistic modelling is that, if data D correspond to samples from a model M, then, in the limit of an infinitely large amount of data, the likelihood takes its maximum value for M, i.e. $P(D|M) > P(D|\tilde{M})$, where $\tilde{M}$ is any other model. In particular, if M has a set of parameters, such as

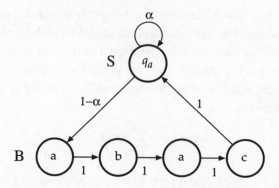

Figure 4.7 *This FSA emits sequences from* S *with probability* q_a, *and strings* abac *from the block* B *of four states below. If the probability of transition to* B *is low, the most probable path will never use* B, *even if the sequence includes the motif* abac.

the transition and emission probabilities of an HMM, the likelihood of the data will be maximised by giving the model the parameter values corresponding to the sample.

As a consequence, if the parameters of a pair HMM describe the statistics of pairs of related sequences well, then we should use that model with those parameter values for searching. If we also have a model, R, that gives a good description of the generation of random sequence, then Bayesian model comparison with M and R is an appropriate procedure (p. 36 in Chapter 2). According to this philosophy, we should be using probabilistic models for searching. However, most currently used algorithms (Chapter 2) fall short of this in two ways. First, they do not compute the full probability $P(x, y|M)$ of the pair of sequences, summing over all alignments, but instead find the best match, or Viterbi path. Second, regarded as FSAs, their parameters may not be readily translated into probabilities.

Consider first the effects of using Viterbi paths. It is easy to show that, in this case, a model whose parameters match the data need not be the best search model. Figure 4.7 shows a simple HMM example. A state S generates symbols with probabilities q_a; S has a transition to itself with probability α and can make a transition with probability $1 - \alpha$ to a sequential block B of states that emits a fixed string abac of length four before returning to the original state. The probability of emitting abac from S is $P_S(\text{abac}) = \alpha^4 q_a q_b q_a q_c$, whereas the probability of emitting abac from B (starting at S) is $1 - \alpha$. If $P_S(\text{abac}) > 1 - \alpha$, the most probable path for any set of data will only use S, because the transition to B is too improbable. Nonetheless, the presence of a greater than expected number of strings abac in the data is what distinguishes the output of the model from that of the random model that emits symbols with probabilities q_a. Model comparison using the best match rather than the total probability, will fail to

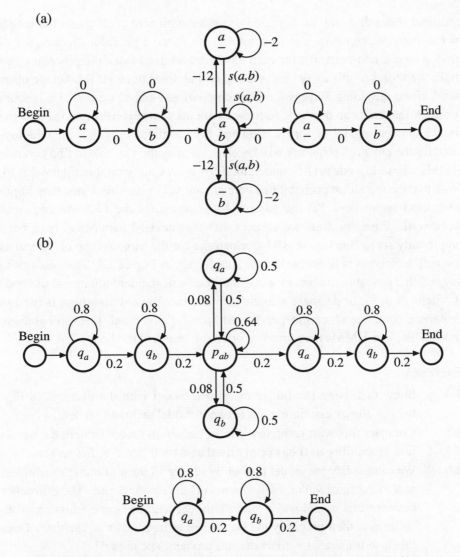

Figure 4.8 *(a) An FSA that computes the local match algorithm. s(a,b) are the scores for the* BLOSUM50 *matrix. (b) Two HMMs, an aligned sequence model (above) and a random model (below) whose log-odds ratio score is the same as the score of the FSA shown in (a). The probabilities p_{ab} and q_a are those used to define the* BLOSUM50 *matrix.*

detect the source of the data, even for very large datasets. We can partially correct for these deficiencies by changing our parameters. For instance, the model will be able to detect these types of sequences if the probability of the transition to B is increased to τ where $\tau > P_S(\text{abac})$. However, then every abac will be classed as coming from B, which is not correct either.

Consider now the problem of turning an FSA for pairwise alignment into a probabilistic model. Figure 4.8(a) shows an FSA for local matches; it has initial and final states that emit an unpaired sequence with zero cost. Since the length of this unpaired sequence can be arbitrary, and since a probabilistic model will always have a non-zero cost for each emission, no fixed rescaling procedure can make the scores of this model into the log probabilities of an HMM. On the other hand, if we are doing Bayesian model comparison, and if we define a random model R that emits an unpaired sequence with the same probabilities used by the local alignment model M for its inital and final unaligned regions, then the log-odds for the unpaired sequence will be zero. We may then be able to find two pair HMMs whose log-odds ratios match the FSA scores, for example Figure 4.8(b). Note that the transition probabilities here are not very plausible, since they imply very short sequences. Yet the parameters assumed for the FSA are known to work well. Based on this, we suspect that the standard parameters have been empirically set to 'unconsciously' compensate for the same failing of Viterbi as a search method as is illustrated in the simple case of Figure 4.8. This leads us to suggest that probabilistic models may underperform standard alignment methods if Viterbi is used for database searching, but if the forward algorithm is used to provide a complete score independent of specific alignment, then probabilistic models like pair HMMs may improve upon the standard methods.

Exercises

4.4   Show that using the full probabilistic model with the example in Figure 4.8 allows discrimination between model and random data.

4.5   Compare this with using the Viterbi path in the model where the transition probability to B has been raised to τ such that $\tau > P_S(\texttt{abac})$.

4.6   We can modify the model further by setting all the emission probabilities at S to the same value, $1/A$, where A is the alphabet size. The difference between this model and a random model with the same emission probabilities is then precisely the number of strings $\texttt{abac}$ in the data. Does this discriminate as well as the full probabilistic model?

4.6   Further reading

Although the explicit formulation of pairwise alignment in terms of pair hidden Markov models that we have given here is not standard, several authors have considered an equivalent full probabilistic model. Bucher & Hofmann [1996] discuss searching with a local probabilistic model normalised via a partition function. Bishop & Thompson [1986] introduced a related model in the context of evolutionary analysis, a strand that has been developed more recently by Thorne, Kishino & Felsenstein [1991; 1992], who have developed parameter estimation methods for probabilistic models of gapped alignment of DNA sequences. We discuss some of these evolutionary motivated models further in Chapter 8.

Zuker [1991] describes a method for finding suboptimal alignments similiar to that of Waterman & Eggert [1987]. Another approach is given in Barton [1993]. Mevissen & Vingron [1996] give an alternative approach to quantifying the reliability of a dynamic programming alignment, and Vingron [1996] provides a good recent review of methods for finding and assessing the significance of suboptimal alignments.

5

Profile HMMs for sequence families

So far we have concentrated on the intrinsic properties of single sequences, such as CpG islands in DNA, or on pairwise alignment of sequences. However, functional biological sequences typically come in families, and many of the most powerful sequence analysis methods are based on identifying the relationship of an individual sequence to a sequence family. Sequences in a family will have diverged from each other in their primary sequence during evolution, having separated either by a duplication in the genome, or by speciation giving rise to corresponding sequences in related organisms. In either case they normally maintain the same or a related function. Therefore, identifying that a sequence belongs to a family, and aligning it to the other members, often allows inferences about its function.

If you already have a set of sequences belonging to a family, you can perform a database search for more members using pairwise alignment with one of the known family members as the query sequence. To be more thorough, you could even search with all the known members one by one. However, pairwise searching with any one of the members may not find sequences distantly related to the ones you have already. An alternative approach is to use statistical features of the whole set of sequences in the search. Similarly, even when family membership is clear, accurate alignment can be often be improved significantly by concentrating on features that are conserved in the whole family.

How, in brief, do we identify such features? Just as a pairwise alignment captures much of the relationship between two sequences, a multiple alignment can show how the sequences in a family relate to each other. Figure 5.1 shows a multiple alignment of seven sequences from the large globin family (hundreds of globin sequences are available in the protein sequence databases). The three dimensional structure has been obtained for each protein in the alignment shown, and the sequences have been aligned on the basis of aligning the eight alpha helices of the conserved globin fold, and also on the basis of aligning certain key residues in the sequences, such as two conserved histidines (H) which are the residues which interact with an oxygen-binding heme prosthetic group in the globin active site.

It is clear that some positions in the globin alignment are more conserved than others. In general the helices are more conserved than the loop regions between

100

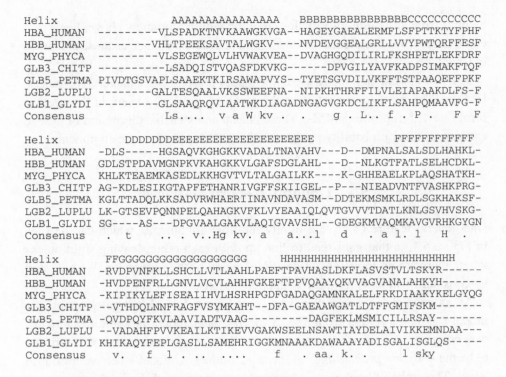

Figure 5.1 *An alignment of seven globins from Bashford, Chothia & Lesk [1987]. To the left is the protein identifier in the* SWISS-PROT *database [Bairoch & Apweiler 1997]. The eight alpha helices are shown as* A–H *above the alignment. A consensus line below the alignment indicates residues that are identical among at least six of the seven sequences in upper case, ones identical in four or five sequences in lower case, and positions where there is a residue identical in three sequences with a dot.*

them, and certain residues are particularly strongly conserved. When identifying a new sequence as a globin, it would be desirable to concentrate on checking that these more conserved features are present. How to obtain and use such information will be the subject of this chapter.

As might be expected, our approach to consensus modelling will be to make a probabilistic model. In particular, we will develop a particular type of hidden Markov model well suited to modelling multiple alignments. We call these *profile HMMs* after standard *profiles*, which are closely related non-probabilistic structures introduced previously for the same purpose by Gribskov, McLachlan & Eisenberg [1987]. Profile HMMs are probably the most popular application of hidden Markov models in molecular biology at the moment [Eddy 1996].

We will assume for the purposes of this chapter that we are given a correct multiple alignment, from which we will build a model that can be used to find and score potential matches to new sequences. The multiple alignment could be built from structural information, like the globin alignment shown here, or it could come from a sequence-based alignment procedure, such as those discussed in Chapter 6.

Much of this chapter makes use of the theory presented in Chapter 3 for general HMMs. The most important algorithms will be presented again in the specific form relevant to profile HMMs. There is also an extensive discussion of how to estimate optimal probability parameters from multiple sequence alignments.

5.1 Ungapped score matrices

One general feature of protein family multiple alignments, which can be seen in Figure 5.1, is that gaps tend to line up with each other, leaving solid blocks where there are no insertions or deletions in any of the sequences. We will start by considering models for these ungapped regions.

As an example, consider the E helix of Figure 5.1. A natural probabilistic model for such a region would be to specify independent probabilities $e_i(a)$ of observing amino acid a in position i (we use letter e because these will turn out to be the *emission probabilities* of the hidden Markov model when we introduce gaps). The probability of a new sequence x according to this model is then

$$P(x|M) = \prod_{i=1}^{L} e_i(x_i),$$

where L is the length of the block, 21 in this case. As usual, we are in fact more interested in the ratio of this probability to the probability of x under a random model, and so to test for membership in the family we evaluate the log-odds ratio

$$S = \sum_{i=1}^{L} \log \frac{e_i(x_i)}{q_{x_i}}.$$

The values $\log \frac{e_i(a)}{q_a}$ behave like elements in a score matrix $s(a,b)$, where the second index is position i, rather than amino acid b. For this reason, such an approach is known as a *position specific score matrix* (PSSM). A PSSM can be used to search for a match in a longer sequence x of length N by evaluating the score S_j for each starting point j in x from 1 to $N - L + 1$, where L is the length of the PSSM.

5.2 Adding insert and delete states to obtain profile HMMs

Although a PSSM captures some conservation information, it is clearly an inadequate representation of all the information in a multiple alignment of a protein family. We have to find some way to take account of gaps. It is possible to combine the scores of multiple ungapped block models, and this is the approach taken by Henikoff & Henikoff [1991] in the BLOCKS database. However, we will pursue here the aim of developing a single probabilistic model for the whole extent of the alignment.

One approach is to allow gaps at each position in the alignment, using the same gap score $\gamma(g)$ at each position, as in pairwise alignment. However, this is also ignoring information, because the alignment gives us explicit indications of where gaps are more and less likely. We want to capture this information to give us position sensitive gap scores, just as the emission probabilities gave us position sensitive substitution scores.

The approach we take is to build a hidden Markov model (HMM), with a repetitive structure of states, but different probabilities in each position. This will provide a full probabilistic model for sequences in the sequence family. We start off by observing that the PSSM can be viewed as a trivial HMM with a series of identical states that we will call *match* states, separated by transitions of probability 1.

Alignment is trivial because there is no choice of transitions. We rename the emission probabilities for the match states to $e_{M_i}(a)$.

The next step is to deal with gaps. We must treat insertions and deletions separately. To handle insertions, i.e. portions of x that do not match anything in the model, we introduce a set of new states I_i, where I_i will be used to match insertions after the residue matching the ith column of the multiple alignment. The I_i have emission distribution $e_{I_i}(a)$, but these are normally set to the background distribution q_a, just as for seeing an unaligned inserted residue in a pairwise alignment. We need transitions from M_i to I_i, a loop transition from I_i to itself, to accommodate multi-residue insertions, and a transition back from I_i to M_{i+1}. Here is a single insert state of this kind:

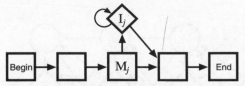

We denote insert states in our diagrams by diamonds. The log-odds cost of an insert is the sum of the costs of the relevant transitions and emissions. Assuming that $e_{I_i}(a) = q_a$ as described above, there is no log-odds contribution from the emission, and the score of a gap of length k is

$$\log a_{M_j I_j} + \log a_{I_j M_{j+1}} + (k-1)\log a_{I_j I_j}.$$

From this you can see that the type of insert state shown corresponds to an affine gap scoring model.

Deletions, i.e. segments of the multiple alignment that are not matched by any residue in x, could be handled by forward 'jump' transitions between non-neighbouring match states:

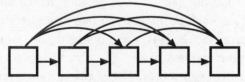

However, to allow arbitrarily long gaps in a long model this way would require a lot of transitions. Instead we introduce silent states D_j as described in Section 3.4:

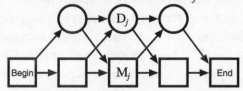

Because the silent states do not emit any residues, it is possible to use a sequence of them to get from any match state to any later one, between two residues in the sequence. The cost of a deletion will then be the sum of the costs of an $M \to D$ transition followed by a number of $D \to D$ transitions, then a $D \to M$ transition. This is at first sight exactly analogous to the cost of an insert, although the path through the model looks different. In detail, it is possible that the $D \to D$ transitions will have different probabilities, and hence contribute differently to the score, whereas all the $I \to I$ transitions for one insert involve the same state, and so are guaranteed to have the same cost.

The full resulting HMM has the structure shown in Figure 5.2. This form of model, which we call a profile HMM, was first introduced in Haussler *et al.* [1993] and Krogh *et al.* [1994]. We have added transitions between insert and delete states, as they did, although these are usually very improbable. Leaving them out has negligible effect on scoring a match, but can create problems when building the model.

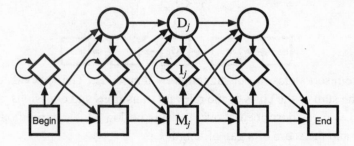

Figure 5.2 *The transition structure of a profile HMM. We use diamonds to indicate the insert states and and circles for the delete states.*

Profile HMMs generalise pairwise alignment

We have seen how the costs of using gap states in a profile HMM mirror those used in pairwise alignment with affine gaps. To help make clear the relationship, it is useful to consider the degenerate case where the multiple alignment from which we build the HMM contains just one sequence.

Let us compare Figure 5.2 with Figure 4.2. If we call the example sequence y, then Figure 5.2 is an unrolled version of Figure 4.2, with the y_j emissions each coming from a separate copy of the pair HMM. The states M_j correspond to a sequence of match states M, the I_j to corresponding incarnations of X, and the D_j to incarnations of Y. To achieve as close a correspondence as possible, the natural values for the match emission probabilities $e_{M_i}(a)$ are p_{y_ia}/q_{y_i}, the conditional probabilities of seeing a given y_i in a pairwise alignment, and for the transition probabilities $a_{M_iI_i} = a_{M_iD_{i+1}} = \delta$ and $a_{I_iI_i} = a_{D_iD_{i+1}} = \varepsilon$ for all i.

In formal terms our profile HMM is effectively the hidden Markov model obtained by conditioning the pair HMM of Figure 4.2 on emitting sequence y as one of the sequences in its alignment. Because of this, the Viterbi equations for finding the most probable alignment of x to our profile HMM are essentially the same as those for the most probable alignment of x and y to the pair HMM described in Chapter 4. If we convert them into log-odds ratio form we recover our standard affine gap cost pairwise alignment equations of (2.16), as we will see below. Any differences are due to slightly different Begin and End arrangements.

5.3 Deriving profile HMMs from multiple alignments

Although it is nice to see that the profile HMM is doing the same sort of dynamic programming as we have used before for pairwise alignment, this is not why we introduced them. The key idea behind profile HMMs is that we can use the same structure as shown in Figure 5.2, but set the transition and emission probabilities to capture specific information about each position in the multiple alignment of the whole family. Essentially, we want to build a model representing the consensus sequence for the family, not the sequence of any particular member.

There are a number of different ways to derive the parameter values from a multiple alignment of the sequences in the family. To provide an example for illustrating these methods, Figure 5.3 shows a short section of the globin alignment shown in Figure 5.1.

Non-probabilistic profiles

A model similar to the profile HMM was first introduced by Gribskov, McLachlan & Eisenberg [1987] who coined the name 'profile' (see also Gribskov, Lüthy & Eisenberg [1990]). However, they did not have an underlying probabilistic model,

```
HBA_HUMAN    ...VGA--HAGEY...
HBB_HUMAN    ...V----NVDEV...
MYG_PHYCA    ...VEA--DVAGH...
GLB3_CHITP   ...VKG------D...
GLB5_PETMA   ...VYS--TYETS...
LGB2_LUPLU   ...FNA--NIPKH...
GLB1_GLYDI   ...IAGADNGAGV...
             *** *****
```

Figure 5.3 *Ten columns from the multiple alignment of seven globin protein sequences shown in Figure 5.1. The starred columns are ones that will be treated as 'matches' in the profile HMM.*

but rather directly assigned position specific scores for each match state and gap penalty, for use in standard 'best match' dynamic programming. They set the scores for each consensus position to the averages of the standard substitution scores from all the residues seen in the corresponding multiple alignment column. For example, they would set the score for residue a in column 1 of our example to be

$$\tfrac{5}{7}s(V,a) + \tfrac{1}{7}s(F,a) + \tfrac{1}{7}s(I,a)$$

where $s(a,b)$ is the standard substitution matrix. They also set gap penalties for each column using a heuristic equation that decreased the cost of a gap (either insertion or deletion) according to the length of the longest gap observed in the multiple alignment spanning the column.

Although this seems an intuitively obvious way to combine information, and it has been used effectively by many people for finding new members of families, it does produce anomalies. For example, column 1 is much more strongly conserved than column 2 in the example shown in Figure 5.3, but the information in column 1 will be smeared out just as much by the substitution matrix as that in column 2. If we had an alignment with 100 sequences, all with a cysteine (C) at some position, then the implicit probability distribution for that column for an 'average' profile would be exactly the same as would be derived from a single sequence. This does not correspond to our expectation that the likelihood of a cysteine should go up as we see more confirming examples.

In addition to these observations about substitution scores, the scores for gaps do not behave as expected. For example, from the alignment in Figure 5.3 the score for a deletion would be set to be the same in column 2, where there is a deletion in one sequence, HBB_HUMAN, as in column 4, where there is a deletion opening in five of the seven sequences. It would be more reasonable to set the probability of a new gap opening to be higher in column 4.

Changes have been made to non-probabilistic profiles to address these and other problems [Thompson, Higgins & Gibson 1994b; Gribskov & Veretnik 1996], and we shall return to some of these later.

Basic profile HMM parameterisation

Let us turn back to hidden Markov model profiles. Like all HMMs, these have emission and transition probabilities. Assuming that these probabilities are non-zero, a profile HMM can model any possible sequence of residues from the given alphabet. It therefore defines a probability distribution over the whole space of sequences. The aim of the parameterisation process it to make this distribution peak around members of the family.

The parameters we have available to control the shape of the distribution are the values of the probabilities, and also the length of the model. There is a lot to say about setting these optimally. We give here the basic methods from Krogh *et al.* [1994]. After sections on database searching and variants for local alignment, we will return to an extended discussion of alternative parameter estimation techniques.

The choice of length of the model corresponds more precisely to a decision on which multiple alignment columns to assign to match states, and which to assign to insert states. The profile HMM we derived above from the single sequence y had a match state for each residue y_i. However, looking at Figure 5.3 it seems clear that the consensus sequence for this region should only have eight residues, and that the two non-starred residues in GLB1_GLYDI should be treated as an insertion with respect to the consensus. For the time being we will use a heuristic rule to decide which columns should correspond to match states, and which to inserts. A simple rule that works well is that columns that are more than half gap characters should be modelled by inserts.

The second problem is how to assign the probability parameters. We regard the alignment as providing a set of independent samples of alignments of sequences x to our HMM. Since the alignments are given, we can estimate the parameters directly using equations (3.18) from Section 3.3. We just count up the number of times each transition or emission is used, and assign probabilities according to

$$a_{kl} = \frac{A_{kl}}{\sum_{l'} A_{kl'}} \quad \text{and} \quad e_k(a) = \frac{E_k(a)}{\sum_{a'} E_k(a')}$$

where k and l are indices over states, and a_{kl} and e_k are the transition and emission probabilities, and A_{kl} and E_k are the corresponding frequencies.

In the limit of having a very large number of sequences in our training alignment, this will give an accurate and consistent estimate of the probabilities. However, it has problems when there are only a few sequences. A major difficulty is that some transitions or emissions may not be seen in the training alignment, and so would acquire zero probability, which would mean they would never be

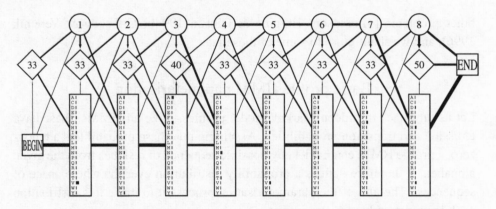

Figure 5.4 *A hidden Markov model derived from the small alignment shown in Figure 5.3 using Laplace's rule. Emission probabilities are shown as bars opposite the different amino acids for each match state, and transition probabilities are indicated by the thickness of the lines. The* $I \to I$ *transition probabilities times 100 are shown in the insert states. (Figure generated automatically using the* SAM *package.)*

allowed in the future. More broadly, we are not using any previous knowledge about protein alignments, as the earlier non-probabilistic methods did implicitly, by using an independently derived substitution matrix. As a minimal approach to avoid zero probabilities, we can add pseudocounts to the observed frequencies (as in Chapters 1 and 3). The simplest pseudocount method is Laplace's rule: to add one to each frequency. We discuss better ways to choose the pseudocount values, and other approaches to estimating the parameters, at greater length below in Section 5.6.

Example: Parameters for an HMM based on Figure 5.3

Let us assume that we use Laplace's rule to obtain parameters for an HMM corresponding to the alignment in Figure 5.3. Then $e_{M_1}(V) = 6/27, e_{M_1}(I) = e_{M_1}(F) = 2/27$, and $e_{M_1}(a) = 1/27$ for all residue types a other than V, I, F. Similarly, $a_{M_1 M_2} = 7/10, a_{M_1 D_2} = 2/10$ and $a_{M_1 I_1} = 1/10$ (following column 1 there are six transitions from match to match, one transition to a delete state, in HBB_HUMAN, and no insertions). Figure 5.4 shows the complete set of parameters for the HMM in diagrammatic form. □

5.4 Searching with profile HMMs

One of the main purposes of developing profile HMMs is to use them to detect potential membership in a family by obtaining significant matches of a sequence to the profile HMM. We will assume for now that we are looking for global matches.

In practice, as for pairwise alignment, one of the local alignment methods may be more sensitive for finding distant matches. We discuss these in the next section.

We have a choice of ways to score a match to a hidden Markov model. We can either use the Viterbi equations to give the most probable alignment π^* of a sequence x together with its probability $P(x, \pi^*|M)$, or the forward equations to calculate the full probability of x summed over all possible paths $P(x|M)$.

In either case, for practical purposes the result we want to consider when evaluating potential matches is the log-odds ratio of the resulting probability to the probability of x given our standard random model

$$P(x|R) = \prod_i q_{x_i}.$$

We therefore show here versions of the Viterbi and forward algorithms that are designed specifically for profile HMMs, and which result directly in the desired log-odds values. Note that changing to log-odds does not change the result; we could have subtracted the random model log score afterwards. However, it is cleaner and more efficient. Another practical reason for working in log-odds units is to avoid problems of underflow when working with raw probabilities, as we discussed in Section 3.6.

Viterbi equations

Let $V_j^M(i)$ be the log-odds score of the best path matching subsequence $x_{1...i}$ to the submodel up to state j, ending with x_i being emitted by state M_j. Similarly $V_j^I(i)$ is the score of the best path ending in x_i being emitted by I_j, and $V_j^D(i)$ for the best path ending in state D_j. Then we can write

$$V_j^M(i) = \log \frac{e_{M_j}(x_i)}{q_{x_i}} + \max \begin{cases} V_{j-1}^M(i-1) + \log a_{M_{j-1}M_j}, \\ V_{j-1}^I(i-1) + \log a_{I_{j-1}M_j}, \\ V_{j-1}^D(i-1) + \log a_{D_{j-1}M_j}; \end{cases}$$

$$V_j^I(i) = \log \frac{e_{I_j}(x_i)}{q_{x_i}} + \max \begin{cases} V_j^M(i-1) + \log a_{M_j I_j}, \\ V_j^I(i-1) + \log a_{I_j I_j}, \\ V_j^D(i-1) + \log a_{D_j I_j}; \end{cases} \tag{5.1}$$

$$V_j^D(i) = \max \begin{cases} V_{j-1}^M(i) + \log a_{M_{j-1}D_j}, \\ V_{j-1}^I(i) + \log a_{I_{j-1}D_j}, \\ V_{j-1}^D(i) + \log a_{D_{j-1}D_j}. \end{cases}$$

These are the general equations. In a typical case, there is no emission score $e_{I_j}(x_i)$ in the equation for $V_j^I(i)$ because we assume that the emission distribution from the insert states I_j is the same as the background distribution, so the probabilities cancel in the log-odds form. Also, the D → I and I → D transition terms may not be present, as discussed above.

We need to take a little care over initialisation and termination of the dynamic programming. We want to allow the alignment to start and end in a delete or insert state, in case the beginning or end of the sequence does not match the first or the last match state of the model. The simplest way to ensure this mechanistically is to rename the Begin state as M_0 and set $V_0^M(0) = 0$ (as we did in Chapter 3). We then allow transitions to I_0 and D_1. Similarly, at the end we can collect together possible paths ending in insert and delete states by renaming the End state to M_{L+1} and using the top relation without the emission term to calculate $V_{L+1}^M(n)$ as the final score.

If these recurrence equations are compared with those for standard gapped dynamic programming in (2.16), it can be seen that apart from renaming of variables this is the same algorithm, but with the substitution, gap-open and gap-extend scores all depending on position in the model, j.

Forward algorithm

The recurrence equations for the forward algorithm are similar to the Viterbi equations, but with the max() operation replaced by addition. We define variables $F_j^M(i)$, $F_j^I(i)$ and $F_j^D(i)$ for the partial full log-odds ratios, corresponding to $V_j^M(i)$, $V_j^I(i)$ and $V_j^D(i)$. The recurrence equations are then:

$$F_j^M(i) = \log \frac{e_{M_j}(x_i)}{q_{x_i}} + \log \left[a_{M_{j-1}M_j} \exp \left(F_{j-1}^M(i-1) \right) \right.$$
$$\left. + a_{I_{j-1}M_j} \exp \left(F_{j-1}^I(i-1) \right) + a_{D_{j-1}M_j} \exp \left(F_{j-1}^D(i-1) \right) \right];$$

$$F_j^I(i) = \log \frac{e_{I_j}(x_i)}{q_{x_i}} + \log \left[a_{M_j I_j} \exp \left(F_j^M(i-1) \right) \right.$$
$$\left. + a_{I_j I_j} \exp \left(F_j^I(i-1) \right) + a_{D_j I_j} \exp \left(F_j^D(i-1) \right) \right];$$

$$F_j^D(i) = \log \left[a_{M_{j-1}D_j} \exp \left(F_{j-1}^M(i) \right) + a_{I_{j-1}D_j} \exp \left(F_{j-1}^I(i) \right) \right.$$
$$\left. + a_{D_{j-1}D_j} \exp \left(F_{j-1}^D(i) \right) \right].$$

Initialisation and termination conditions are handled as for the Viterbi case, with $F_0^M(0)$ being initialised to 0.

Although these appear a little complicated, in a practical implementation the operation $\log(e^x + e^y)$ can be performed efficiently to adequate accuracy by function lookup and interpolation; see Section 3.6.

Alternatives to log-odds scoring

In some of the earlier papers on HMMs, rather than calculating the log-odds score relative to a random model, the logarithm of the probability of the sequence given the model was used directly. This was called the LL score for 'log likelihood': $LL(x) = \log P(x|M)$. The LL score is strongly length dependent, so for searching

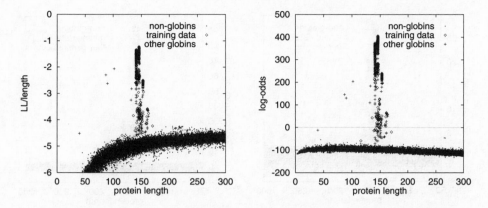

Figure 5.5 *To the left the length-normalized LL score is shown as a function of sequence length. The right plot shows the same for the log-odds score.*

it is not good enough to use a simple threshold. It is better to use LL divided by the sequence length, but even that is not always perfect, because the dependence between LL and sequence length is not linear (see example below).

A way to get around this is to estimate an average score and a standard deviation as a function of length and then use the number of standard deviations each sequence is away from the average. This is called the Z-score, and is also illustrated in the example below.

Example: Modelling and searching for globins

From 300 randomly picked globin sequences a profile HMM was estimated from scratch, i.e. starting from unaligned sequences using procedures we will explain in Chapter 6. A simple pseudocount regulariser was used. The estimation was done several times and the model with the highest overall LL score was picked. (We used the default settings of the SAM package, version 1.2; Hughey & Krogh [1996]).

With this model a database of about 60 000 proteins (SWISS-PROT release 34; Bairoch & Apweiler [1997]) was searched using the forward algorithm. The LL and log-odds scores were found for each sequence. For the null model we used the amino acid frequencies of the 300 sequences in the training set. In Figure 5.5 the length-normalised scores are shown for all the globins in the training set, all the other globins in the database and all the rest of the proteins with lengths up to 300 amino acids.[1] The globin sequences are clearly separated from the non-globins apart from a few in the 'twilight zone.'

The main difference between the two is in the variance of the score for non-globins, which is lower for the log-odds score, and therefore the separation is clearer. However, just choosing a cut-off of zero for the log-odds would miss a

---

[1] A few dubious globins and other strange sequences were removed from these data.

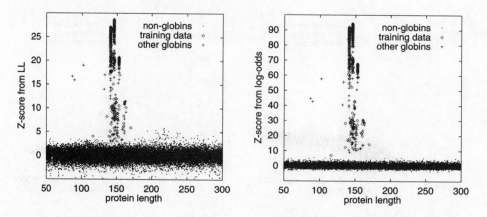

Figure 5.6 *The Z-score calculated from the LL scores (left) and the log-odds (right).*

lot of real globins in the search. This is because the profile HMM is not broad enough: it is too concentrated on a subset of the globins. Although there are ways to address this problem directly that we will return to later in the chapter, it is also possible to take a pragmatic approach to the separation of signal from noise given the results of the search, and calculate Z-scores for each hit.

To calculate Z-scores, a smooth curve is fitted to the LL or log-odds score of the non-globin sequences (a method is outlined in Krogh *et al.* [1994]). A standard deviation is then estimated for each length (or rather a little interval around it), and for each score the distance from the smooth curve is calculated in units of the standard deviation. This is the Z-score. The result (still as a function of sequence length) is shown in Figure 5.6.[2]

It is evident that it is now possible to find a threshold which will separate most globins from all other sequences. It is also clear that the score based on log-odds is much better for discrimination, with approximately three times the signal to noise ratio of the LL score. The reason for this is that dividing by the probability of the random model adjusts for the residue composition of the sequence. Without doing that, sequences with similar residue compositions as globins will tend to score more highly than sequences containing different residues, increasing the variance of the noise.   □

Alignment

Aside from finding matches, the other principal use of profile HMMs is to give an alignment of a sequence to the family, or more precisely to add it into the multiple alignment of the family. This is primarily the subject of the next chapter,

---

[2] There is no analytical result about the shape of these score distributions. The global alignment distribution is probably not exactly a Gaussian [Waterman 1995], but it appears to be a good approximation. For local alignments the extreme value distribution may be more reasonable, as discussed in Chapter 2.

on multiple alignment methods, which covers alignment with profile HMMs at length. For now, we will just point out that the natural solution is to take the highest scoring, or Viterbi, alignment. This is obtained by tracing back on the Viterbi variables $V_j^\bullet(i)$, exactly as with pairwise alignment. Beyond this, all the methods of Chapter 4 can be applied, to explore variants, and to assess the reliability of the alignment.

5.5 Profile HMM variants for non-global alignments

We have seen that there is a very close relationship between the Viterbi alignment of a sequence to a profile HMM and the global dynamic programming comparison between two sequences using affine gap penalties, which we described in Chapter 2. It is therefore possible to generalise all the variations of dynamic programming, such as those that find local, repeat and overlap matches, to use profile HMMs.

However, we have developed probabilistic models much more fully since Chapter 2, and this time we want to take more care to ensure that the result of converting to a local algorithm remains a proper probabilistic model, i.e. that we assign each sequence a true probability so that the sum over all sequences $\sum_x P(x|M) = 1$. Our approach to doing this is to specify a new model for the complete sequence x, which incorporates the original profile HMM together with one or more copies of a simple self-looping model that is used to account for the regions of unaligned sequence. These behave very like the insert states that we added to the profile itself. We call them *flanking model* states, because they are used to model the flanking sequences to the actual profile match itself.

The model for local (Smith–Waterman style) alignment is shown here:

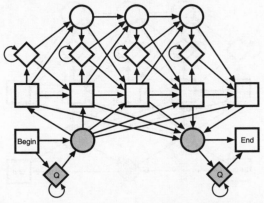

The flanking model states are shown as shaded diamonds. Notice that as well as specifying the emission probabilities of the new states, which will normally of course be q_a, we must specify a number of new transition probabilities. The looping probability on the flanking states should be close to 1, since they must account for long stretches of sequence. Let us set these to $(1 - \eta)$. Note also that we have made use of silent states, shown as shaded circles, as 'switching points' to reduce the total number of transitions.

The next issue is how to set all the transition probabilities from the left flanking state to different start points in the model. One option is to give them equal probabilities, η/L. Another is to assign more probability to starting at the beginning of the model. The default option in the HMMER package for profile HMMs [Eddy 1996] assigns probability $\eta/2$ to the start of the profile, and $\eta/(2(L-1))$ to the other positions, favouring matches that start at the beginning of the model.

If all the probability is assigned to the first model state, then it forces this model to match only complete copies of the profile in the searched sequence, ensuring a type of 'overlap' match constraint. This can be appropriate when, for example, the HMM represents a protein domain that you expect to find either present as a whole or absent. However, to allow for rare cases where the first residue might be missing, it may be wise in such cases to allow a direct transition from the flanking state into a delete state, as shown here:

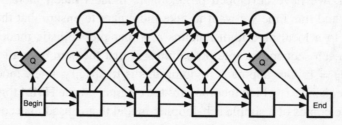

It is clear that by tinkering with the transition connections and probabilities a wide variety of different models can be produced, each potentially useful in different circumstances. A final example similar to the first model for local matches is

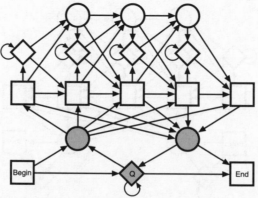

which allows repeat matches to subsections of the profile model, like the repeat algorithm variant in Chapter 2.

Note that all these variants of transition connectivity and probability assignment affect not only the types of match that are allowed, but also the score. More restrictive transition distributions will give higher match scores if a good match is found, so are preferable if they can be designed to represent the types of correct matches that are expected.

Exercises

5.1 Show that if the random model is the same as that described in Chapter 4 (a succession of two states looping on themselves with probability $(1 - \eta)$), with η the same as in the flanking models, the local alignment model gives update equations like those of equation (2.9).

5.2 Explain the reasons for any differences.

5.6 More on estimation of probabilities

As promised above, we now return to the subject of parameter estimation at greater length. Although our discussion for most of this section will be focused on the emission probabilities, analogous methods can be used for the transition probabilities. The aim here is to introduce methods that can be used. A more detailed mathematical discussion about the estimation of probabilites from sample counts is given in Chapter 11 (p. 311).

The most straightforward approach to parameter estimation would be to give the maximum likelihood estimates for the parameters. We will change notation slightly from that used before. Given observed frequencies c_{ja} of residue a in position j of the alignment, maximum likelihood estimates for $e_{M_j}(a)$, the corresponding model parameters, are

$$e_{M_j}(a) = \frac{c_{ja}}{\sum_{a'} c_{ja'}}. \tag{5.2}$$

As we described above, a clear problem with this is that if there are no observed examples of a particular outcome then its probability is estimated as zero. This will frequently occur. For example, in the alignment of Figure 5.3 only V, I and F are present in the first column. However, it is quite likely that other amino acids will occur in that position amongst all the other globin sequences in biology. The easiest way to deal with this problem is to add pseudocounts to the observed counts c_{ja}. Below, we first discuss the pseudocount approach at greater length, then give some more complex alternatives.

Simple pseudocounts

A very simple and much-used pseudocount method is to add a constant to all the counts, which prevents the problem with zero probabilities. When the constant is one, as we used above in our example, this is called 'Laplace's rule'. A slightly more sophisticated method is to add a quantity proportional to the background distribution, giving

$$e_{M_j}(a) = \frac{c_{ja} + Aq_a}{\sum_{a'} c_{ja'} + A},$$

(5.3)

where c_{ja} are the real counts, and A is the weight put on the pseudocounts as compared to the real counts. Values of A of around twenty seem to work well for protein alignments.

This form of regularisation has the appealing feature that $e_{M_j}(a)$ is approximately equal to q_a if very little data is available, i.e. all the real counts are very small compared to A. At the other extreme, where a large amount of data is available, the effect of the regulariser becomes insignificant and $e_{M_j}(a)$ is essentially equal to the maximum likelihood solution. So, at this intuitive level, pseudocounts make a lot of sense.

Adding pseudocounts amounts to adding some fake imagined data into the alignment, based on our general knowledge of proteins, to represent all the other things that might happen. They thus correspond to prior information about protein families, before having seen the specific data for the family in the form of the alignment. This statement can be formalised in a Bayesian framework. Bayes' equation tells us how to combine data, D, with a prior probability distribution over the parameters $P(\theta)$ to give a posterior distribution over θ, from which we can take either the maximum or the mean as our best estimate,

$$P(\theta|D) = \frac{P(D|\theta)P(\theta)}{P(D)}.$$

In our case the parameters θ are our model probabilities. The pseudocount method given above corresponds in this Bayesian framework to assuming a Dirichlet prior distribution with parameters $\alpha_a = Aq_a$ over the probabilities; see Chapter 11 for mathematical details.

Dirichlet mixtures

The problem with the simple pseudocounts, as compared to the substitution matrix based methods, is that only the most rudimentary prior knowledge can be contained in a single pseudocount vector. For this reason we need a lot of example data in the alignment to get good estimates of the parameters. Experience suggests that to achieve good discrimination typically fifty or more examples are desirable when modelling proteins.

In order to include better prior information, it was therefore suggested by Brown *et al.* [1993] that one should use a *mixture* of Dirichlet distributions as the prior. The idea is that there might be several different sets of pseudocount priors $\alpha_{\bullet}^1, \ldots, \alpha_{\bullet}^K$ corresponding to different types of alignment environments, where

α_a^k corresponds to Aq_a in the example above. One set might be relevant for exposed loop environments, one for buried small residue environments, etc. Given our counts c_{ja} we first estimate how likely each prior distribution k is (based on how well it fits the observed data), then combine their effects according to these posterior probabilities:

$$e_{M_j}(a) = \sum_k P(k|c_j) \frac{c_{ja} + \alpha_a^k}{\sum_{a'}(c_{ja'} + \alpha_{a'}^k)},$$

where the $P(k|c_i)$ are the *posterior mixture coefficients*. We calculate these by Bayes' rule,

$$P(k|c_j) = \frac{p_k P(c_j|k)}{\sum_{k'} p'_k P(c_j|k')}$$

where the p_k are the prior probabilities of each mixture component, and $P(c_j|k)$ is the probability of the data according to Dirichlet mixture k. The equation for $P(c_j|k)$ has a frightening looking form, which is in fact fairly simple to calculate:

$$P(c_j|k) = \frac{(\sum_a c_{ja})!}{\prod_a c_{ja}!} \frac{\Gamma\left(\sum_a c_{ja} + \alpha_a^k\right)}{\prod_a \Gamma(c_{ja} + \alpha_a^k)} \frac{\Gamma\left(\sum_a \alpha_a^k\right)}{\prod_a \Gamma(\alpha_a^k)},$$

where $\Gamma(x)$ is the gamma function, a standard function over the reals related to the factorial function on the integers. For further details and an explanation of this equation, see Chapter 11, where we also describe how the mixture component distributions $\alpha_{\bullet}^k$ are obtained.

Using this type of approach, it seems that good profile HMMs can be fit to alignments with as few as ten or twenty examples [Sjölander *et al.* 1996].

Substitution matrix mixtures

An alternative approach to using a mixture of Dirichlets is to adjust the pseudo-counts in a single Dirichlet formulation, using information from the observed counts and a substitution matrix. This is not a theoretically well-founded approach, but it makes intuitive sense as a heuristic, combining features of the non-probabilistic profile methods and the Dirichlet pseudocount methods.

The first step is to convert the matrix entries $s(a,b)$ into conditional probabilities $P(b|a)$. If we assume that the substitution matrix entries are derived as log-odds ratios, as in Chapter 2, then $s(a,b) = \log(P(a,b)/q_a q_b)$, which is the same as $\log(P(b|a)/P(b))$, so $P(b|a) = q_b e^{s(a,b)}$. We can in fact derive $P(b|a)$ values from an arbitrary score matrix $s(a,b)$ given background probabilities q_a; see below.

Given conditional probabilities $P(b|a)$ we can generate pseudocounts as follows. Let f_{ja} be the maximum likelihood probabilities derived from the counts, so $f_{ja} = c_{ja} / \sum_a' c_{ja'}$. Using these we set pseudocount values with

$$\alpha_{ja} = A \sum_b f_{jb} P(a|b),$$

where A is a positive constant comparable to the one we used with simple pseudo-counts [Tatusov, Altschul & Koonin 1994; Claverie 1994; Henikoff & Henikoff 1996]. We then use essentially the same equation as (5.3) to obtain the model parameters:

$$e_{M_j}(a) = \frac{c_{ja} + \alpha_{ja}}{\sum_{a'} c_{ja'} + \alpha_{ja'}}.$$

There is no obvious statistical interpretation for this type of pseudocount, but the idea is quite natural: amino acid i contributes to pseudocount j in proportion to its abundance in the column and the probability of its changing to amino acid j. The formula interpolates between the treatment of pairwise alignments and the maximum likelihood solution. The substitution matrix term dominates if there are small numbers of sequences (especially if $A \gg 1$), and values close to the maximum likelihood estimate are obtained when the number of counts is large (more precisely when the total number of counts $C_j \gg A$).

There are various choices for the scaling constant A of the pseudocounts. For instance $A = 1$ was used in Lawrence *et al.* [1993], but this appears to be too weak in practice. Claverie [1994] suggests $A = \min(20, N)$, and Henikoff & Henikoff [1996] suggest $A = 5R$, where R is the number of different residue types observed in the column (i.e. the number of a for which $c_{ja} > 0$).

Deriving $P(b|a)$ from an arbitrary matrix

Even if a score matrix $s(a,b)$ was not derived as a log-odds matrix, as long as certain conditions are fulfilled it is possible to find a scale factor λ such that $\lambda s(a,b)$ will behave correctly when interpreted as a log-odds matrix [Altschul 1991]. The conditions are that the matrix is negatively biased, i.e. $\sum_{ab} q_a q_b s(a,b) < 0$, and that it contains at least one positive entry.

What we want is a set of values r_{ij} for which

$$s(a,b) = \frac{1}{\lambda} \log \frac{r_{ab}}{q_a q_b},$$

where r_{ab} can be interpreted as the probability for the pair a,b. This equation is easily inverted, so we get the pair probabilities expressed in terms of the substitution matrix $r_{ab} = q_a q_b \exp(\lambda s(a,b))$. To be legitimate probabilities the r_{ab} have to sum to one. We therefore need to find a λ such that

$$f(\lambda) = \sum_{a,b} q_a q_b e^{\lambda s(a,b)} = 1. \tag{5.4}$$

One such value is $\lambda = 0$, but clearly this is not what we want. The two conditions we gave above turn out to be sufficient to ensure there is another, positive solution to this equation; see the exercises below.

The resulting value of λ is called the natural scaling factor of the substitution matrix. This probabilistic interpretation of the substitution matrix leads to an entropy measure for the matrix of $\sum_{ab} r_{ab} \log(r_{ab}/q_a q_b)$, which is a useful quantity for characterising and comparing substitution matrices [Altschul 1991].

Exercises

5.3   Use the negative bias condition to show that $f(\lambda)$ is negative for small enough λ. Hint: calculate $f'(0)$, the derivative of $f(\lambda)$ at $\lambda = 0$.

5.4   Use the second condition, that there is at least one positive $s(a,b)$, to show that $f(\lambda)$ becomes positive for large enough λ.

5.5   Finally, show that the second derivative of $f(\lambda)$ is positive, and from this and the results of the previous two exercises that there is one and only one positive value of λ satisfying (5.4).

Estimation based on an ancestor

There is a more principled and direct way to use the information in substitution matrices for estimating the HMM probabilities than that described above. This approach does not use pseudocounts. Instead, it assumes that all the observed sequences have been derived independently from a common ancestor, and generates an estimate of the residue present in a given position in that common ancestor (or rather a posterior probability distribution for what that residue was). From this we can estimate the probability of seeing each residue in a new descendant of the ancestor, different from those in the sample.

Assume we have example sequences x^k with residues x_j^k in column j of the alignment (we have adjusted our notation slightly; this x_j^k is not the jth residue in sequence x^k if there are gaps, but it is a convenient notation for what we need here). Once again, we need the conditional probabilities $P(b|a)$ derived from the substitution matrix. Let the residue in the common ancestor be y_j. Then we can use Bayes rule to calculate the posterior probability that $y_j = a$

$$P(y_j = a | \text{alignment}) = \frac{q_a \prod_k P(x_j^k | a)}{\sum_{a'} q_{a'} \prod_k P(x_j^k | a')}. \tag{5.5}$$

Note that we needed a prior distribution for residues at the common ancestor, which we set to q_a because that is our background probability for amino acids in the absence of further information.

We can now calculate the HMM emission probabilities as the predicted probabilities for a new sequence

$$e_{M_j}(a) = \sum_{a'} P(a|a') P(y_j = a' | \text{alignment}). \tag{5.6}$$

One problem with this approach is that, as we noticed above, different columns vary widely in their degree of conservation. Indeed, that is one of the properties that we wanted to exploit when using alignments to estimate profile HMMs. However, using a single substitution matrix implies assuming a fixed degree of conservation. As we discussed in Chapter 2, matrices typically come in families varying in their level of implied conservation. Examples are the PAM [Dayhoff, Schwartz & Orcutt 1978] and the BLOSUM [Henikoff & Henikoff 1992] series of matrices. We can therefore significantly improve the approach in (5.5) and (5.6) if we optimise over choice of matrix from a family. This way, a very conserved column might use a conservative matrix, such as PAM30, and a very varied column would use a divergent matrix, such as PAM500.

How do we choose the optimal matrix? A natural approach is to maximise the likelihood of the observed data

$$P(x_j^1, \ldots, x_j^N | t) = \sum_a q_a \prod_k P(x_j^k | a, t) \qquad (5.7)$$

where t is the matrix family parameter (t for evolutionary *time*). It would also be possible to use a Bayesian approach here, proposing a prior distribution over t, then combining this with (5.7) in Bayes' rule to obtain a posterior distribution for t, and summing over this in (5.6). However, that would require signficantly more computation.

The maximum likelihood time-dependent approach is closely related to the 'evolutionary weights' method in the PROFILE package [Gribskov & Veretnik 1996]. However, that method estimates different evolutionary times t for each possible ancestral amino acid, and also adjusts the resulting weights with respect to a set of baseline probabilities; for details see Gribskov & Veretnik [1996]. There are also strong connections between the methods of this subsection and those discussed later in Chapter 8 when building phylogenetic trees using maximum likelihood methods.

Testing the pseudocount methods

All the methods mentioned above have been tested in various ways. Direct tests, in which profiles were constructed and used for searching, were carried out extensively by Henikoff & Henikoff [1996]. The best method turned out to be the substitution matrix based method (5.6), with $A = 5R$ as described above; the Dirichlet mixture regulariser came a reasonably close second. Other tests gave different results [Tatusov, Altschul & Koonin 1994; Karplus 1995], so it is not clear which method is best, and it is likely that this will depend on the application and the details of the mixture components or substitution matrix used.

An interesting method was for testing various regularisers was given by Karplus [1995]. Instead of performing a huge number of database searches, he asked the following question: How well can an amino acid distribution be approximated from a small sample? Columns were extracted from a large set of deep alignments (the BLOCKS database; Henikoff & Henikoff [1991]). Imagine we take a small sample of size n with counts c_a from a column with complete counts C_a. From the sample counts c_a we can estimate the frequencies $e_c(a)$ of other symbols that might occur in the same column, using one of the methods described above (we use a subscript c to remind ourselves that this estimation is dependent on the sample counts). We can now calculate the log likelihood of the other symbols that actually do occur, as $\sum_a(C_a - c_a)\log e_c(a)$. Note that we do not score the counts that were used in the sample. We can now repeat this process for all samples of size n from all columns and sum all the resulting log likelihoods,

$$\text{LL} = \sum_{\text{columns } C} \sum_{\text{samples } c} \sum_a (C_a - c_a) \log e_c(a). \qquad (5.8)$$

Karplus proposed that a good regulariser should maximise this value. Furthermore, he pointed out that there is clearly an optimal strategy for such as process, which is to tabulate for each possible set of sample counts c_a the maximum likelihood distribution emission distribution $e_c(a)$. This is only practically possible to do explicitly up to $n = 5$.

Karplus showed that several of the more complex regularisers described above resulted in estimators that were very close to optimal, in the sense that the difference in LL values from optimal was very small at $n = 5$. Of course, ultimately we are interested in database searches, and it is not evident that the regulariser obtaining the lowest LL score will actually be best for searching. It is likely that the typical similarities in the source alignment database are not the same as the ones that we will be searching for with our HMM.

The actual averaging over all samples can only be done explicitly for sample sizes up to around $n = 5$, but that is also the most interesting regime, because it is for small sample sizes that regularisation is most crucial. For larger sample sizes we would have to use some form of random sampling method to estimate the average.

As well as evaluating methods, Karplus' approach can also be used to set the free parameters in the various methods described above, for example the total number of pseudocounts A to use in (5.3). For any value of A we can calculate the average log likelihood from our database of columns, either directly or by some sort of random sampling, and in fact we can also calculate the gradient of the relative entropy with respect to A. We can therefore find the value of A that minimises this average relative entropy, using gradient descent methods [Press *et al.* 1992], or by other optimisation methods. Of course, in principle this can be done for any sample size n, yielding parameters dependent on n. However, because the averaging is difficult for $n > 5$, this technique has yet to prove its potential.

5.7 Optimal model construction

When we first discussed the parameterisation of profile HMMs, we pointed out that as well as estimating the probability parameters, it is necessary to decide which columns of the alignment should be assigned to insert states, and which to match states. We call this process model construction. At the time we proposed a simple heuristic, but we can do better than that. There is an efficient dynamic programming algorithm which can find the column assignments that maximise the posterior probability of the mode, at the same time as fitting optimal probability parameters.

In the profile HMM formalism, it is assumed that an aligned column of symbols corresponds either to emissions from the same match state or to emissions from the same insert state. It therefore suffices to mark which columns come from match states to specify a profile HMM architecture and the state paths for all the sequences in the alignment, as shown in Figure 5.7. In a marked column, symbols are assigned to match states and gaps are assigned to delete states. In an unmarked column, symbols are assigned to insert states and gaps are ignored. State transition and symbol emission counts are obtained from the state paths, and these counts can be used to estimate probability parameters by one of the methods in the previous section. In passing, we note that this model estimation procedure implicitly assumes that the multiple alignment is correct, i.e. that the implied state paths have probability one and all other state paths have probability zero, which is akin to a Viterbi assumption. The next chapter addresses issues of simultaneous alignment and model estimation.

There are 2^L combinations of markings for an alignment of L columns, and hence 2^L different profile HMMs to choose from. There are at least three ways to determine the marking. In *manual* construction, the user marks alignment columns by hand. This is perhaps the simplest way to allow users to manually specify the model architecture to use for a given alignment. In *heuristic* construction, a rule is used to decide whether a column should be marked. For instance, a column might be marked when the proportion of gap symbols in it is below a certain threshold. In *MAP* construction, a maximum *a posteriori* choice is determined by dynamic programming. A description of this algorithm follows.

MAP match–insert assignment

The MAP construction algorithm recursively calculates a number S_j, which is the log probability of the optimal model for the alignment up to and including column

(a) Multiple alignment:

```
        x x . . . x
  bat   A G - - - C
  rat   A - A G - C
  cat   A G - A A -
  gnat  - - A A A C
  goat  A G - - - C
        1 2 . . . 3
```

(b) Profile-HMM architecture:

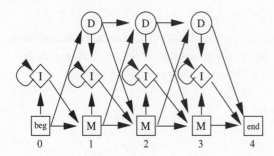

(c) Observed emission/transition counts

		model position			
		0	1	2	3
match emissions	A	-	4	0	0
	C	-	0	0	4
	G	-	0	3	0
	T	-	0	0	0
insert emissions	A	0	0	6	0
	C	0	0	0	0
	G	0	0	1	0
	T	0	0	0	0
state transitions	M-M	4	3	2	4
	M-D	1	1	0	0
	M-I	0	0	1	0
	I-M	0	0	2	0
	I-D	0	0	1	0
	I-I	0	0	4	0
	D-M	-	0	0	1
	D-D	-	1	0	0
	D-I	-	0	2	0

Figure 5.7 *As an example of model construction from an alignment, a small DNA multiple alignment is given (a), with three columns marked above with x's. These three columns are assigned to positions 1–3 in the model architecture (b). The assignment of columns to model positions determines the symbol emission and state transition counts (c) from which probability parameters would be estimated.*

j, assuming that column j is marked. S_j is calculated from smaller subalignments ending at a marked column i ($i < j$) by incrementing S_i with the summed log probability of the transitions and emissions for the columns between i and j. The relevant probability parameters are estimated 'on the fly' from the counts that are implied by marking columns i and j while leaving unmarked the intervening columns (if any).

Transition and emission counts for a section of alignment bounded by marked columns i and j are independent of how columns are marked before i and after j, thus making a dynamic programming recursion possible. Only marked columns are considered in the recursion, because transition and emission counts for unmarked columns are not independent of the assignment of neighbouring columns; a single insert state may account for more than one column in the alignment.

For instance, let $\mathcal{T}_{ij}$ be the summed log probability of all the state transitions between marked columns i and j. We can determine $\mathcal{T}_{ij}$ from the observed state transition counts c_{xy} and the probabilities t_{xy}:

$$\mathcal{T}_{ij} = \sum_{x,y \in M,D,I} c_{xy} \log a_{xy}.$$

Transition counts c_{xy} are obtained from the partial state paths implied by marking i and j. For instance, if in one sequence we see a gap in column i, five residues in columns $i+1$ to $j-1$, and a residue in column j, we would count one delete–insert transition, four insert–insert transitions, and one insert–match transition. The transition probabilities a_{xy} are estimated from the c_{xy} in the usual fashion, possibly including Dirichlet prior terms α_{xy} (or indeed, any form of prior that is independent of the marking outside of $i,\ldots,j$):

$$a_{xy} = \frac{c_{xy}+\alpha_{xy}}{\sum_y c_{xy}+\alpha_{xy}}.$$

Let $\mathcal{M}_j$ be the analogous log probability contribution for match state symbol emissions in column j, and $\mathcal{I}_{i+1,j-1}$ be the same for the insert state emissions for columns $i+1,\ldots,j-1$ (for $j-i > 1$). We can now give the algorithm:

Algorithm: MAP model construction

Initialisation:
$$S_0 = 0,\ \mathcal{M}_{L+1} = 0.$$

Recurrence: for $j = 1,\ldots,L+1$:
$$S_j = \max_{0 \le i < j} S_i + \mathcal{T}_{ij} + \mathcal{M}_j + \mathcal{I}_{i+1,j-1} + \lambda;$$

$$\sigma_j = \operatorname*{argmax}_{0 \le i < j} S_i + \mathcal{T}_{ij} + \mathcal{M}_j + \mathcal{I}_{i+1,j-1} + \lambda.$$

Traceback: From $j = \sigma_{L+1}$, while $j > 0$:
Mark column j as a match column;
$j = \sigma_j$.   ◁

A profile HMM is then built from the marked alignment. The extra term λ is a penalty used to favour models with fewer match states. In Bayesian terms, λ is the log of the prior probability of marking each column, implying a simple but adequate exponentially decreasing prior distribution over model lengths.

With some care in implementation, this algorithm is $O(L)$ in memory and $O(L^2)$ in time for an alignment of L columns.

5.8 Weighting training sequences

One issue that we have avoided completely so far is that of weighting sequences when estimating parameters. In a typical alignment, there are often some sequences that are very closely related to each other. Intuitively, some of the information from these sequences is shared, so we should not give them each the same influence in the estimation process as a single sequence that is more highly diverged from all the others. In the extreme that two sequences are identical, it makes sense that they should each get half the weight of other sequences, so that

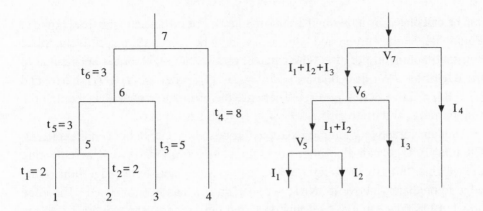

Figure 5.8 *On the left, a tree of sequences with branch lengths. On the right, the corresponding 'current' and 'voltage' values used in the 'Kirchhoff's law' approach to sequence weighting (see text).*

the net effect is of having only one of them. Statistically, the problem is that typically the examples we have do not constitute a good random sample from all the sequences that belong to the family; the assumption of independence is incorrect. To deal with this sort of situation, there have been a large number of proposals for different ways to assign weights to sequences. In principle, any of these can be used in combination with any of the methods of the preceding sections on fitting model parameters and model construction.

Simple weighting schemes derived from a tree

Many weighting approaches are based on building a tree relating the sequences. Since sequences in a family are related by an evolutionary tree, a very natural approach is to try to reconstruct this tree and use it when estimating the independent contribution of each of the observed sequences, downweighting sequences that have only recently diverged. We discuss phylogenetic tree construction at length later in Chapters 7 and 8, as well as in the next chapter on multiple sequence alignment. For our current purposes, the fine details of the method are probably not too important, and we will assume that we are given a tree connecting the sequences, with branch lengths indicating the relative degrees of divergence for each edge in the tree.

One of the intuitively simplest weighting schemes [Thompson, Higgins & Gibson 1994b] can be expressed nicely as follows. We are given a tree made of a conducting wire of constant thickness and apply a voltage V to the root. All the leaves are set to zero potential and the currents flowing from them are measured and taken to be the weights. Clearly, the currents will be smaller in the highly divided parts of the tree so these weights have the right qualitative properties. They can be calculated by applying Kirchhoff's laws. For instance, in the tree shown in Figure 5.8, let the current and voltage at node n be I_n and V_n, respectively. Since constant factors do not affect the calculation, we can set the resistance equal to the edge-time. We then find $V_5 = 2I_1 = 2I_2$, $V_6 = 2I_1 + 3(I_1 + I_2) = 5I_3$, and $V_7 = 8I_4 = 5I_3 + 3(I_1 + I_2 + I_3)$. There are therefore three equations relating the four currents, and these give $I_1 : I_2 : I_3 : I_4 = 20 : 20 : 32 : 47$.

Another attractively simple idea was proposed by Gerstein, Sonnhammer & Chothia [1994]. Their algorithm works up the tree from the leaves, incrementing the weights. Initially the weight of a sequence is set equal to the edge-time of the edge immediately above it. Now, suppose node n has been reached. The edge above n has edge-time t_n, and this is shared out amongst the weights of all the sequences at the leaves below n, incrementing them by a fraction proportional to their current weight values. Formally, the increase Δw_i in a weight w_i is given by

$$\Delta w_i = t_n \frac{w_i}{\sum_{\text{leaves } k \text{ below } n} w_k}. \tag{5.9}$$

The same operation is carried out up to the root.

This is clearly an easy and efficient algorithm. For instance, the weights in the tree of Figure 5.8 are computed as follows: Initially the weights are set to the edge lengths of the leafs, $w_1 = w_2 = 2$, $w_3 = 5$, and $w_4 = 8$. At node 5 the edge length of 3 above node 5 is shared out equally to w_1 and w_2, giving them $3/2$ each, so now $w_1 = w_2 = 2 + 3/2 = 3.5$. At node 6 we find the edge of length 3 above node 6 is shared out to nodes 1, 2 and 3 in the ratio $3.5 : 3.5 : 5$, making $w_1 = w_2 = 3.5 + 3 \times 3.5/12$, and $w_3 = 5 + 3 \times 5/12$. With $w_4 = 8$, this gives $w_1 : w_2 : w_3 : w_4 = 35 : 35 : 50 : 64$. Even though these weights are close to those given by the Kirchhoff rule, the methods are in a sense opposed, for in a tree with two leaves and one edge longer than the other, the longer edge is down weighted by Kirchhoff and up weighted by (5.9).

Root weights from Gaussian parameters

One view of weights is that they should represent the influence of leaves on the root distribution. It is possible to make this idea precise, as Altschul, Carroll & Lipman [1989] showed. They built on the version of Felesenstein's 'pruning' algorithm which applies to continuous parameters [Felsenstein 1973]. Instead of discrete members of an alphabet we have a continuous real-valued variable, like the weight of an organism. In place of a substitution matrix we have a probability density that defines the probability of substituting one value, x, of this variable by another, y. A simple example of such a density is a Gaussian, where the probability of $x \rightarrow y$ along an edge with time t is $\exp(-(x - y)^2/(2\sigma^2 t'))$. The

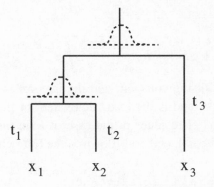

Figure 5.9 *The tree described in the text when deriving Gaussian weights.*

pruning algorithm now proceeds exactly as for a finite alphabet, but with integrals replacing discrete sums [Felsenstein 1973].[3]

Felsenstein's algorithm yields a Gaussian distribution for the parameter in question at the root whose mean μ depends linearly on the values x_i of the parameters at the leaves, so $\mu = \sum w_i x_i$. Altschul, Carroll & Lipman [1989] proposed that these w_i should be used as weights. They represent the influence of each leaf at the root.

Example: Altschul–Carroll–Lipman weights for a three-leaf tree

To illustrate how the weights are derived, consider the simple three-leaf tree shown in Figure 5.9, where leaf i takes the value x_i. The probability distribution at node 4 is given by

$$P(x \text{ at node } 4 \mid L_1, L_2) = K_1 e^{-\frac{(x-x_1)^2}{2t_1}} e^{-\frac{(x-x_2)^2}{2t_2}}$$

where K_1 is a normalising constant. One can rewrite this as

$$P(x \text{ at node } 4 \mid L_1, L_2) = K_1 e^{-\frac{(x-v_1 x_1 - v_2 x_2)^2}{2t_{12}}}$$

where $v_1 = t_2/(t_1 + t_2)$, $v_2 = t_1/(t_1 + t_2)$ and $t_{12} = t_1 t_2/(t_1 + t_2)$. If we were considering only the two-leaf tree with root at node 4, the mean of the root distribution would be given by $\mu = v_1 x_1 + v_2 x_2$, and the weights would be v_1 and v_2. Continuing with our three-leaf tree, however, we find next that the distribution at node 5

---

[3] Historically, the continuous case came first, and Felsenstein defined the pruning algorithm for Gaussian distributions of real-valued parameters. In the cited paper he takes account of the distribution of the parameters at each leaf, e.g. the mean and variance of the weight of an organism. Puzzlingly, he also introduces covariances between values for different leaves. It is not clear how to calculate a covariance between, say, the weights of cows and cats. For proteins, having multiple corresponding sites in an alignment would allow correlations to be considered in principle.

is given by

$$P(y \text{ at node } 5 \mid L_1, L_2, L_3) = K_2 e^{-\frac{(y-x_3)^2}{2t_3}} \int e^{-\frac{(x-v_1x_1-v_2x_2)^2}{2t_{12}}} e^{-\frac{(x-y)^2}{2t_4}} dx$$

where K_2 is a normalising constant, and the integral is taken over all possible values of x at node 4 (and is the exact equivalent of the sum over all possible ancestral assignments of residues in the case of a discrete alphabet). This is a standard Gaussian integral, and boils down to the following

$$P(y \text{ at node } 5 \mid L_1, L_2, L_3) = K_3 e^{-\frac{(y-w_1x_1-w_2x_2-w_3x_3)^2}{2t_{123}}}$$

where K_3 is a new normalising constant and $t_{123} = t_3\{t_1t_2 + t_4(t_1 + t_2)\}/\Omega$, with $\Omega = t_1t_2 + (t_3 + t_4)(t_1 + t_2)$. The mean of the distribution of y, i.e. of the root distribution, is given by

$$\mu = w_1x_1 + w_2x_2 + w_3x_3$$

with $w_1 = t_2t_3/\Omega$, $w_2 = t_1t_3/\Omega$, and $w_3 = \{t_1t_2 + t_4(t_1 + t_2)\}/\Omega$. These are therefore the Altschul–Carroll–Lipman weights for a tree with three leaves.   □

Voronoi weights

There are also weighting schemes not based on trees. One approach is based on an image of the sequences from a family lying in 'sequence space'. In general, some will lie in clusters and others will be widely separated. The philosophy of the Voronoi scheme [Sibbald & Argos 1990] is to assume that this unevenness represents effects of sampling, including the 'sampling' performed by natural selection in favouring certain phyla. A more thorough trawl through all eligible sequences of the protein family, or perhaps a multitude of reruns of evolution, should produce a flat distribution within some region. To compensate for the gaps, we want to give sequences a weight proportional to the volume of empty space around them.

If sequence space were two-dimensional, or even low-dimensional, we could use standard methods from computational geometry to divide up space into regions around each example point. The standard approach is to take lines joining neighbouring pairs of points and draw their perpendicular bisectors, extending them till they join up. This produces a partitioning into polygons (in two dimensions) called a *Voronoi diagram* [Preparata & Shamos 1985], which has the property that the polygon around each point is the set of all points closer to that point than any other.

Sequence space is of course a high-dimensional construct in which the Voronoi geometry is hard to picture or calculate. However, we can implement the underlying principle of it by sampling sequences randomly from sequence space and testing to see which of the family sequences each sequence lies closest to. The trick is in the sampling. This is accomplished by choosing, at each position of the alignment, uniformly from those residues which occur at that position in any sequence. If we count n_i such sample sequences closest to the ith family member (dividing up the counts if there is a tie), then we can define the ith weight to be $n_i / \sum_k n_k$.

Maximum discrimination weights

Another approach to weighting comes indirectly, from focusing initially on a reformulation of the primary goal in building the model [Eddy, Mitchison & Durbin 1995]. Rather than maximising the likelihood of sequences in the family, or even their posterior probability derived from Bayesian priors, we are normally interested in making the correct decision on whether sequences are members of the family or not. We are therefore interested in the probability

$$P(M|x) = \frac{P(x|M)P(M)}{P(x|M)P(M) + P(x|R)(1 - P(M))},$$

where x is a sequence from the family, M is the model for the family that we are fitting, R is our alternative, random model for sequences not in the family, and $P(M)$ is the prior probability of a new sequence belonging to the family. Given example training sequences x^k, we would like to maximise the probability of classifying them all correctly, which is

$$D = \prod_k P(M|x^k),$$

not $\prod P(x^k|M)$ as usual with maximum likelihood based approaches. We call D the *discrimination* of the model on the set of sequences x^k. Maximising D will have the effect of emphasising performance on distant or difficult members of the family. Sequences that are easily classified will have $P(M|x)$ values very close to one; changing parameters to increase their likelihood $P(x|M)$ will have very little effect on D. On the other hand, increasing the likelihood of sequences for which $P(M|x)$ is small can potentially have a big effect.

It turns out that the parameter values that maximise D can be shown to be the ones that maximise a weighted version of the likelihood, where the weights are proportional to $1 - P(M|x_i)$, i.e. the probability of misclassifying sequence i. This can be seen from the observation that if $y = e^x / (K + e^x)$, then

$$\frac{\partial \log y}{\partial x} = \frac{1}{K + e^x} = K(1 - y).$$

If we set $x = \log \left(\frac{P(x|M)}{P(x|R)} \right)$, which is the log likelihood ratio for sequence x, then $y = P(M|x)$. So at a maximum of $\log D$ we will also be at a maximum of the weighted sum of log likelihood ratios, with weights $1 - P(M|x_i)$, and since the random model is fixed this is equivalent to a maximum of the weighted log likeli-hood of the model M. The maximum discrimination criterion therefore amounts to another sequence weighting system.

One difference from previous systems, however, is that these weights are de-fined in a somewhat circular fashion; they depend upon the model that is being fit. When using maximum discrimination weighting as a method, an iterative ap-proach must be used; an initial set of weights gives rise to a model, from which posterior probabilities $P(M|x)$ can be calculated, giving rise to new weights, and hence a new model, and so on until convergence is achieved. This iterative re-estimation procedure is analogous to the versions of the EM algorithm used to fit HMM parameters to sets of unlabelled sequences (p. 64 and p. 323).

Maximum discrimination training has a big advantage in that it is directly op-timising performance on the type of operation that the model will be used for, ensuring that the most effort is applied to recognising the most distant sequences. On the other hand, exactly the same point can lead to problems. If there is any training sequence that has been misclassified, then the distortion needed to give it a good score can damage performance for correct members of the class. To some extent, though, this same problem occurs with all weighting schemes: incorrectly assigned sequences will be the most distant ones in any tree that gets built from the examples.

Maximum entropy weights

Finally, we describe two weighting methods based on the idea of trying to make the statistical spread of the model as broad as possible.

Assume column i of a multiple alignment has k_{ia} residues of type a and a total of m_i different types of residues. To make a distribution as uniform as possible from these counts by weighting each sequence, we can choose a weight for sequence k of $1/(m_i k_{ix_i^k})$. Maximum likelihood estimation will then yield a distribution $p_{ia} = k_{ia}/(m_i k_{ia}) = 1/m_i$, i.e. all the residues appearing in the column will have the same probability. To illustrate the idea, suppose we have ten sequences with residue A at a site, and one sequence with a B, so the unweighted frequencies of A and B are $c_A = \frac{10}{11}$, $c_B = \frac{1}{11}$. The weights of the ten sequences are $w_1 = w_2 = \ldots = w_{10} = 1/(2 \times 10) = 0.05$, and $w_{11} = 1/(2 \times 1) = 0.5$, which have the effect of making the overall weighting for each of A and B equal.

The preceding paragraph only considered one column. With just one weight per sequence, it is of course not possible to make the distribution uniform for all columns in an alignment. However, by averaging over all columns, one may hope to obtain reasonable weights. That is, the weights are calculated as

$$w_k = \sum_i \frac{1}{m_i k_{ix_i^k}},$$

and then normalised to sum to one. This weighting scheme was proposed by [Henikoff & Henikoff 1994].

Instead of averaging, there is another approach to combining the information from the different columns that has a simple theoretical justification. A standard measure of the 'uniformity' of a distribution is the entropy (11.8), which is larger the more uniform the distribution is. Indeed, it is easy to see that the weights chosen above based on a single column maximise the entropy of the distribution p_{ia} for that column. An HMM defines a probability distribution over sequences, and therefore a natural extension of the single column weighting to full sequences is to maximise the entropy of the complete HMM distribution [Krogh & Mitchison 1995]. We will see that, perhaps surprisingly, this is closely related to maximum discrimination weighting.

Let us consider all the sites in an alignment with no gaps. We then sum the entropies from each site, and choose the weights to maximise this sum; that is we maximise $\sum_i H_i(w\bullet) + \lambda \sum_k w_k$, where $H_i(w\bullet) = \sum_a p_{ia} \log p_{ia}$, and p_{ia} is the weighted frequency of residue a at the ith site, computed as above.

Suppose for instance that we have the sequences $x^1 = \text{AFA}$, $x^2 = \text{AAC}$, and $x^3 = \text{DAC}$. Giving them weights w_1, w_2 and w_3, respectively, the entropies at each site are

$$
\begin{aligned}
H_1(w\bullet) &= -(w_1 + w_2)\log(w_1 + w_2) - w_3 \log w_3, \\
H_2(w\bullet) &= -w_1 \log w_1 - (w_2 + w_3)\log(w_2 + w_3), \\
H_3(w\bullet) &= -w_1 \log w_1 - (w_2 + w_3)\log(w_2 + w_3).
\end{aligned}
$$

We assume that the weights sum to one, and therefore we have to use a Lagrange multiplier term $\lambda \sum_k w_k$, when differentiating and finding the maximum of the entropy. Setting the derivatives of $H_1(w\bullet) + H_2(w\bullet) + H_3(w\bullet) + \lambda \sum_k w_k$ to zero gives $(w_1 + w_2)w_1^2 = (w_1 + w_2)(w_2 + w_3)^2 = w_3(w_2 + w_3)^2$, which implies $w_1 = w_3 = 0.5, w_2 = 0$. This makes the frequencies in each column equal, which was our goal. If it seems odd to give a sequence zero weight, note that the residue at each site in x^2 is always present in one of the other two sequences. Intuitively, x^2 lies 'between' x^1 and x^3, (in fact, it would be a possible ancestral sequence of x^1 and x^3 in an evolutionary reconstruction based on parsimony; see Chapter 7).

Another way to view to the result of this example is that if we set the model probabilities to be the weighted counts frequencies, as a weighted maximum likelihood procedure would, the resulting model assigns an equal probability to all of the original sequences, x^1, x^2 and x^3. This seems very reasonable, according to the view that all the example sequences should be treated as equally good members of the family for which we are building the model. In fact, Krogh & Mitchison [1995] show that the maximum entropy procedure assigns weights to the example sequences so that some subset of the sequences (perhaps all of them) have non-zero weight and equal probabilities under the resulting model, or they have a higher probability, in which case they have zero weight. The former can be thought of as boundary points for the region of sequence space occupied by the whole sequence set, while the latter are internal points.

Furthermore, empirical tests indicate that the maximum entropy weights are optimal in the sense that they maximise the minimum score assigned to any of the example sequences [Krogh & Mitchison 1995]. This is an absolute version of the criterion specified in the previous section on maximum discrimination weights; rather than simply weighting the weakest match most strongly, all the parameter-fitting effort is applied to increasing its score, until it reaches that of the other non-zero-weighted sequences. Although satisfying an attractive goal, maximum entropy weighting suffers from the same problems as maximum discrimination: if a sequence is an outlier that should not be a full member of the family, the method will force it in, possibly at a substantial cost in performance on all other sequences. In addition, the rejection of all information from some of the sequences may seem intuitively undesirable.

Exercise

5.6   Compute the weights for the following sequence set, using each of the weighting methods described above except Voronoi weights (which requires random sampling of sequences): AGAA, CCTC, AGTC.

5.9   Further reading

PSSM methods were introduced during the 1980s for finding new members of sequence families, although the matrix values were not always obtained using an explicit probability-based derivation. They are also known by other names, such as *weight matrices* [Staden 1988]. More recent papers using related methods include those by Stormo [1990]; Henikoff & Henikoff [1994]; Tatusov, Altschul & Koonin [1994].

The non-probabilistic versions of profiles already have a long history, and many variants of the profile method have been suggested and tested. Thompson, Higgins & Gibson [1994b] and Luthy, Xenarios & Bucher [1994] report an improvement when the sequences are weighted using one of the BLOSUM matrices [Henikoff & Henikoff 1992] instead of a PAM matrix. In Thompson, Higgins & Gibson [1994b] the treatment of gaps is also improved.

Several ways have been suggested for incorporating structural information into profiles. In Luthy, McLachlan & Eisenberg [1991] substitution matrices were estimated for six different structural environments: the three secondary structure elements α-helix, β-sheet, and 'other' combined with an outside/inside classification, which was based on the exposure of an amino acid to solvent. Other variations of structural profiles can be found in Bowie, Luthy & Eisenberg [1991]; Wilmanns & Eisenberg [1993].

Early on, profile HMMs were adopted by Baldi *et al.* [1994], who used them to model globins, immunoglobulins and kinases. In this work a different estimation method was also introduced, which was based on gradient descent, see also Baldi & Chauvin [1994]. The same basic structure of profile HMMs has since been used in several different areas. A library of HMMs for all the big protein families has been established under the name of PFAM [Sonnhammer, Eddy & Durbin 1997]. The library of regular expressions called PROSITE [Bairoch, Bucher & Hofmann 1997] is being extended to something essentially like profile HMMs [Bucher *et al.* 1996]. Profile HMMs also have several uses for DNA. For instance they can be used to find DNA repeat family members in large-scale genomic sequence.

6

Multiple sequence alignment methods

In Chapter 5, we assumed that a reasonable multiple sequence alignment was already known and provided the starting point for constructing a profile HMM. We now look at what a 'reasonable' multiple alignment is, and at ways to construct one automatically from unaligned sequences.

Multiple alignments must usually be inferred from primary sequences alone. Biologists produce high quality multiple sequence alignments by hand using expert knowledge of protein sequence evolution. This knowledge comes from experience. Important factors include: specific sorts of columns in alignments, such as highly conserved residues or buried hydrophobic residues; the influence of secondary and tertiary structure, such as the alternation of hydrophobic and hydrophilic columns in exposed beta sheet; and expected patterns of insertions and deletions, that tend to alternate with blocks of conserved sequence. Furthermore, the phylogenetic relationships between sequences dictate constraints on the changes that occur in columns and in the patterns of gaps. RNA alignments involve similar knowledge but additionally they are often strongly constrained by a secondary structure model that in many cases has also been inferred from primary sequence data (Chapter 10).

Manual multiple alignment is tedious. Automatic multiple sequence alignment methods are a topic of extensive research in computational biology. In general, an automatic method must have a way to assign a score so that better multiple alignments get better scores. We should carefully distinguish the problem of scoring a multiple alignment from the problem of searching over possible multiple alignments to find the best one. Descriptions of multiple sequence alignment programs tend to emphasise the alignment algorithm rather than the scoring function. However, by now it should be clear that the scoring function is our primary concern in probabilistic modelling, and algorithms, though important, are secondary. One of our goals in probabilistic modelling is to incorporate as many of an expert's evaluation criteria as possible into our scoring procedure.

We therefore start our discussion of automatic multiple alignment by considering carefully what we want to do. We look at what a multiple sequence alignment means, structurally and evolutionarily. Then we consider the question of how best to turn the biological criteria into a numerical scoring scheme, so that a program will recognise a good multiple alignment. We examine various approaches taken by different multiple alignment programs. We conclude by describing full probabilistic multiple alignment approaches based on the profile HMMs we introduced in Chapter 5 and comparing the strengths and weaknesses of profile HMM alignment to other methods. We will focus primarily on protein alignments, though most of the discussion applies to DNA alignments as well. (Alignment of RNA is complicated by long-range correlations due to base pairing and is not treated until Chapter 10.)

6.1 What a multiple alignment means

In a multiple sequence alignment, homologous residues among a set of sequences are aligned together in columns. 'Homologous' is meant in both the structural and evolutionary sense. Ideally, a column of aligned residues occupy similar three-dimensional structural positions and all diverge from a common ancestral residue. For example, in Figure 6.1, a manually generated multiple alignment of ten immunoglobulin superfamily sequences is shown. A crystal structure of one of the sequences (1tlk, telokin) is known. The telokin structure and alignments to other related sequences reveal conserved characteristics of the I-set immunoglobulin superfamily fold, including eight conserved β-strands and certain key residues in the sequences, such as two completely conserved cysteines in the b and f strands which form a disulfide bond in the core of the folded structure. The other nine sequences, from various neural cell adhesion molecules, have been manually aligned to 1tlk based on this expert structural knowledge.

Except for trivial cases of highly identical sequences, it is not possible to unambiguously identify structurally or evolutionarily homologous positions and create a single 'correct' multiple alignment. Since protein structures also evolve (though more slowly than protein sequences), we do not expect two protein structures with different sequences to be entirely superposable. Chothia & Lesk [1986] examined pairwise structural alignments in several different protein families and found that for a given pair of divergent but clearly homologous (30% identical) protein sequences, usually only about 50% of the individual residues were superposable in the two structures (Figure 6.2). The globin family, often used as a 'typical' protein family in computational work, is in fact exceptional: almost the entire structure is conserved among divergent sequences. Even the definition of 'structurally superposable' is subjective and can be expected to vary among experts.

In principle, there is always an unambiguously correct evolutionary alignment even if the structures diverge. In practice, however, an evolutionarily correct alignment can be even more difficult to infer than a structural alignment. While structural alignment has an independent point of reference (superposition of crystal or NMR structures), the evolutionary history of the residues of a sequence

```
structure:    ...aaaaa...bbbbbbbbbb.....cccccccCCC..C........ddd
1tlk          ILDMDVVEGSAARFDCKVEGY--PDPEVMWFKDDNP--VKESR----HFQ
AXO1_RAT      RDPVKTHEGWGVMLPCNPPAHY-PGLSYRWLLNEFPNFIPTDGR---HFV
AXO1_RAT      ISDTEADIGSNLRWGCAAAGK--PRPMVRWLRNGEP--LASQN----RVE
AXO1_RAT      RRLIPAARGGEISILCQPRAA--PKATILWSKGTEI--LGNST----RVT
AXO1_RAT      ----DINVGDNLTLQCHASHDPTMDLTFTWTLDDFPIDFDKPGGHYRRAS
NCA2_HUMAN    PTPQEFREGEDAVIVCDVVSS--LPPTIIWKHKGRD--VILKKDV--RFI
NCA2_HUMAN    PSQGEISVGESKFFLCQVAGDA-KDKDISWFSPNGEK-LTPNQQ---RIS
NCA2_HUMAN    IVNATANLGQSVTLVCDAEGF--PEPTMSWTKDGEQ--IEQEEDDE-KYI
NRG_DROME     RRQSLALRGKRMELFCIYGGT--PLPQTVWSKDGQR--IQWSD----RIT
NRG_DROME     PQNYEVAAGQSATFRCNEAHDDTLEIEIDWWKDGQS--IDFEAQP--RFV
consensus:    .......G..+.+.C.+.........+.W.........+........++

structure:    ddd.....eeeeee.......ffffffffff.......gggggggggggg.
1tlk          IDYDEEGNCSLTISEVCGDDDAKYTCKAVNSL-----GEATCTAELLVET
AXO1_RAT      SQTT----GNLYIARTNASDLGNYSCLATSHMDFSTKSVFSKFAQLNLAA
AXO1_RAT      VLA-----GDLRFSKLSLEDSGMYQCVAENKH-----GTIYASAELAVQA
AXO1_RAT      VTSD----GTLIIRNISRSDEGKYTCFAENFM-----GKANSTGILSVRD
AXO1_RAT      AKETI---GDLTILNAHVRHGGKYTCMAQTVV-----DGTSKEATVLVRG
NCA2_HUMAN    VLSN----NYLQIRGIKKTDEGTYRCEGRILARG---EINFKDIQVIVNV
NCA2_HUMAN    VVWNDDSSSTLTIYNANIDDAGIYKCVVTGEDG----SESEATVNVKIFQ
NCA2_HUMAN    FSDDSS---QLTIKKVDKNDEAEYICIAENKA-----GEQDATIHLKVFA
NRG_DROME     QGHYG---KSLVIRQTNFDDAGTYTCDVSNGVG----NAQSFSIILNVNS
NRG_DROME     KTND----NSLTIAKTMELDSGEYTCVARTRL-----DEATARANLIVQD
consensus:    ..........L.+..+...+.+.Y.C................+.+.+..
```

Figure 6.1 *A multiple alignment of ten I-set immunoglobulin superfamily domains, adapted from Harpaz & Chothia [1994]. To the left are sequence identifiers from the* PDB *or* SWISS-PROT *databases. The eight β-strands of the telokin structure, 1tlk, are annotated at the top (a-g; C represents the c' strand). Aligned columns are annotated at the bottom if all residues are identical (letter) or highly conservative (+).*

family is not independently known from any source; it must itself be inferred from sequence alignment. Since sequence tends to diverge more rapidly than structure, parts of proteins which are structurally unalignable are typically not alignable by sequence either.

Thus, our ability to define a single 'correct' alignment will vary with the relatedness of the sequences being aligned. An alignment of very similar sequences will generally be unambiguous, but these alignments are not of great interest to us; a simple program can get the alignment right. For cases of interest (e.g. for a family of proteins sharing perhaps only 30% average pairwise sequence identity), we must keep in mind that there is no objective way to define an unambiguously correct alignment. Usually, a small subset of key residues will be identifiable which can be aligned unambiguously for all the sequences in a family almost regardless of sequence divergence [Harpaz & Chothia 1994]; core structural elements will also tend to be conserved and meaningfully alignable; but other regions may not be meaningfully alignable because of structural evolution and sequence divergence.

Assessments of multiple alignment quality must keep these considerations in mind. Asking a sequence alignment program to produce *exactly* the same align-

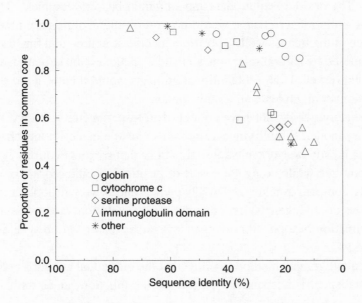

Figure 6.2 *Proportion of structurally superposable residues in pairwise alignments as a function of sequence identity; redrawn from data in Chothia & Lesk [1986]. 'Other' structural alignments include pairwise alignments of two dihydrofolate reductases, two lysozymes, plastocyanin/azurin, and papain/actinidin.*

ment as a manual structural alignment, for instance, means building in the same meaningless biases about how to 'align' structurally unalignable regions. Instead, we should focus attention on the subset of columns corresponding to key residues and core structural elements that can be aligned with confidence [McClure, Vasi & Fitch 1994].

6.2 Scoring a multiple alignment

Our scoring system should take into account at least two important features of multiple alignments: (1) the fact that some positions are more conserved than others, e.g. position-specific scoring; and (2) the fact that the sequences are not independent, but instead are related by a phylogenetic tree. An idealised way to score a multiple alignment would therefore be to specify a complete probabilistic model of molecular sequence evolution. Given the correct phylogenetic tree for the sequences, the probability of a multiple alignment is the product of the probabilities of all the evolutionary events necessary to produce that alignment via ancestral intermediate sequences times the prior probability of the root ancestral sequence. The desired evolutionary model would be very complex. The probabilities of evolutionary change would depend on the evolutionary times along each branch of the tree, as well as position-specific structural and functional constraints imposed by natural selection, so that key residues and structural elements would be conserved. High-probability alignments would then be good structural and evolutionary alignments under this model.

Unfortunately, we do not have enough data to parameterise such a complex evolutionary model. Simplifying assumptions must be made. In this chapter, we concentrate mostly on workable approximations that partly or entirely ignore the phylogenetic tree while doing some sort of position-specific scoring of aligning structurally compatible residues. In Chapters 7 and 8 we will look at more explicit models of phylogenetic trees and molecular evolution, most of which make an approximation of a position-independent rather than position-specific evolutionary model.

Almost all alignment methods assume that the individual columns of an alignment are statistically independent. Such a scoring function can be written as

$$S(m) = G + \sum_i S(m_i) \tag{6.1}$$

where m_i is column i of the multiple alignment m, $S(m_i)$ is the score for column i, and G is a function for scoring the gaps that occur in the alignment.

We write G as an unspecified function because methods of scoring gaps in multiple alignments differ greatly. The simplest method is to treat a gap symbol as an extra residue type, which then just gives $S(m) = \sum_i S(m_i)$. However, most multiple alignment methods use affine scoring functions that pay a higher cost for opening the gap than extending it, so successive gap residues are not treated independently. For simplicity, we will focus in the next several paragraphs on definitions of $S(m_i)$ for scoring a column of aligned residues with no gaps.

Minimum entropy

We now define some notation. As above, m is a multiple alignment. Let m_i^j be the symbol in column i for sequence j. Let c_{ia} be the observed counts for residue a in column i; $c_{ia} = \sum_j \delta(m_i^j = a)$, where $\delta(m_i^j = a)$ is 1 if $m_i^j = a$ and 0 otherwise. Let m_i be the column $m_i^1, \ldots, m_i^N$ of aligned symbols in column i, and let c_i be the count vector $c_{i1}, \ldots, c_{iK}$ of observed symbols in column i for an alphabet of K different residues.

If the phylogenetic tree for the sequences has many intermediate ancestors, then the statistical dependence between sequences is complex (see Chapter 7). The scoring problem is greatly simplified if we assume that sequences have all been generated independently. If we assume that residues *within* the column are independent, as well as being independent *between* columns, then the probability of a column m_i is

$$P(m_i) = \prod_a p_{ia}^{c_{ia}}, \qquad (6.2)$$

where p_{ia} is the probability of residue a in column i. We can define a column score as the negative logarithm of this probability:

$$S(m_i) = -\sum_a c_{ia} \log p_{ia}. \qquad (6.3)$$

This is an *entropy* measure directly related to the equation for Shannon entropy in information theory (Chapter 11). It is a convenient measure of the variability observed in an aligned column of residues. The more variable the column is, the higher the entropy. A completely conserved column would score 0. We could define a good alignment to be one which minimises the total entropy of the alignment (e.g. $\sum_i S(m_i)$).

As we have seen before (Chapter 5), the parameters p_{ia} can be estimated from counts c_{ia}; for instance, the maximum likelihood estimate is just

$$p_{ia} = \frac{c_{ia}}{\sum_{a'} c_{ia'}}. \qquad (6.4)$$

In practice we would normally regularise this probability estimate with pseudocounts or Dirichlet priors.

This is obviously near to the HMM formulation of the problem. Profile HMMs go further and also model insertions and deletions in the alignment probabilistically. In return for giving up the evolutionary tree and assuming independence between sequences, we gain the ability to straightforwardly estimate a position-specific model of both residue probabilities in columns and insertions and deletions. Standard profiles make a similar assumption.

The assumption that the sequences are independent can be reasonable if representative sequences of a sequence family are carefully chosen. It is often the case, though, that the sample of sequences is biased and certain evolutionary sub-families are under- or over-represented relative to others. A variety of tree-based weighting schemes have been proposed to deal with this problem to partially compensate for the defects of the sequence independence assumption (see Chapter 5).

Sum of pairs: SP scores

The standard method of scoring multiple alignments is not the HMM formulation, but is similar in that it does not use a phylogenetic tree and it assumes statistical independence for the columns. Columns are scored by a 'sum of pairs' (SP) function using a substitution scoring matrix. The SP score for a column is defined as:

$$S(m_i) = \sum_{k<l} s(m_i^k, m_i^l),$$ \hfill (6.5)

where scores $s(a,b)$ come from a substitution scoring matrix such as a PAM or BLOSUM matrix. For simple linear gap costs, gaps are handled by defining $s(a,-)$ and $s(-,a)$ to be the gap cost. Otherwise gaps are scored separately (e.g. for affine gap costs).

Summing all the pairwise substitution scores in the column might seem to be a natural thing to do. However, substitution scores are usually derived as log-odds scores for pairwise comparisons. The correct extension to multiple alignments would be, for instance, $\log(p_{abc}/q_a q_b q_c)$ for a three-way alignment, rather than the SP score $\log(p_{ab}/q_a q_b) + \log(p_{bc}/q_b q_c) + \log(p_{ac}/q_a q_c)$. There is no probabilistic justification of the SP score; each sequence is scored as if it descended from the $N-1$ other sequences instead of a single ancestor. Evolutionary events are over-counted, a problem which increases as the number of sequences increases. Altschul, Carroll & Lipman [1989] recognised the problem and proposed a weighting scheme designed to partially compensate for this defect in SP scores (Chapter 5).

Example: A problem with SP scores

As an intuitive, concrete example of a problem with the standard SP multiple alignment scoring system, consider an alignment of N sequences which all have leucine (L) at a certain position for some important functional reason. The score of an L aligned to L according to the BLOSUM50 substitution matrix (Figure 2.2) is 5, so the SP score of the column is $5 \times N(N-1)/2$, where $N(N-1)/2$ is the number of symbol pairs in the column. If instead there were one glycine (G) in the column and $N-1$ Ls, the score for the column would be $9 \times (N-1)$ less, because a G-L pair scores -4 instead of $+5$, and $N-1$ pairs are affected. That is, the SP score for a column with one G is worse than the score for a column of all Ls by a fraction of

$$\frac{9(N-1)}{5N(N-1)/2} = \frac{18}{5N}.$$

Notice the inverse dependence on N; the relative difference in score between the correct alignment and the incorrect alignment *decreases* with the number of sequences in the alignment. This is clearly counter-intuitive. The relative difference ought to increase with the more evidence we have for a conserved leucine. See p. 105 for another example. □

6.3 Multidimensional dynamic programming

With some appreciation of scoring issues in mind, we turn to algorithms for constructing multiple alignments.

It is possible to generalise pairwise dynamic programming alignment (Chapter 2) to the alignment of N sequences. However, this turns out to be impractical for more than a few sequences, as we will see shortly. We assume the columns of an alignment are statistically independent, and for now we also assume that gaps are scored with a linear gap cost $\gamma(g) = gd$ for a gap of length g and some gap cost d. Thus we can calculate the overall score $S(m)$ for an alignment as a sum of the scores $S(m_i)$ for each column i:

$$S(m) = \sum_i S(m_i). \tag{6.6}$$

Multidimensional dynamic programming with affine gap costs and multiple states is possible, using methods like those in Chapter 2, but the formalism becomes tedious in many dimensions.

Define $\alpha_{i_1,i_2,...,i_N}$ as the maximum score of an alignment up to the subsequences ending with $x_{i_1}^1, x_{i_2}^2, \ldots, x_{i_N}^N$. The dynamic programming algorithm is

$$\alpha_{i_1,i_2,...,i_N} = \max \begin{cases} \alpha_{i_1-1,i_2-1,...,i_N-1} & + \quad S(x_{i_1}^1, x_{i_2}^2, \ldots, x_{i_N}^N), \\ \alpha_{i_1,i_2-1,...,i_N-1} & + \quad S(-, x_{i_2}^2, \ldots, x_{i_N}^N), \\ \alpha_{i_1-1,i_2,i_3-1,...,i_N-1} & + \quad S(x_{i_1}^1, -, \ldots, x_{i_N}^N), \\ \quad \vdots & \\ \alpha_{i_1-1,i_2-1,...,i_N} & + \quad S(x_{i_1}^1, x_{i_2}^2, \ldots, -), \\ \alpha_{i_1,i_2,i_3-1,...,i_N-1} & + \quad S(-, -, \ldots, x_{i_N}^N), \\ \quad \vdots & \\ \alpha_{i_1,i_2-1,...,i_{N-1}-1,i_N} & + \quad S(-, x_{i_2}^2, \ldots, -), \\ \quad \vdots & \end{cases} \tag{6.7}$$

where all combinations of gaps appear except the one where all residues are replaced by gaps. There are $2^N - 1$ such combinations. Initialisation, termination, and traceback steps for the algorithm are not shown here, but also follow analogously from the pairwise dynamic programming algorithm.

It is possible to simplify the notation by introducing Δ_i which is 0 or 1 and define the 'product'

$$\Delta_i \cdot x = \begin{cases} x & \text{if } \Delta_i = 1, \\ - & \text{if } \Delta_i = 0. \end{cases} \tag{6.8}$$

Now the recursion can be written as follows [Sankoff & Cedergren 1983; Waterman 1995]:

$$\alpha_{i_1,i_2,\ldots,i_N} = \max_{\Delta_1+\ldots+\Delta_N>0} \left\{\alpha_{i_1-\Delta_1,i_2-\Delta_2,\ldots,i_N-\Delta_N} + S(\Delta_1\cdot x_{i_1}^1,\Delta_2\cdot x_{i_2}^2,\ldots,\Delta_N\cdot x_{i_N}^N)\right\}.$$

(6.9)

The algorithm requires the computation of the whole dynamic programming matrix with $L_1L_2\cdots L_N$ entries. To calculate each entry we need to maximise over all 2^N-1 combinations of gaps in a column, excluding the case where all the Δ_k are zero. Assuming the sequences are of roughly the same length $\bar{L}$, the memory complexity of the multidimensional dynamic programming algorithm is $O(\bar{L}^N)$ and the time complexity is $O(2^N\bar{L}^N)$.

Note that we did not specify the functional form of the column score $S(m_i)$. The only assumption necessary to make multidimensional dynamic programming work is that column scores are independent. In principle, $S(m_i)$ could be calculated using an evolutionary model [Sankoff 1975].

Exercise

6.1   Assume we have a number of sequences that are 50 residues long, and that a pairwise comparison of two such sequences takes one second of CPU time on our computer. An alignment of four sequences takes $(2L)^{N-2} = 10^{2N-4} = 10^4$ seconds (a few hours). If we had unlimited memory and we were willing to wait for the answer until just before the sun burns out in five billion years, how many sequences could our computer align?

MSA

A clever algorithm for reducing the volume of the multidimensional dynamic programming matrix that needs to be examined was described by Carrillo & Lipman [1988]. This algorithm was implemented in the multiple alignment program MSA [Lipman, Altschul & Kececioglu 1989]. MSA can optimally align up to five to seven protein sequences of reasonable length (200–300 residues).

Carrillo & Lipman assume an SP scoring system for both residues and gaps. We assume here that the score of a multiple alignment is the sum of the scores of all pairwise alignments defined by the multiple alignment; a somewhat broader definition of the score is possible [Altschul 1989]. Let a^{kl} denote the pairwise alignment between sequences k and l. Then the score of the complete alignment is given by

$$S(a) = \sum_{k<l} S(a^{kl}).$$

(6.10)

Let $\hat{a}^{kl}$ be the optimal *pairwise* alignment of k,l, which we can calculate in $O(\bar{L}^2)$ time by standard dynamic programming. Obviously $S(a^{kl}) \le S(\hat{a}^{kl})$.

Combining this simple observation and the definition of the SP scoring system, we can obtain a lower bound on the score of any pairwise alignment that can occur in the optimal multiple alignment. Assume for the moment that we have a lower bound $\sigma(a)$ on the score of the optimal multiple alignment, so $\sigma(a) \leq S(a)$. From the above and the SP score definition it must be true that, for the optimal multiple alignment a,

$$\sigma(a) \;\leq\; S(a^{kl}) - S(\hat{a}^{kl}) + \sum_{k'<l'} S(\hat{a}^{k'l'})$$

and thus

$$S(a^{kl}) \;\geq\; \beta^{kl}$$
$$\text{where} \quad \beta^{kl} \;=\; \sigma(a) + S(\hat{a}^{kl}) - \sum_{k'<l'} S(\hat{a}^{k'l'}).$$

Therefore, we know we only need to consider pairwise alignments of k and l that score better than β^{kl}. This lower bound β^{kl} is easily calculated. We can obtain a good bound $\sigma(a)$ by any fast heuristic multiple alignment algorithm (for example one of the progressive alignment algorithms given below). The $N(N-1)/2$ optimum pairwise alignments $\hat{a}^{kl}$ are each calculated and scored by standard pairwise alignment. The higher these bounds are, the smaller the volume of dynamic programming matrix that must be calculated and the faster the algorithm will run. (Indeed, by default MSA heuristically picks a higher β^{kl} and so does not guarantee an optimal multiple alignment.)

Now, for each pair k,l we can find the complete set B^{kl} of coordinate pairs (i_k,i_l) such that the best alignment of x^k to x^l through (i_k,i_l) scores more than β^{kl}. This set is calculated in $O(\bar{L}^2)$ time by summing the forward and backward Viterbi scores for each cell of the complete pairwise dynamic programming table, and testing if the result is greater than β^{kl}. The costly multidimensional dynamic programming algorithm can then be restricted to evaluate only cells in the intersection of all these sets: i.e. cells $(i_1,i_2,\ldots,i_N)$ for which (i_k,i_l) is in B^{kl} for all k,l (see Figure 6.3). It is tricky to manage the intersection matrix and perform the dynamic programming calculation efficiently. Details are given in Gupta, Kececioglu & Schaffer [1995].

Altschul & Lipman [1989] extended the theory of the Carrillo–Lipman algorithm to more realistic scoring systems based on evolutionary stars and trees instead of SP scores, but we are not aware of any implementations of those ideas.

6.4 Progressive alignment methods

Probably the most commonly used approach to multiple sequence alignment is *progressive alignment*. This works by constructing a succession of pairwise alignments. Initially, two sequences are chosen and aligned by standard pairwise alignment; this alignment is fixed. Then, a third sequence is chosen and aligned

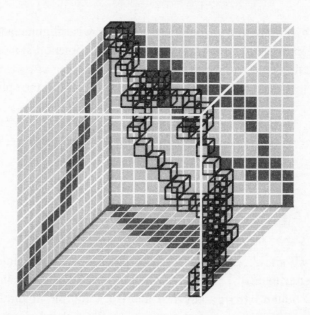

Figure 6.3 *Carrillo & Lipman's algorithm allows the search for optimal alignments to be restricted to a subset of the multidimensional programming matrix, shown here as three-dimensional. The sets B^{kl} are shown in dark grey, and the cells in the matrix to which the search can be confined are outlined in black.*

to the first alignment, and this process is iterated until all sequences have been aligned.

Progressive alignment strategies were introduced by a number of authors [Hogeweg & Hesper 1984; Waterman & Perlwitz 1984; Feng & Doolittle 1987; Taylor 1987; Barton & Sternberg 1987; Higgins & Sharp 1989]. The algorithms differ in several ways: (1) in the way that they choose the order to do the alignment; (2) in whether the progression involves only alignment of sequences to a single growing alignment or whether subfamilies are built up on a tree structure and, at certain points, alignments are aligned to alignments; and (3) in the procedure used to align and score sequences or alignments against existing alignments.

Progressive alignment is heuristic: it does not separate the process of scoring an alignment from the optimisation algorithm. It does not directly optimise any global scoring function of alignment correctness. The advantage of progressive alignment is that it is fast and efficient, and in many cases the resulting alignments are reasonable.

The most important heuristic of progressive alignment algorithms is to align the most similar pairs of sequences first. These are the most reliable alignments. Most algorithms build a 'guide tree'. This is a binary tree whose leaves represent sequences and whose interior nodes represent alignments. The root node represents a complete multiple alignment. The nodes furthest from the root represent the most similar pairs. The methods used to construct guide trees are similar to the methods used to construct phylogenetic trees (Chapter 7), but guide trees are typically 'quick and dirty' trees unsuitable for serious phylogenetic inference.

Feng–Doolittle progressive multiple alignment

The Feng–Doolittle algorithm was one of the first progressive alignment algorithms [Feng & Doolittle 1987]. In overview, it is as follows:

Algorithm: Feng–Doolittle progressive alignment

(i) Calculate a diagonal matrix of $N(N-1)/2$ distances between all pairs of N sequences by standard pairwise alignment, converting raw alignment scores to approximate pairwise 'distances'.

(ii) Construct a guide tree from the distance matrix using the clustering algorithm by Fitch & Margoliash [1967a].

(iii) Starting from the first node added to the tree, align the child nodes (which may be two sequences, a sequence and an alignment, or two alignments). Repeat for all other nodes in the order that they were added to the tree (i.e. from most similar pairs to least similar pairs) until all sequences have been aligned. ◁

The method for converting alignment scores to distances does not need to be especially accurate, as the goal is only to create an approximate guide tree, not an evolutionary tree. Feng & Doolittle calculate the distance D as

$$D = -\log S_{\text{eff}} = -\log \frac{S_{\text{obs}} - S_{\text{rand}}}{S_{\text{max}} - S_{\text{rand}}}, \qquad (6.11)$$

where S_{obs} is the observed pairwise alignment score; S_{max} is the maximum score, the average of the score of aligning either sequence to itself; and S_{rand} is the expected score for aligning two random sequences of the same length and residue composition. The last one, S_{rand}, may either be calculated by random shuffling of the two sequences, or by an approximate calculation given in Feng & Doolittle [1996]. The effective score S_{eff} can thus be viewed as a normalised percentage similarity; it is expected to roughly decay exponentially towards zero with increasing evolutionary distance, hence the $-\log$ to make the measure more approximately linear with evolutionary distance. In phylogenetic tree construction, more care must be taken in calculating distances from alignments.

The Fitch–Margoliash algorithm is one of the fast clustering algorithms that build evolutionary trees from distance matrices. Clustering algorithms are described in Chapter 7.

Sequence–sequence alignments are done with the usual pairwise dynamic programming algorithm. A sequence is added to an existing group by aligning it pairwise to each sequence in the group in turn. The highest scoring pairwise alignment determines how the sequence will be aligned to the group. For aligning a group to a group, all sequence pairs between the two groups are tried; the best pairwise sequence alignment determines the alignment of the two groups. Thus, the scoring system is essentially the standard pairwise PAM score with an affine gap penalty. After an alignment is completed, gap symbols are replaced with a neutral X character. Feng & Doolittle call this the rule of 'once a gap, always a gap'. The rule allows pairwise sequence alignments to be used to guide the alignment of sequences to groups or groups to groups; otherwise, any given pairwise sequence alignment would not necessarily be consistent with the preexisting alignment of a group. Since there is no cost for aligning an X with anything (including a gap symbol), the rule has a desirable side effect of encouraging gaps to occur in the same columns in subsequent pairwise alignments. The X rewriting is not needed in profile-based progressive alignment algorithms (see below).

Profile alignment

A problem with the Feng–Doolittle approach is that all alignments are determined by pairwise sequence alignments. Once an aligned group has been built up, it is advantageous to use position-specific information from the group's multiple alignment to align a new sequence to it. The degree of sequence conservation at each position should be taken into account and mismatches at highly conserved positions penalised more stringently than mismatches at variable positions. Gap penalties in positions might be reduced where lots of gaps occur in the cluster alignment, and increased where no gaps occur. This is the same argument that motivated the development of sequence profiles for database searching (Chapter 5). It also makes sense to apply profiles in progressive multiple sequence alignment.

Many progressive alignment methods use pairwise alignment of sequences to profiles [Thompson, Higgins & Gibson 1994a; Gribskov, McLachlan & Eisenberg 1987] or of profiles to profiles (see e.g. Gotoh [1993]) as a subroutine which is used many times in the process. The exact definition of the scoring function used in profile–sequence or profile–profile alignment varies. Aligned residues are usually scored by some form of a sum-of-pairs score, but the handling of gaps varies substantially between different methods.

For linear gap scoring, profile alignment is simple, because the gap scores can be included in the SP score (6.5) by setting $s(-,a) = s(a,-) = -g$ and $s(-,-) = 0$. Assume we have two multiple alignments (or 'profiles'), one containing sequence 1 to n, and the other containing sequence $n + 1$ to N. An alignment of these two profiles means that gaps are inserted in whole columns, so the alignment within one of the profiles is not changed. The score (6.5) of the global alignment is then

$$\sum_i S(m_i) = \sum_i \sum_{k<l\leq N} s(m_i^k, m_i^l)$$

$$= \sum_i \sum_{k<l\leq n} s(m_i^k, m_i^l) + \sum_i \sum_{n<k<l\leq N} s(m_i^k, m_i^l) + \sum_i \sum_{k\leq n, n<l\leq N} s(m_i^k, m_i^l).$$

All we did was to split up the sum into two sums only concerning the two profiles and one sum containing all the cross terms. The first two sums are unaffected by the global alignment, because adding columns of gap characters to a profile adds zero to the score ($s(-,-) = 0$). Therefore the *optimal alignment* of the two profiles can be obtained by only optimising the last sum with the cross terms. This can be done exactly like a standard pairwise alignment, where columns are scored against columns by adding the pair scores. Obviously one of the profiles can consist of a single sequence only, which corresponds to aligning a single sequence to a profile.

CLUSTALW

One widely used implementation of profile-based progressive multiple alignment is the CLUSTALW program [Thompson, Higgins & Gibson 1994a], which succeeded an earlier popular program, CLUSTALV [Higgins, Bleasby & Fuchs 1992]. CLUSTALW works in much the same way as the Feng–Doolittle method except for its carefully tuned use of profile alignment methods. In overview, the CLUSTALW algorithm is as follows:

Algorithm: CLUSTALW **progressive alignment**

(i) Construct a distance matrix of all $N(N-1)/2$ pairs by pairwise dynamic programming alignment followed by approximate conversion of similarity scores to evolutionary distances using the model of Kimura [1983].

(ii) Construct a guide tree by a neighbour-joining clustering algorithm by Saitou & Nei [1987].

(iii) Progressively align at nodes in order of decreasing similarity, using sequence–sequence, sequence–profile, and profile–profile alignment. ◁

CLUSTALW is unabashedly *ad hoc* in its alignment construction and scoring stage. In addition to the usual methods of profile construction and alignment, various additional heuristics of CLUSTALW contribute to its accuracy:

• Sequences are weighted to compensate for biased representation in large subfamilies. The profile scoring function in CLUSTALW is fundamentally sum-of-pairs. As with Carrillo–Lipman, sequence weighting is important to compensate for the defects of the sum-of-pairs.

- The substitution matrix used to score an alignment is chosen on the basis of the similarity expected of the alignment; closely related sequences are aligned with 'hard' matrices (e.g. BLOSUM80), and distant sequences are aligned with 'soft' matrices (e.g. BLOSUM50).
- Position-specific gap-open profile penalties are multiplied by a modifier that is a function of the residues observed at the position. These penalties were obtained from gap frequencies observed in a large number of structurally based alignments. In general, hydrophobic residues (which are more likely to be buried) give higher gap penalties than hydrophilic or flexible residues (which are more likely to be surface-accessible).
- Gap-open penalties are also decreased if the position is spanned by a consecutive stretch of five or more hydrophilic residues.
- Both gap-open and gap-extend penalties are increased if there are no gaps in a column but gaps occur nearby in the alignment. This rule tries to force all the gaps to occur in the same places in an alignment.
- In the progressive alignment stage, if the score of an alignment is low, the guide tree may be adjusted on the fly to defer the low-scoring alignment until later in the progressive alignment phase when more profile information has been accumulated.

From the standpoint of probabilistic modelling, it is of interest to study such carefully crafted heuristics. It might be good to co-opt the heuristics into more formal probabilistic models, bringing to bear the ability of full probabilistic models to optimise large sets of free parameters.

Iterative refinement methods

One problem with progressive alignment algorithms is that the subalignments are 'frozen'. That is, once a group of sequences has been aligned, their alignment to each other cannot be changed at a later stage as more data arrive. Iterative refinement algorithms attempt to circumvent this problem [Barton & Sternberg 1987; Berger & Munson 1991; Gotoh 1993].

In iterative refinement, an initial alignment is generated, for instance as outlined above; then one sequence (or a set of sequences) is taken out and realigned to a profile of the remaining aligned sequences. If a meaningful score is being optimised, this either increases the overall score or results in the same score. Another sequence is chosen and realigned, and so on, until the alignment does not change. The procedure is guaranteed to converge to a local maximum of the score provided that all the sequences are tried and a maximum score exists, simply because the sequence space is finite.

The method of Barton & Sternberg [1987] is an example of how some of the methods mentioned so far can be combined. It works as follows:

Algorithm: Barton–Sternberg multiple alignment

  (i) Find the two sequences with the highest pairwise similarity and align them using standard pairwise dynamic programming alignment.

 (ii) Find the sequence that is most similar to a profile of the alignment of the first two, and align it to the first two by profile–sequence alignment. Repeat until all sequences have been included in the multiple alignment.

(iii) Remove sequence x^1 and realign it to a profile of the other aligned sequences $x^2, \ldots, x^N$ by profile–sequence alignment. Repeat for sequences $x^2, \ldots, x^N$.

 (iv) Repeat the previous realignment step a fixed number of times, or until the alignment score converges. ◁

The ideas of profile alignment and iterative refinement come quite close to the formulation of probabilistic hidden Markov model approaches for the multiple alignment problem. We turn to HMM methods now.

6.5 Multiple alignment by profile HMM training

In Chapter 5 it was shown that sequence profiles could be recast in probabilistic form as profile HMMs. Thus, profile HMMs could simply be used in place of standard profiles in progressive or iterative alignment methods. The use of profile HMM formalisms may have certain advantages. In particular, the essentially *ad hoc* SP scoring scheme can be replaced by the more explicit profile HMM assumption that the sequences are generated independently from a single 'root' probability distribution.

Profile HMMs can also be trained from initially unaligned sequences using the Baum–Welch expectation maximisation algorithm from Chapter 3. These sorts of approaches, drawn from the HMM literature, were in fact the first HMM-based multiple alignment approaches to be applied. If the trained model is used for a final step of Viterbi alignment of each individual sequence, training generates a multiple alignment in addition to a model [Krogh *et al.* 1994].

Multiple alignment with a known profile HMM

Before tackling the problem of estimating a model and a multiple alignment simultaneously from initially unaligned training sequences, we consider the simpler problem of obtaining a multiple alignment from a known model. This problem often arises in sequence analysis, for instance when we have a multiple alignment and a model of a small representative set of sequences in a family, and we wish to use that model to align a large number of other family members together.

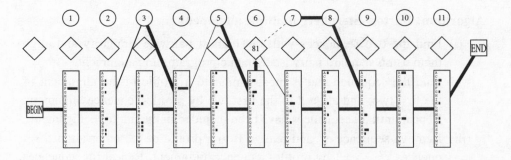

```
FPHF-DLS-----HGSAQ
FESFGDLSTPDAVMGNPK
FDRFKHLKTEAEMKASED
FTQFAG-KDLESIKGTAP
FPKFKGLTTADQLKKSAD
FS-FLK-GTSEVPQNNPE
FG-FSG----AS---DPG
```

Figure 6.4 *A model (top) estimated from an alignment (bottom). The residues in the shaded area of the alignment were treated as inserts. See Figure 5.4 for a description of the model drawing.*

We have seen how to align a sequence to a profile HMM: the most probable path through the model is found by the Viterbi algorithm. Constructing a multiple alignment just requires calculating a Viterbi alignment for each individual sequence. Residues aligned to the same profile HMM match state are aligned in columns. This implies an important difference between profile HMM multiple alignments and traditional multiple alignments which will be clearer by example.

Figure 6.4 shows a small profile HMM and the multiple alignment it was derived from. The shaded residues were arbitrarily defined to be insertions for the purposes of this example, and the other ten columns correspond to ten profile HMM match states. The same seven sequences were realigned to the model, giving the optimal Viterbi paths shown in Figure 6.5. These paths result in the multiple alignment shown in Figure 6.6, left, where lower-case residues were assigned to an insert state and upper-case residues were assigned to a match state.

The important observation here is that the original alignment (Figure 6.4) and the new alignment (Figure 6.6, left) are the same alignment. A profile HMM does not attempt to align the lower-case residues assigned to insert states. The choice of how to put the insert residues in the alignment is arbitrary; some profile HMM implementations simply left-justify insert regions, as shown in Figure 6.6. The insert state residues usually represent parts of the sequences which are atypical, unconserved, and not meaningfully alignable. As we discussed earlier, this is a biologically realistic view of multiple alignment. For instance, we expect loops of homologous protein structures often to be structurally different and unalignable.

Position	1	2	3	4	5	6	insert	7	8	9	10	11
	F	P	H	F	–	D	LS	H	G	S	A	Q
	F	E	S	F	G	D	LSTPDAV	M	G	N	P	K
	F	D	R	F	K	H	LKTEAEM	K	A	S	E	D
	F	T	Q	F	A	G	KDLESI	K	G	T	A	P
	F	P	K	F	K	G	LTTADQL	K	K	S	A	D
	F	S	–	F	L	K	GTSEVP	Q	N	N	P	E
	F	G	–	F	S	G	AS	–	–	D	P	G

Figure 6.5 *The most probable paths of the seven sequences through the model. If the path goes through a match state in position i of the model, the corresponding residue is placed in the column labelled i. If it goes through a delete state, a '–' is placed in the table instead, and when it goes through the insert state in position 6 the corresponding residue is placed in the column labelled 'insert'.*

```
                        FS-FLKngvdptaai--NPK
FPHF-Dls.....HGSAQ      FPHF-Dls.......HGSAQ
FESFGDlstpdavMGNPK      FESFGDlstpdav..MGNPK
FDRFKHlkteaemKASED      FDRFKHlkteaem..KASED
FTQFAGkdlesi.KGTAP      FTQFAGkdlesi...KGTAP
FPKFKGlttadqlKKSAD      FPKFKGlttadql..KKSAD
FS-FLKgtsevp.QNNPE      FS-FLKgtsevp...QNNPE
FG-FSGas.....--DPG      FG-FSGas.......--DPG
```

Figure 6.6 *Left: the alignment of the seven sequences is shown with lower-case letters meaning inserts. The dots are just space-filling characters to make the matches line up correctly. Right: the alignment is shown after a new sequence was added to the set. The new sequence is shown at the top, and because it has more inserts more space-filling dots were added.*

In contrast, many other multiple alignment algorithms align the whole sequences, regardless of what parts of the sequence are meaningfully alignable or not.

The alignment on the right in Figure 6.6 shows a new sequence aligned to the same model. This sequence has more inserted residues than any of the other seven sequences in the shaded area assigned to insert state 6, so the alignment of the other seven sequences must be adjusted to allow space for these two new residues. In an implementation, we typically look at all the Viterbi paths and find the maximum number of inserted residues for each insert state before building the multiple alignment, so we know up front how much room we need to leave to accommodate insertions.

Overview of profile HMM training from unaligned sequences

Now we turn to the harder problem of estimating both a model and a multiple alignment from initially unaligned sequences. The method is summarised as follows:

Algorithm: Multiple alignment using profile HMMs

Initialisation: Choose the length of the profile HMM and initialise parameters.

Training: Estimate the model using the Baum–Welch algorithm (p. 64) or the Viterbi alternative (p. 65). It is usually necessary to use a heuristic method for avoiding local optima (see below).

Multiple alignment: Align all sequences to the final model using the Viterbi algorithm (p. 55) and build a multiple alignment as described in the previous section.   ◁

We now consider the problems of initialisation and training in detail.

Initial model

A profile HMM is a repeating linear structure of three states (match, delete, and insert). The only decision that must be made in choosing an initial architecture for Baum–Welch estimation is the length of the model M. Here M is the number of *match* states in the profile HMM rather than the total number of states, which is $3M + 3$ for the profile HMM architecture of Chapter 5. A commonly used rule is to set M to be the average length of the training sequences (or to set it based on prior knowledge).

Since Baum–Welch estimation finds local optima, not global, it is important to choose initial models carefully. The model should be encouraged to use 'sensible' transitions; for instance, transitions into match states should be large compared to other transition probabilities. At the same time, we want to start Baum–Welch from multiple different points to see if all converge to approximately the same optimum, so we want some randomness in the choice of initial model parameters.

One reasonable approach is to sample the model's initial parameters from the model's Dirichlet prior over parameters (Chapter 11). Alternatively, we can initialise the model with frequencies derived from the prior, use this model to generate a small number of random sequences, and then use these counts as 'data' to estimate an initial model. A further possibility is to estimate the initial model by model construction from an existing guess at the multiple alignment of some or all of the sequences.

Baum–Welch expectation maximisation

The basic parameter estimation is done by a straightforward application of the Baum–Welch algorithm from Chapter 3. Below we give the algorithms in the notation of Chapter 5 for reference.

Algorithm: Forward algorithm for profile HMMs

Initialisation: $f_{M_0}(0) = 1$.

Recursion:
$$f_{M_k}(i) = e_{M_k}(i)[f_{M_{k-1}}(i-1)a_{M_{k-1}M_k} + f_{I_{k-1}}(i-1)a_{I_{k-1}M_k}$$
$$+ f_{D_{k-1}}(i-1)a_{D_{k-1}M_k}];$$
$$f_{I_k}(i) = e_{I_k}(i)[f_{M_k}(i-1)a_{M_kI_k} + f_{I_k}(i-1)a_{I_kI_k}$$
$$+ f_{D_k}(i-1)a_{D_kI_k}];$$
$$f_{D_k}(i) = f_{M_{k-1}}(i)a_{M_{k-1}D_k} + f_{I_{k-1}}(i)a_{I_{k-1}D_k} + f_{D_{k-1}}(i)a_{D_{k-1}D_k}.$$

Termination:
$$f_{M_{M+1}}(L+1) = f_{M_M}(L)a_{M_MM_{M+1}} + f_{I_M}(L)a_{I_MM_{M+1}}$$
$$+ f_{D_M}(L)a_{D_MM_{M+1}}. \qquad \triangleleft$$

Algorithm: Backward algorithm for profile HMMs

Initialisation:
$$b_{M_{M+1}}(L+1) = 1;$$
$$b_{M_M}(L) = a_{M_MM_{M+1}};$$
$$b_{I_M}(L) = a_{I_MM_{M+1}};$$
$$b_{D_M}(L) = a_{D_MM_{M+1}}.$$

Recursion:
$$b_{M_k}(i) = b_{M_{k+1}}(i+1)a_{M_kM_{k+1}}e_{M_{k+1}}(x_{i+1})$$
$$+ b_{I_k}(i+1)a_{M_kI_k}e_{I_k}(x_{i+1}) + b_{D_{k+1}}(i)a_{M_kD_{k+1}};$$
$$b_{I_k}(i) = b_{M_{k+1}}(i+1)a_{I_kM_{k+1}}e_{M_{k+1}}(x_{i+1})$$
$$+ b_{I_k}(i+1)a_{I_kI_k}e_{I_k}(x_{i+1}) + b_{D_{k+1}}(i)a_{I_kD_{k+1}};$$
$$b_{D_k}(i) = b_{M_{k+1}}(i+1)a_{D_kM_{k+1}}e_{M_{k+1}}(x_{i+1})$$
$$+ b_{I_k}(i+1)a_{D_kI_k}e_{I_k}(x_{i+1}) + b_{D_{k+1}}(i)a_{D_kD_{k+1}}. \qquad \triangleleft$$

The forward and backward variables can then be combined to re-estimate emission and transition probability parameters as follows:

Algorithm: Baum–Welch re-estimation equations for profile HMMs

Expected emission counts from sequence x:

$$E_{M_k}(a) = \frac{1}{P(x)} \sum_{i|x_i=a} f_{M_k}(i)b_{M_k}(i);$$

$$E_{I_k}(a) = \frac{1}{P(x)} \sum_{i|x_i=a} f_{I_k}(i)b_{I_k}(i).$$

Expected transition counts from sequence x:

$$A_{X_k M_{k+1}} = \frac{1}{P(x)} \sum_i f_{X_k}(i) a_{X_k M_{k+1}} e_{M_{k+1}}(x_{i+1}) b_{M_{k+1}}(i+1);$$

$$A_{X_k I_k} = \frac{1}{P(x)} \sum_i f_{X_k}(i) a_{X_k I_k} e_{I_k}(x_{i+1}) b_{I_k}(i+1);$$

$$A_{X_k D_{k+1}} = \frac{1}{P(x)} \sum_i f_{X_k}(i) a_{X_k D_{k+1}} b_{D_{k+1}}(i).$$

$\triangleleft$

As usual the Baum–Welch re-estimation procedure can be replaced by the Viterbi alternative described on p. 65 (see below also). Other types of estimation have also been used for estimation of profile HMMs, such as gradient descent [Baldi *et al.* 1994].

Avoiding local maxima

The Baum–Welch algorithm is guaranteed to find a local maximum on the probability 'surface' but there is no guarantee that this local optimum is anywhere near the global optimum nor a biologically reasonable solution. Much the same is true for any practical score optimising multiple alignment method (multidimensional dynamic programming finds global optima but is not practical). Part of the reason is that these models are usually quite long, and thus there are many opportunities to get stuck in a wrong solution. For instance, two variations of the same conserved motif may end up being modelled as two different motifs or a conserved region is squeezed in between two other regions and ends up as being modelled as an insert. One way to search the parameter space is simply to start again many times from different (random) initial models and keep the best scoring final one.

A more involved approach is to use some form of stochastic search algorithm that 'bumps' Baum–Welch off from local maxima. (The two approaches can be combined, and usually are.) The most common stochastic algorithm is *simulated annealing* [Kirkpatrick, Gelatt & Vecchi 1983]. We describe what simulated annealing does, and then discuss a profile HMM training algorithm inspired by simulated annealing.

Theoretical basis of simulated annealing

Some compounds only crystallise if they are slowly annealed from high temperature to low temperature. If the temperature is lowered too fast the structure ends up in a local free energy minimum and is disordered. In an optimisation problem we have some function to minimise, which we can call the 'energy' $E(x)$, where x represents all the variables in which it has to be minimised. (Maximising a function is identical to minimising the negative value of the function.) Inspired

by the physics example, one can introduce an artificial 'temperature' T, and by
the laws of statistical physics the probability of a configuration (or 'state') x is
given by the Gibbs distribution:[1]

$$P(x) = \frac{1}{Z} \exp\left(-\frac{1}{T}E(x)\right). \qquad (6.12)$$

The normalising term $Z = \int \exp(-\frac{1}{T}E(x))dx$ is called the *partition function* in
statistical physics. Since x is usually multidimensional, this is a complicated
integral and often it is impossible to calculate Z.

In the limit of $T \to 0$, all configurations except the one(s) with the lowest energy have probability 0 (the system is 'frozen'). In the limit of $T \to \infty$, all configurations are equiprobable (the system is 'molten'). By analogy with crystalli-
sation, the minimum (or minima) can be found by sampling this probability distribution at high temperature first, and then at gradually decreasing temperatures.
This is called simulated annealing. In other applications, not considered here,
simulated annealing is usually done by so-called Monte Carlo methods [Binder
& Heerman 1988].

For an HMM, a natural energy function is the negative logarithm of the likeli-
hood, $-\log P(\text{data}|\theta)$, so the probability (6.12) is

$$\frac{\exp\left(-\frac{1}{T}[-\log P(\text{data}|\theta)]\right)}{Z} = \frac{P(\text{data}|\theta)^{1/T}}{Z} = \frac{P(\text{data}|\theta)^{1/T}}{\int P(\text{data}|\theta')^{1/T}d\theta'} \qquad (6.13)$$

To pick a model from this distribution appears distinctly non-trivial. The two
methods we mention below are approximations.

Noise injection during Baum–Welch HMM re-estimation

An *ad hoc* approach inspired by simulated annealing was introduced in Krogh
et al. [1994]. The important point of simulated annealing is that it is possible
to escape local minima because of the stochastic choice of configuration (as opposed to algorithms that seek to always lower the energy). A similar effect can be
obtained by adding noise to the counts estimated in the forward–backward procedure, and to let the size of this noise decrease slowly, just as the temperature
decreases in simulated annealing. In Krogh *et al.* [1994] the noise was generated
by a random walk in the initial model. Some systematic studies of the effective-
ness of the method were presented in Hughey & Krogh [1996].

---

[1] In physics, the temperature is multiplied by Boltzmann's constant, but the temperature here
is not a real physical temperature, and therefore it is not needed here.

Simulated annealing Viterbi estimation of HMMs

A second approach was introduced by Eddy [1995], in which a model is trained by a simulated annealing variant of the Viterbi approximation to Baum–Welch estimation. A similar algorithm was described by Allison & Wallace [1993], but in the context of finite state automata rather than HMMs.

Recall that in Viterbi estimation (p. 65) the most probable path for each sequence is used to obtain the counts from which the new model is estimated, rather than summing over all paths to obtain expectations for the counts. If there are N sequences, there is an exact translation from the N paths $\pi^1, \ldots, \pi^N$ to the parameters of the model. Therefore, we can treat the paths as the fundamental parameters in which to maximise the likelihood, so the simulated annealing can be done in these (discrete) variables instead of the (continuous) model parameters, θ.

The key difference between Viterbi estimation and the simulated annealing variant is that while Viterbi selects the highest probability path π for each sequence x, simulated annealing samples each path π according to the likelihood of the path given the current model as modified by a temperature T:

$$\text{Prob}(\pi) = \frac{P(\pi, x|\theta)^{1/T}}{\sum_{\pi'} P(\pi', x|\theta)^{1/T}}$$

The denominator is Z, the partition function. However, it is just a sum over all paths and therefore can be obtained by a modified forward algorithm using *exponentiated* transition and emission parameters. Exponentiated parameters are pre-calculated for computational efficiency: $\hat{a}_{ij} = a_{ij}{}^{1/T}$, and $\hat{e}_j(x) = e_j(x)^{1/T}$ and used in place of the unmodified probability parameters in the forward calculation described on p. 58. The partition function Z is then the result of the forward algorithm, which would be $P(x)$ when the unexponentiated parameters are used.

A suboptimal path π is then selected from the forward dynamic programming lattice by a *stochastic* traceback. The alignment consists of a series of states π_i which are recursively chosen with a probability determined from the forward variable. As this algorithm applies to any HMM, we use the same general notation as used for the forward algorithm as described in Chapter 3:

Algorithm: Stochastic sampling traceback algorithm for HMMs

Initialisation: $\pi_{L+1} = \text{End}$.

Recursion: for $L + 1 \geq i \geq 1$,

$$\text{Prob}(\pi_{i-1}|\pi_i) = f_{i-1, \pi_{i-1}} \hat{a}_{\pi_{i-1}, \pi_i} / \left(\sum_k f_{i-1, k} \hat{a}_{i, \pi_i} \right).$$

◁

In other words, for each state reached in the traceback, a previous state is chosen based on its share of the (exponentiated) path sum probability coming into the current state.[2]

This suboptimal alignment algorithm is then used to implement a simulated annealing variant of Viterbi training. Instead of determining an optimal multiple alignment with respect to the current model at each step of each iteration, a suboptimal multiple alignment is sampled. The degree of the suboptimality is controlled by a temperature factor T, which is started high (giving very randomised alignments) and slowly reduced. Since the new alignment is chosen from the probability distribution over alignments given the *previous* model (as in the expectation step of expectation maximisation), not by the probability of the alignment according to its optimal model, the procedure is not entirely correct according to the formal statistical mechanical basis of simulated annealing [Kirkpatrick, Gelatt & Vecchi 1983].

Finding the best 'schedule' for how fast to lower the temperature is a whole science (or perhaps art) in itself. There is a theoretical result for simulated annealing saying that if the temperature is lowered slowly enough, finding the optimum is guaranteed, but the time required for this is prohibitive. In practice a simple exponentially or linearly decreasing temperature schedule is often used, where each step amounts to either multiplying T by some number less than 1 or to reducing it by some small constant amount.

Comparison to Gibbs sampling

The 'Gibbs sampler' algorithm described by Lawrence *et al.* [1993] has substantial similarities. The statistical model used by Lawrence *et al.* is a short ungapped motif model which is essentially a profile HMM with no insert or delete states (though they do not refer to it as an HMM). The training data consist of a set of sequences which contain (in the simplest case) exactly one instance of some motif, such as a specific protein-binding site on DNA, where the position of the motif is initially unknown. The problem is to simultaneously find the motif positions and to estimate the parameters for a consensus statistical model of them (by knowing one, we can find the other, just as an alignment implies an HMM and vice versa). It is a natural problem for expectation maximisation (EM; Chapter 11), where the missing data are the positions of the motifs, that can simply be specified by their start points in the sequences; these correspond to the alignments which are the missing data we are trying to infer in HMM training. Indeed, earlier algorithms [Lawrence & Reilly 1990] applied expectation maximisation to the problem, but these approaches proved prone to poor local optima.

In an HMM framework, both the above simulated annealing algorithm and the

---

[2] An algebraic proof that this algorithm correctly calculates the partition function and correctly samples a path from the Boltzmann/Gibbs distribution over all possible paths (given the current model θ) follows fairly straightforwardly from the fact that exponentiation is distributive over multiplication, e.g. $(a_{1,2}a_{2,3}a_{3,4})^{1/T} = (a_{1,2})^{1/T}(a_{2,3})^{1/T}(a_{3,4})^{1/T} = \hat{a}_{1,2}\hat{a}_{2,3}\hat{a}_{3,4}$.

Gibbs sampler are stochastic sampling variants of the Viterbi approximation of EM. At each iteration of Gibbs sampling, a sequence is removed from the alignment; an HMM is built of the remaining aligned sequences; and then a new alignment of the sequence to the rest is sampled probabilistically using the stochastic sampling algorithm at $T = 1$. This iteration is repeated until the model reaches a region of high probability. The Gibbs sampler is thus like running the above simulated annealing Viterbi algorithm at a constant $T = 1$, where alignments are sampled from a probability distribution unmodified by any effect of a temperature factor. For a general description of Gibbs sampling, see Chapter 11.

Adaptively modifying model architecture; model surgery

After (or even during) training a model, we can look at the alignment it produces and decide that: (a) some of the match states are redundant and should be absorbed in an insert state; or (b) it seems like one or more insert states absorb too much sequence, in which case they should be expanded (i.e. more match modules can be inserted before or after the insert state). This can happen both because the initial choice of model length was not as good as it could have been, and because of local optima encountered during training. It is advantageous to devise procedures to adaptively modify the model's architecture during training and just after training has been completed.

In Krogh *et al.* [1994] a method called *model surgery* was described. From the 'counts' estimated by the forward–backward procedure (or the Viterbi analogue) we can see how much a certain transition is used by the training sequences. The usage of a match state is the sum of counts for all letters in the state. If a certain match state is used by less than half the sequences (or some other predefined fraction) the corresponding module is deleted. Similarly if more than half (or some other predefined fraction) of the sequences use the transitions into a certain insert state, this is expanded to some number of new modules. The number of new modules is determined by the average length of the insertions. Though it is *ad hoc*, it works well.

Another approach is to re-estimate both a model architecture and model parameters using the maximum *a posteriori* (MAP) model construction algorithm given in Chapter 5. As this procedure requires an alignment, not expected counts, it cannot be applied during the usual Baum–Welch expectation maximisation procedure. It can be applied correctly during training by the Viterbi approximation to Baum–Welch, and in fact can completely replace the usual parameter re-estimation process, leading to a (locally) convergent optimisation algorithm that simultaneously optimises both the architecture and parameters of the HMM. It can also be applied periodically (much like model surgery is applied) during full Baum–Welch estimation by inserting an iteration of Viterbi alignment and MAP model construction. In this use, it is not necessarily guaranteed to improve the overall likelihood of the data, but, like model surgery, is a pretty good heuristic.

6.6 Further reading

Surveys of the voluminous multiple alignment algorithm literature include Carrillo & Lipman [1988], Chan, Wong & Chiu [1992], and Gotoh [1996].

A class of multiple alignment algorithms we have not discussed here are simulated annealing algorithms that define 'moves' (small changes in a candidate alignment) and an objective function for determining the probability of whether a proposed move should be accepted or not. These sampling algorithms are Monte Carlo-style simulated annealing algorithms that are quite distinct from the simulated annealing variant of Viterbi HMM estimation we have discussed [Lukashin, Engelbrecht & Brunak 1992; Hirosawa *et al.* 1993; Kim & Pramanik 1994; Kim, Pramanik & Chung 1994].

Consensus motif finding algorithms like the Gibbs sampler are closely akin to multiple alignment algorithms, as we briefly discussed. Other examples of motif finders besides the Gibbs sampler include Stormo & Hartzell [1989], Hertz, Hartzell & Stormo [1990], Bailey & Elkan [1994], and Bailey & Elkan [1995]. The problem of multiply aligning three-dimensional structures is also related to the multiple sequence alignment problem (but even harder) [Russell & Barton 1992; Holm & Sander 1993; Gerstein & Levitt 1996].

Several papers have systematically tested the accuracy of different multiple alignment algorithms against structurally or manually generated alignments, including McClure, Vasi & Fitch [1994] and Gotoh [1996].

7

Building phylogenetic trees

In the previous chapter, we considered the problem of multiple alignment of sets of sequences. One can argue [Sankoff, Morel & Cedergren 1973] that alignment of sequences should take account of their evolutionary relationship. For example, an alignment that implies many substitutions between closely related sequences is less plausible than one that makes most of its changes over large evolutionary distances.

Some multiple alignment algorithms use a tree; for instance, we have seen that several progressive alignment algorithms use a 'guide tree'. As the name suggests, this tree is meant to guide the clustering process rather than satisfy a taxonomist. In this chapter we shift emphasis, and begin to take a serious interest in building trees. However, we do not lose sight of alignment: the last section describes methods for simultaneous alignment and tree building.

We concentrate here on two general approaches to tree building: distance methods and parsimony; the next chapter formulates phylogeny probabilistically.

7.1   The tree of life

The similarity of molecular mechanisms of the organisms that have been studied strongly suggests that all organisms on Earth had a common ancestor. Thus any set of species is related, and this relationship is called a *phylogeny*. Usually the relationship can be represented by a *phylogenetic tree*. The task of phylogenetics is to infer this tree from observations upon the existing organisms.

Traditionally, morphological characters (both from living and fossilised organisms) have been used for inferring phylogenies. Zuckerkandel & Pauling's pioneering paper [1962] showed that molecular sequences provide sets of characters that can carry a large amount of information. If we have a set of sequences from different species, therefore, we may be able to use them to infer a likely phylogeny of the species in question. This assumes that the sequences have descended from some common ancestral gene in a common ancestral species.

The widespread occurrence of gene duplication means that the foregoing assumption needs to be checked carefully. The phylogenetic tree of a group of sequences does not necessarily reflect the phylogenetic tree of their host species, because gene duplication is another mechanism, in addition to speciation, by which two sequences can be separated and diverge from a common ancestor. Genes which diverged because of speciation are called *orthologues*. Genes which diverged by gene duplication are called *paralogues*. If we are interested in inferring the phylogenetic tree of the species carrying the genes, we must use orthologous sequences. But, of course, we might be interested in the phylogeny of duplication events, in which case we might construct a phylogeny of paralogues, even the paralogues within a single species. The distinction between paralogues and orthologues is illustrated by Figure 7.1.

7.2 Background on trees

In this chapter, all trees will be assumed to be binary, meaning that an edge that branches splits into two daughter edges (Figure 7.2). This is equivalent to saying that three edges meet at every branch node, a *node* being an endpoint of an edge. The assumption that the tree is binary is not a serious limitation, because any other branching pattern can be approximated by a binary tree in which some of the branches are very short.

Each edge of the tree has a certain amount of evolutionary divergence associated to it, defined by some measure of distance between sequences, or from a model of substitution of residues over the course of evolution. We adopt the general term 'length' or 'edge length' here, and represent this by the lengths of edges in the figures we draw. The relationship between phylogenetically determined lengths and palaeontological time periods was examined by Langley & Fitch [1974], who found that different proteins can change at very different rates, and the same sequence can evolve much faster in some organisms than others. However, averaging over larger sets of proteins does demonstrate a broad correspondence between lengths and evolutionary time periods [Doolittle *et al.* 1996; Wray, Levinto & Shapiro 1996].

A true biological phylogeny has a 'root', or ultimate ancestor of all the sequences. Some algorithms provide information, or at least a conjecture, about the location of the root. Others, like parsimony and the probabilistic models in the next chapter, are completely uninformative about its position, and other criteria have to be used for rooting the tree. We consider here how to represent both rooted and unrooted trees.

Figure 7.2 shows an unrooted tree and a rooted version of it. Note that, in the latter, we have drawn the root at the top, with the *leaves*, the terminal nodes corresponding to the observed sequences, at the base.

The leaves of trees have names or numbers. Sometimes these can be swopped without altering the phylogeny (e.g. numbers 4 and 5 in Figure 7.2), but they often cannot (e.g. swopping 1 and 2 in the figure changes the phylogeny). A tree

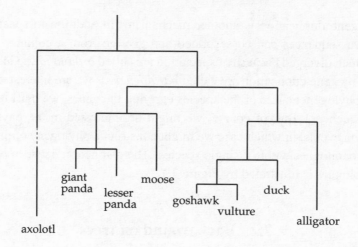

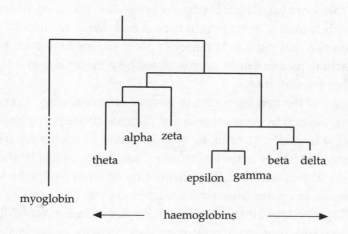

Figure 7.1 *Above: a tree of orthologues based on a set of alpha haemoglobins. Below: a tree of paralogues, the alpha, beta, gamma, delta, epsilon, zeta and theta chains of human haemoglobins, and human myoglobin. The orthologues are the alpha haemoglobins with* SWISS-PROT *identifiers* HBA_ACCGE, HBA_AEGMO, HBA_AILFU, HBA_AILME, HBA_ALCAA, HBA_ALLMI, HBA_AMBME, *and* HBA_ANAPL, *chosen because they were the alphabetically the first eight alpha globins in* PFAM *[Sonnhammer, Eddy & Durbin 1997] (http://genome.wustl.edu/Pfam/). The paralogues are globins with* SWISS-PROT *identifiers* HBAT_HUMAN, HBAZ_HUMAN, HBA_HUMAN, HBB_HUMAN, HBD_HUMAN, HBE_HUMAN, HBG_HUMAN, *and* MYG_HUMAN. *The trees were made by neighbour-joining, Section 7.3, using J. Felsenstein's package* PHYLIP *(http://evolution.genetics.washington.edu/phylip.html). The distances used for neighbour-joining were the* PAM-*based ML distances (see p. 228) determined by the program* PROTDIST *in* PHYLIP.

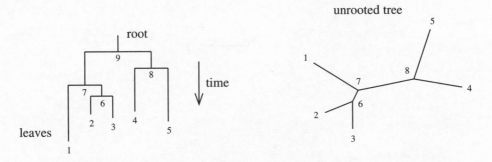

Figure 7.2 *An example of a binary tree, showing the root and leaves, and the direction of evolutionary time (the most recent time being at the bottom of the figure). The corresponding unrooted tree is also shown; the direction of time here is undetermined.*

with a given labelling will be called a *labelled branching pattern*. More loosely, we refer to this as the tree *topology*[1] and denote it by the symbol T. To complete the definition of a phylogenetic tree, one must also define the lengths of its edges; these will generally be denoted[2] by t_i with a suitable numbering scheme for the is.

Counting and labelling trees

The nodes and edges of a rooted tree can be counted as follows: Suppose there are n leaves. As we move up the tree, the edges coalesce as each new node is reached. Each time this happens, the number of edges is reduced by one. So there must be $(n-1)$ nodes in addition to the n leaves, giving $(2n-1)$ nodes in all, and one fewer edges, i.e. $(2n-2)$, discounting the edge above the root node. We shall label the leaves using the numbers 1 to n, and assign the branch nodes the numbers $n+1$ to $2n-1$, reserving $2n-1$ for the root node. The lengths of edges will be labelled by the node at the bottom of the edge, so d_1 is the length associated to the edge above node 1, and so on.

An unrooted tree with n leaves has $2n-2$ nodes altogether and $2n-3$ edges. A root can be added to it at any of its edges, thereby producing $(2n-3)$ rooted trees from it. Figure 7.3 shows this for $n=3$; the three positions for the root yield three rooted trees. There are therefore $(2n-3)$ times as many rooted trees as unrooted trees, for a given number n of leaves.

Instead of the root, we can add an extra edge or 'branch' with a distinct label at its leaf (i.e. a '4') to the unrooted tree with three leaves in Figure 7.3, thereby obtaining an unrooted tree with four leaves. There are three such trees, with

---

[1] A topologist would reserve this term for the unlabelled branching patterns, i.e. the distinct classes of tree that cannot be rearranged into each other by permutation of edges at nodes or shrinking or extending of edges.

[2] A deliberate echo of 'time', the variable we are ultimately interested in.

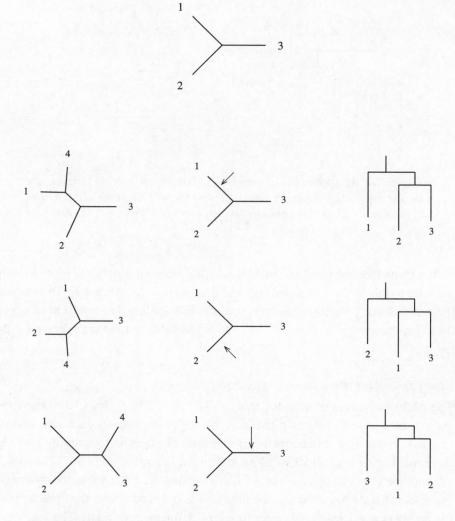

Figure 7.3 *The rooted trees (right-hand column) derived from the unrooted tree for three sequences by picking different edges as positions for the root (arrows).*

$(2n - 3) = 5$ edges, and it is easy to see that they are distinct labelled branching patterns. There are then five ways of adding a further branch labelled with a distinct label ('5'), giving in all $3 \times 5 = 15$ unrooted trees with five leaves. Continuing this, we see that there are $(3) \cdot (5) \cdot \ldots \cdot (2n - 5)$ unrooted trees with n leaves; this number is also written $(2n - 5)!!$. From what was said above, it follows that there are $(2n - 3)!!$ rooted trees. The number of trees grows very rapidly with n; for $n = 10$ there are about two million unrooted trees, and for $n = 20$, 2.2×10^{20} of them. For further information on tree counting, see Felsenstein [1978b].

Exercises

7.1   Draw the rooted trees obtained by adding the root in all seven possible positions to the unrooted tree in Figure 7.2.

7.2   The trees with three and four leaves in Figure 7.3 all have the same *unlabelled* branching pattern. How many leaves do there have to be to obtain more than one unlabelled branching pattern, and how does this number grow up to 10 leaves? Can you conjecture a recurrence relation for this number? (The answer is given in Edwards & Cavalli-Sforza [1964].)

7.3   All trees considered so far have been binary, but one can envisage ternary trees that, in their rooted form, have *three* branches descending from a branch node. The unrooted trees therefore have four edges radiating from every branch node. If there are m branch nodes in an unrooted ternary tree, how many leaves are there and how many edges?

7.4   Consider next a composite unrooted tree with m ternary branch nodes and n binary branch nodes. How many leaves are there, and how many edges? Let $N_{m,n}$ denote the number of distinct labelled branching patterns of this tree. Extend the counting argument for binary trees to show that

$$N_{m,n} = (3m + 2n - 1)N_{m,n-1} + (n+1)N_{m-1,n+1}$$

(Hint: the first term after the '=' counts the number of ways that a new edge can be added to an existing edge, thereby creating an additional binary node; the second term corresponds to edges added at binary nodes, thereby producing ternary nodes.)

7.5   Use the above recurrence relationship to calculate $N_{m,0}$, the number of distinct pure ternary trees with m branch nodes, for small values of m. (Hint: We know that $N_{0,i} = (2i - 1)!!$, and the recurrence relationship allows one to express $N_{m,0}$ in terms of $N_{0,i}$, for $i \leq n$. Programmers will enjoy writing a recursive program that carries out this operation.) Check that the calculated numbers satisfy $N_{m,0} = \prod_{i=1}^{m} (1 + 9i(i-1)/2)$. Can you prove this formula?

7.3   Making a tree from pairwise distances

Some of the more intuitively accessible methods of tree building begin with a set of distances d_{ij} between each pair i, j of sequences in the given dataset. There are many different ways of defining distance. For example, one can take d_{ij} to be the fraction f of sites u where residues x_u^i and x_u^j differ (presupposing an alignment of the two sequences). This gives a sensible definition for small fractions f. For two unrelated sequences, however, random substitutions will cause f to approach the fraction of differences expected by chance, and we would like the distance to become large as f tends to this value. Markov models of residue substitution, such as the Jukes–Cantor model for DNA (p. 195), can be used to define distances that behave this way. Thus the Jukes–Cantor distance is $d_{ij} = -\frac{3}{4}\log(1 - 4f/3)$, which tends to infinity as the equilibrium value of f (75% of residues different) is approached. We return to the definition of distances in Section 8.6.

Clustering methods: UPGMA

We begin with a clustering procedure [Sokal & Michener 1958] called UPGMA, which stands for unweighted pair group method using arithmetic averages. Despite its formidable acronym, the method is simple and intuitively appealing. It works by clustering the sequences, at each stage amalgamating two clusters and at the same time creating a new node on a tree. The tree can be imagined as being assembled upwards, each node being added above the others, and the edge lengths being determined by the difference in the heights of the nodes at the top and bottom of an edge.

First we define the distance d_{ij} between two clusters C_i and C_j to be the average distance between pairs of sequences from each cluster:

$$d_{ij} = \frac{1}{|C_i||C_j|} \sum_{p \text{ in } C_i, q \text{ in } C_j} d_{pq}, \tag{7.1}$$

where $|C_i|$ and $|C_j|$ denote the number of sequences in clusters i and j, respectively. Note that, if C_k is the union of the two clusters C_i and C_j, i.e. if $C_k = C_i \cup C_j$, and if C_l is any other cluster, then (Exercise 7.6):

$$d_{kl} = \frac{d_{il}|C_i| + d_{jl}|C_j|}{|C_i| + |C_j|}. \tag{7.2}$$

The clustering procedure is:

Algorithm: UPGMA

Initialisation:

    Assign each sequence i to its own cluster C_i,

    Define one leaf of T for each sequence, and place at height zero.

Iteration:

    Determine the two clusters i, j for which d_{ij} is minimal. (If there are
        several equidistant minimal pairs, pick one randomly.)

    Define a new cluster k by $C_k = C_i \cup C_j$, and define d_{kl} for all l by (7.2).

    Define a node k with daughter nodes i and j, and place it at height $d_{ij}/2$.

    Add k to the current clusters and remove i and j.

Termination:

    When only two clusters i, j remain, place the root at height $d_{ij}/2$.   ◁

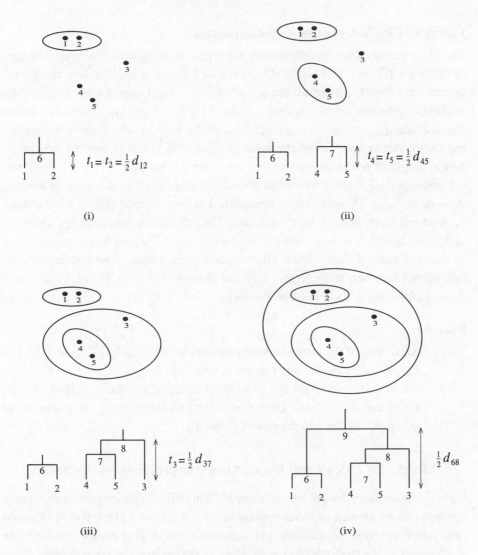

Figure 7.4 *An example of how UPGMA produces a rooted tree by successively clustering sequences, in this case a set of five sequences whose distances can be represented by points in the plane (this will not generally be true of a set of distances).*

To check that this procedure produces well-defined edge lengths, we have to show that a parent node always lies above its daughters (see Exercise 7.7). There are variants of UPGMA that define the distance between clusters as the minimum or maximum of the distances between constituent sequences, rather than the average, but UPGMA seems to have the best performance record.

Example: UPGMA applied to five sequences

The distances between five sequences are represented schematically as distances
in the plane (Figure 7.4). UPGMA works as follows: First, the two closest sequences are found; suppose these are x^1 and x^2. Their parent node is given the
number 6 and edge lengths t_1 and t_2 defined by $t_1 = t_2 = \frac{1}{2}d_{12}$. Next, we define
the distance d_{i6} between a sequence x^i and the new branch node 6, represent-
ing the cluster $\{x^1, x^2\}$, to be the average $\frac{1}{2}(d_{1i} + d_{2i})$, and search for the closest
pair amongst all remaining sequences and node 6. This pair is $\{x^4, x^5\}$; their parent node, node 7, is constructed as above and edge lengths t_4 and t_5 defined by
$d_4 = d_5 = \frac{1}{2}d_{45}$. This process is repeated. The next closest pair is x^3 and node
7. A parent node, node 8, to x^3 and node 7 is introduced, and the edge above x^3
assigned a length $t_3 = \frac{1}{2}d_{37}$, and the edge above node 7 a length $t_7 = \frac{1}{2}d_{37} - \frac{1}{2}d_{45}$,
so that the sum of times down all branches is the same. The last amalgama-
tion occurs between node 6 ($\{x^1, x^2\}$) and node 8 ($\{x^3, x^4, x^5\}$), with a distance
$d_{68} = \frac{1}{6}(d_{13} + d_{14} + d_{15} + d_{23} + d_{24} + d_{25})$.   □

Exercises

7.6   Show that, if distances between clusters are defined by (7.1), and if $C_k =
        C_i \cup C_j$, then d_{kl} for any l is given by (7.2).

7.7   Show that a node always lies above its daughter nodes. (Hint: if not,
        show that an incorrect choice of closest clusters would have been made
        when one of the daughters was formed.)

Molecular clocks and the ultrametric property of distances

UPGMA produces a rooted tree of a special kind. The edge lengths in the resulting tree can be viewed as times measured by a *molecular clock* with a constant
rate. The divergence of sequences is assumed to occur at the same constant rate
at all points in the tree, which is equivalent to saying that the sum of times down
a path to the leaves from any node is the same, whatever the choice of path. If our
distance data are derived by adding up edge lengths in a tree T with a molecular
clock, then UPGMA will reconstruct T correctly. To see this, imagine a horizontal line rising through the tree T starting from the level of the leaves: each time it
crosses a node, the distances of all the leaves in the left branch from that node to
the leaves in the right branch will be the current minimum distance, and a node
will therefore be added precisely where the node is encountered in the original
tree T.

If the original tree is not well-behaved in this way, but has different length
routes to its leaves, as in Figure 7.5 (left), then it may be reconstructed incorrectly by UPGMA (Figure 7.5 right). What goes wrong in this case is that the
closest leaves are not neighbouring leaves: they do not have a common parent
node. A test of whether reconstruction is likely to be correct is the *ultrametric*

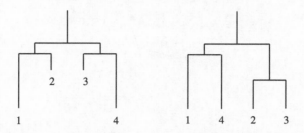

Figure 7.5 *A tree (left) that is reconstructed incorrectly by UPGMA (right).*

condition. The distances d_{ij} are said to be ultrametric if, for any triplet of sequences, x^i, x^j, x^k, the distances d_{ij}, d_{jk}, d_{ik} are either all equal, or two are equal and the remaining one is smaller. This condition holds for distances derived from a tree with a molecular clock.

Exercise

7.8   It can be shown that, if the distances d_{ij} are ultrametric, and if a tree is constructed from these distances by UPGMA, then the distances obtained from this tree by taking twice the height of the node on the path between i and j are identical to the d_{ij}. Check that this is true in the example of UPGMA applied to five sequences if the distances are ultrametric. (Hint: Show that, when two clusters C_k and C_l are amalgamated, the ultrametric condition implies that the distances between any leaf in C_k and any leaf in C_l are the same.)

Additivity and neighbour-joining

In describing the molecular clock property of the trees produced by UPGMA, we implicitly assumed another important property: additivity. Given a tree, its edge lengths are said to be *additive* if the distance between any pair of leaves is the sum of the lengths of the edges on the path connecting them. This property is built in automatically as the UPGMA tree is constructed. However, it is possible for the molecular clock property to fail but for additivity to hold, and in that case there are algorithms that can be used to reconstruct the tree correctly.

Given a tree T with additive lengths $\{d_\bullet\}$, we can try to reconstruct it from the pairwise distances of its leaves $\{d_{ij}\}$ as follows: Find a pair of *neighbouring leaves*, i.e. leaves that have the same parent node, k. Suppose their numbers are i, j. Remove them from the list of leaf nodes and add k to the current list of nodes, defining its distance to leaf m by

$$d_{km} = \tfrac{1}{2}(d_{im} + d_{jm} - d_{ij}). \tag{7.3}$$

By additivity, the distances d_{km} just defined are precisely those between the equiv-

170   7 *Building phylogenetic trees*

Figure 7.6 *For any three leaves i, j and m there is a node, k here, where the branches to them meet. By additivity, $d_{im} = d_{ik} + d_{km}$, $d_{jm} = d_{jk} + d_{km}$ and $d_{ij} = d_{ik} + d_{jk}$, from which it follows that $d_{km} = \frac{1}{2}(d_{im} + d_{jm} - d_{ij})$, which is equation (7.3).*

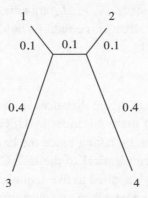

Figure 7.7 *A tree whose closest pair of leaves are not neighbouring leaves. The lengths of edges are shown. If the lengths are additive, we find $d_{12} = 0.3$ and $d_{13} = 0.5$, so the neighbouring pair 1,3 are further apart than the non-neighbouring pair 1,2.*

alent nodes in the original tree (see Figure 7.6). In this way we can strip away leaves, reducing the number by one at each operation, until we get down to a pair of leaves.

If we could determine from distances alone a pair of neighbouring leaves, therefore, we could reconstruct a tree with additive lengths exactly. The remarkable fact is that we can pick neighbouring leaves, using a procedure proposed by Saitou & Nei [1987] and modified by Studier & Keppler [1988].

First, note that it does *not* suffice to pick simply the two closest leaves, i.e. the pair i,j with d_{ij} minimal. Figure 7.7 shows why. If one of a pair of neighbours has a short edge and the other a long edge, the one with the short edge may be closer to another leaf than its true neighbour, as happens in the illustrated tree. To avoid this, the trick is to subtract the averaged distances to all other leaves; in effect, this compensates for long edges. We define

$$D_{ij} = d_{ij} - (r_i + r_j),$$

where

$$r_i = \frac{1}{|L|-2} \sum_{k \in L} d_{ik}, \tag{7.4}$$

and $|L|$ denotes the size of the set L of leaves. The claim now is that a pair of leaves i, j for which D_{ij} is minimal will be neighbouring leaves; we give a proof at the end of this chapter. It is instructive to check that this is true of the tree in Figure 7.7 (see Exercise 7.9).

The complete algorithm for neighbour-joining works by constructing a tree T by steps, keeping a list L of active nodes in this tree. If there were a pre-existing additive tree, L would be the current remaining set of leaf nodes as neighbouring pairs were stripped away, and T would be the tree built up from these stripped-off nodes.

Algorithm: Neighbour-joining

Initialisation:

    Define T to be the set of leaf nodes, one for each given sequence, and put $L = T$.

Iteration:

    Pick a pair i, j in L for which D_{ij}, defined by (7.4), is minimal.

    Define a new node k and set $d_{km} = \frac{1}{2}(d_{im} + d_{jm} - d_{ij})$, for all m in L.

    Add k to T with edges of lengths $d_{ik} = \frac{1}{2}(d_{ij} + r_i - r_j)$, $d_{jk} = d_{ij} - d_{jk}$, joining k to i and j, respectively.

    Remove i and j from L and add k.

Termination:

    When L consists of two leaves i and j add the remaining edge between i and j, with length d_{ij}.

                                                      ◁

The definition of the length d_{ik} by $\frac{1}{2}(d_{ij} + r_i - r_j)$ gives the correct length if additivity holds, since this expression is the average of $\frac{1}{2}(d_{ij} + d_{im} - d_{jm})$ over all leaves m, and each such term is just d_{ik} (compare (7.3)).

Additivity is a property that depends on the distance measure used: a tree may be additive with respect to one distance measure and not with respect to another. In Section 8.6 we shall see that a certain type of maximum likelihood distance measure would be expected to give additivity, in the limit of a large amount of data, if the underlying model assumptions were correct. Real data, of course, will only be at best approximately additive.

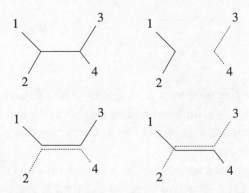

Figure 7.8 *Additivity means that two of the summed lengths $d_{12} + d_{34}$, $d_{13} + d_{24}$, $d_{14} + d_{23}$ must be larger than the third and equal in size. This holds if the pairwise distances are obtained by summing edge lengths, as the diagrams show.*

We can use neighbour-joining even if lengths are not additive, but reconstruction of the correct tree is no longer guaranteed. Just as the ultrametric condition provided a test for the molecular clock property, so we can use the following property of distances as a test for additivity: For every set of four leaves i, j, k and l, two of the distances $d_{ij} + d_{kl}$, $d_{ik} + d_{jl}$ and $d_{il} + d_{jk}$ must be equal and larger than the third. This *four-point condition* is a consequence of additivity, because two of the sums include the length of the 'bridge' connecting pairs of leaves (see Figure 7.8).

Exercises

7.9   Show that the smallest distances D_{ij} in the tree in Figure 7.7 correspond to neighbouring leaves.

7.10   Show that, for a tree with four leaves, D_{ij} for a pair of neighbours is less than D_{ij} for all other pairs by twice the 'bridge length', i.e. the length of the edge joining the two branch nodes in the tree.

Rooting trees

Neighbour-joining, unlike UPGMA, produces unrooted trees. Finding the root is a secondary task, which can be accomplished by adding an *outgroup*, or species that is known to be more distantly related to each of the remaining species than they are to each other. The point in the tree where the edge to the outgroup joins is therefore the best candidate for the root position. In the top tree in Figure 7.1, for instance, the axolotl can be treated as an outgroup, as it is an amphibian whereas all the other species are amniotes. It is therefore reasonable to place the divide between the axolotl and the other species earlier than any of the other branches.

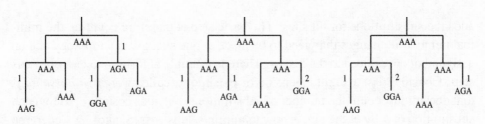

Figure 7.9 *Building a tree by parsimony.*

In the absence of a convenient outgroup, there are somewhat *ad hoc* strategies, such as picking the midpoint of the longest chain of consecutive edges, which would be expected to identify the root if deviations from a molecular clock were not too great.

7.4 Parsimony

We come now to what is probably the most widely used of all tree building algorithms: *parsimony*. It works by finding the tree which can explain the observed sequences with a minimal number of substitutions. It uses a different general strategy from the distance-based algorithms considered so far. Instead of building a tree, it assigns a cost to a given tree, and it is necessary to search through all topologies, or to pursue a more efficient search strategy that achieves this effect (see p. 176), in order to identify the 'best' tree. We can therefore distinguish two components to the algorithm:

(1) the computation of a cost for a given tree T;

(2) a search through all trees, to find the overall minimum of this cost.

We begin with an example. Suppose we have the following four aligned nucleotide sequences:

AAG

AAA

GGA

AGA

We can try out different trees for these four sequences and count the number of substitutions needed in each tree, summing over all sites. Figure 7.9 shows three possible trees for the above four sequences; they differ in the order in which the sequences are assigned to the leaves. In each tree, hypothetical sequences have been assigned to the ancestral nodes so as to minimise the number of changes needed in the whole tree. We shall see shortly how this is done. The leftmost tree needs fewer changes (a total of three) than the two shown to its right (which need four each).

As we see in this example, parsimony treats each site independently, and then adds the substitutions for all sites. The basic step is therefore counting the minimal number of changes that need to be made at one site, given a topology and an assignment of residues to the leaves. There is a simple algorithm to carry out this step. Consider first a slight extension of parsimony, called *weighted parsimony*, that doesn't just count the number of substitutions, but adds costs $S(a,b)$ for each substitution of a by b; the aim is now to minimise this cost [Sankoff & Cedergren 1983]. Weighted parsimony reduces to traditional parsimony when $S(a,a) = 0$ for all a, and $S(a,b) = 1$ for all $a \neq b$.

To compute the minimal cost at site u we proceed as follows: Let $S_k(a)$ denote the minimal cost for the assignment of a to node k.

Algorithm: Weighted parsimony

Initialisation:

Set $k = 2n - 1$, the number of the root node.

Recursion: Compute $S_k(a)$ for all a as follows:

If k is leaf node:

Set $S_k(a) = 0$ for $a = x_u^k$, $S_k(a) = \infty$, otherwise.

If k is not a leaf node:

Compute $S_i(a)$, $S_j(a)$ for all a at the daughter nodes i, j, and

define $S_k(a) = \min_b(S_i(b) + S(a,b)) + \min_b(S_j(b) + S(a,b))$.

Termination:

Minimal cost of tree $= \min_a S_{2n-1}(a)$.

◁

Note that the steps under 'Recursion' require that S_i and S_j are computed for the daughter nodes i, j of k, and this is achieved by returning to 'Recursion' for both i and j. The effect is that the algorithm starts at the leaves and works its way up to the root. This way of passing through a tree is called *post-order traversal*, and plays an important part in many computer implementations of tree algorithms.

It is sometimes of interest to find the ancestral assignments of residues that give the minimal cost. For instance, one way of defining a length for an edge is to count the number of mismatches along that edge that occur in all possible minimal-cost ancestral assignments to the tree. This can be achieved by keeping pointers from each residue a at node k to those residues b and c at daughter nodes i and j, respectively, that were the minimising choices in the equation defining $S_k(a)$ in the weighted parsimony algorithm. We define pointers $l_k(a)$, $r_k(a)$ to left and right daughters of node k, respectively (these pointers perhaps having more than one target if there are several possible minimising residues), and add the steps

$$\text{Set } l_k(a) = \text{argmin}_b(S_i(b) + S(a,b)),$$
$$\text{and } r_k(a) = \text{argmin}_b(S_j(b) + S(a,b)). \qquad (7.5)$$

at the end of the 'Recursion' block of the weighted parsimony algorithm. To obtain an assignment of ancestral residues, we pick a residue a at the root that gives the minimal cost $S_{2n-1}(a)$ and trace back to the leaves using the pointers, choosing arbitrarily whenever the pointers have several possible targets.

In the case of traditional parsimony, where we just count the number of substitutions, all that is needed to obtain the cost of the tree is to keep a list of minimal cost residues at each node, together with the current cost C.

Algorithm: Traditional parsimony [Fitch 1971]

> Initialization:
>> Set $C = 0$ and $k = 2n - 1$.
>
> Recursion: To obtain the set R_k:
>> If k is leaf node:
>>> Set $R_k = x_u^k$.
>>
>> If k is not a leaf node:
>>> Compute R_i, R_j for the daughter nodes i, j of k, and set
>>> $R_k = R_i \cap R_j$ if this intersection is not empty, or else
>>> set $R_k = R_i \cup R_j$ and increment C.
>
> Termination:
>> Minimal cost of tree$= C$.

$\triangleleft$

There is a traceback procedure for finding ancestral assignments in traditional parsimony: We choose a residue from R_{2n-1}, then proceed down the tree. Having chosen a residue from the set R_k, we pick the same residue from the daughter set R_i if possible, and otherwise pick a residue at random from R_i (and similarly for the other daughter set R_j).

The tree T in Figure 7.10 shows two possible sets of ancestral residues obtained by this traceback procedure (the two middle figures). The bottom left figure shows another assignment that cannot be obtained this way. The reason for this failure is that keeping a list of minimal cost residues at each node neglects the possibility that a mismatch cost can be paid at some level in the tree and recouped higher up. This is automatically taken care of with the algorithm for weighted parsimony. The bottom right figure shows the minimal costs for A, B at each node, and the particular choices of back-pointers defined by (7.5) that lead to the assignments in the tree on the bottom left. Note that the left pointer from the top node goes to the residue B whose cost is one more than the minimum. The difference of one is recouped because a B does not have to pay a mismatch penalty for this transition. In general, the assignments not obtained by traceback with R_k can be found by keeping a set Q_k of residues at node k whose cost is one more than that of the residues in R_k. The traditional parsimony algorithm can readily be extended so the Qs are computed at the same time as the Rs.

We have formulated parsimony in the context of rooted trees. However, the minimum cost for a tree in traditional parsimony is independent of where the root is located. In fact, the two edges below the root cannot both have substitutions in them in an optimal tree, for otherwise the assignment at the root could be changed to the assignment of one of the nodes below, with a reduction in cost. This means that, in principle, the root can be removed and costs counted along the edges of the unrooted tree. As it happens, it is easiest to count costs in a rooted tree, because the root defines a direction and hence a unique parent–daughter relationship for applying the ñparsimony algorithm. But the independence of root position means that the number of trees that have to be searched over is reduced.

Exercises

7.11   Show that the tradition parsimony algorithm gives the same cost as that for weighted parsimony using weights $S(a,a) = 0$ for all a, and $S(a,b) = 1$ for all $a \neq b$. (Hint: Show that, to obtain the minimal cost residues a of $S_k(a)$ at each node k, it suffices to keep the list R_k at each node.)

7.12   Show that the minimal cost with weighted parsimony is also independent of the position of the root, provided the substitution cost is a metric, i.e. satisfies $S(a,a) = 0$, symmetry $S(a,b) = S(b,a)$, and the triangle inequality $S(a,c) \leq S(a,b) + S(b,c)$, for all a,b,c.[3] (Hint: If there is a residue A at the root, and different residues B and C at the two daughters, show that the triangle inequality implies the cost cannot be minimal. Use the other two properties of a metric to show that the root can be moved to either of the daughter nodes without increase of cost.)

Selecting labelled branching patterns by branch and bound

We have seen that the search of trees with parsimony can be reduced because only unrooted trees need be considered. Nonetheless, the number of topologies swiftly becomes large as the number of leaves increases. For this reason some more efficient search strategy is needed than simple enumeration.

There are tree-searching methods that proceed stochastically; for instance, we can swop randomly chosen branches on a tree and choose the altered tree if it

[3] Sankoff & Cedergren [1983] assume their cost is a metric, but this is the only place where this property is needed.

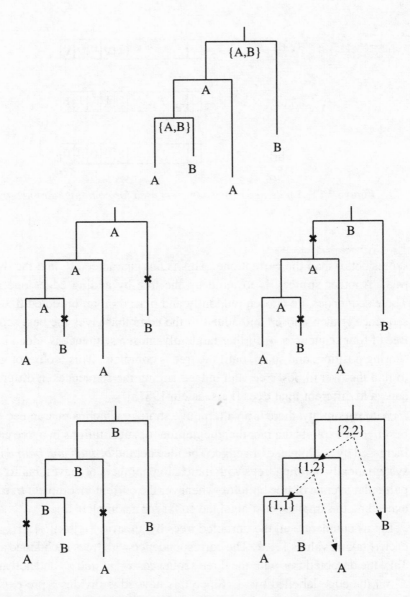

Figure 7.10 *Traditional parsimony with a cost of one for a substitution (marked by an 'X' on the edge), and zero cost otherwise. The sets R_k are shown in the top tree, the two middle trees show assignments of ancestral residues obtained by traceback using the sets R_k. The bottom left tree shows a further eligible set of ancestral residues that cannot be obtained this way, and the bottom right tree shows how this assignment would be obtained using the traceback for weighted parsimony (7.5).*

(i)

(ii)

(iii)

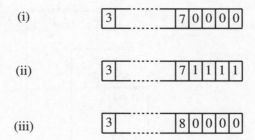

Figure 7.11 *A milometer (or odometer) used for counting unrooted trees.*

scores better than the current one. This is not guaranteed to find the overall best tree. Another strategy is to build up the tree by adding edges one at a time. Three sequences are chosen randomly and placed on an unrooted tree. Another sequence is then chosen and added to the edge that gives the best score for the tree of four sequences. A further randomly chosen sequence is added in the best-scoring position, and so on, until the tree is complete. This too is not guaranteed to find the overall best tree, and indeed adding the sequences in different orders can yield different final trees [Felsenstein 1981a].

With parsimony, there is an alternative strategy which is guaranteed to find the best tree; it exploits the fact that the number of substitutions in a tree can only be increased by adding an extra edge. The idea behind *branch and bound* is to begin systematically building trees with increasing numbers of leaves, but to abandon a particular avenue of tree building whenever the current incomplete tree has a cost exceeding the smallest cost obtained so far for a complete tree.

Let us enumerate all the unrooted trees by an array $[i_3][i_5][i_7]\ldots[i_{2n-5}]$, with each i_k taking values $1\ldots k$. The correspondence with trees is obtained as follows: Take the unrooted tree with the three sequences x^1, x^2 and x^3 and add an edge for x^4 on the edge labelled by i_3. Since this new edge divides a pre-existing edge in two, the total number of edges is now $3 + 2 = 5$. The value of i_5 determines which of these we add x^5 to, giving $5 + 2 = 7$ edges. And so on, up to x^n, which has $(2n - 5)$ choices of position.

Now think of $[i_3][i_5][i_7]\ldots[i_{2n-5}]$ as a milometer (or odometer in the USA) on a car's dashboard. The rightmost numbers advance till they reach $2n - 5$, when they go back to 1 and the next-to-rightmost array index clicks forward by 1. When the next-to-rightmost array index reaches $2n - 7$ it starts again at 1 and the second-to-right array index clicks forward by 1. And so on.

This enumerates all trees with n leaves in a specified sequence, but we also want to count trees with fewer than n leaves, since we are going to build trees of varying sizes. We therefore add a '0' to each counter, meaning that there is no edge of the order specified by the counter, and we let each index cycle from 0 to i_k. However, this will produce some meaningless values, because we cannot add an edge to a non-existent edge, i.e. we cannot have a non-zero counter to the right of a 0. Therefore, when we reach a situation with a row of 0s on the right, we have to advance them all simultaneously to '1' to make the next step (e.g. going from (i) to (ii) in Figure 7.11).

The process starts from the milometer setting [1][0][0]...[0]. Let the smallest cost so far for a complete tree be C. Whenever the cost of our current tree T is more than C, we know that T is not the optimal tree. But (here's the trick) if this happens when all the counters to the right of a given non-zero counter are 0, instead of advancing them all to '1' we can click the rightmost non-zero counter one forward. The reason for this is that the rightmost non-zero counter defines a tree with $k < n$ leaves, and adding more leaves can only increase the cost. We can therefore proceed to the next tree with k leaves (e.g. go from (i) to (iii) in Figure 7.11). This method can save a great deal of searching.

7.5 Assessing the trees: the bootstrap

The tree building algorithms we have described present us with a tree, or perhaps, in the case of parsimony, several optimal trees, but with no measure of how much they should be trusted. Felsenstein [1985] suggested using the bootstrap [Efron & Tibshirani 1993] as a method of assessing the significance of some phylogenetic feature, such as the segregation of a particular set of species on their own branch (a 'clade').

The bootstrap works as follows: Given a dataset consisting of an alignment of sequences, an artificial dataset of the same size is generated by picking columns from the alignment at random with replacement. (A given column in the original dataset can therefore appear several times in the artificial dataset.) The tree building algorithm is then applied to this new dataset, and the whole selection and tree building procedure is repeated some number of times, typically of the order of 1000 times. The frequency with which a chosen phylogenetic feature appears is taken to be a measure of the confidence we can have in this feature.

For certain probabilistic models, the bootstrap frequency of a phylogenetic feature F can be shown to approximate the posterior distribution $P(F|\text{data})$ (see p. 212). When the bootstrap is applied to a non-probabilistically formulated model, such as parsimony, it can be interpreted in terms of statistical hypothesis testing, though a rather more elaborate procedure than that given above may be needed to make the bootstrap conform to standard notions of confidence intervals [Efron, Halloran & Holmes 1996].

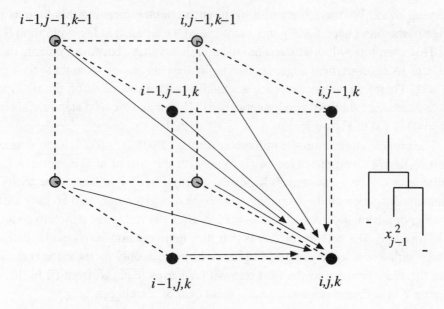

Figure 7.12 *Dynamic programming matrix for Sankoff & Cedergren's algorithm for 3 sequences. Each transition in the matrix is shown by an arrow. For 3 sequences there are 7 transitions at each point. Each has a cost assigned to it, which is the minimal cost, derived by the parsimony algorithm, of a tree whose leaves are defined by the transition, as follows: If 1 is subtracted from a coordinate, the relevant leaf is assigned the preceding character in the input sequence; if the coordinate is unchanged, its leaf is assigned a '-'. For instance, the transition from $(i, j-1, k)$ to (i, j, k) is assigned the tree shown in the figure.*

7.6 Simultaneous alignment and phylogeny

We turn now to the problem of simultaneously aligning sequences and finding a plausible phylogeny for them. There are two parsimony-type algorithms that tackle this problem, the first using a character-subsitution model of gaps, the second using affine gap penalties. Both find an optimal alignment given a tree; it is necessary to search over trees to find the overall optimum.

Sankoff & Cedergren's gap-substitution algorithm

Sankoff & Cedergren's algorithm is guaranteed to find ancestral sequences, and alignments of them and the leaf sequences, that together minimise a tree-based, parsimony-type cost [Sankoff & Cedergren 1983]. The algorithm is, in fact, a combination of two methods already introduced in this book (Figure 7.12). In Chapter 6, p. 141, a dynamic programming method was described for aligning a set of N sequences $x^1, x^2, \ldots, x^N$ [Sankoff & Cedergren 1983; Waterman 1995].

Replacing the max by a min (we use costs here rather than scores), the minimum cost $\alpha_{i_1,i_2,...,i_N}$ of an alignment ending with $x_{i_1}^1, x_{i_2}^2, \ldots, x_{i_N}^N$ is

$$\alpha_{i_1,i_2,...,i_N} = \min_{\Delta_1+...+\Delta_N>0} \{\alpha_{i_1-\Delta_1,i_2-\Delta_2,...,i_N-\Delta_N} + \sigma(\Delta_1 \cdot x_{i_1}^1, \Delta_2 \cdot x_{i_2}^2, \ldots, \Delta_N \cdot x_{i_N}^N)\}.$$
(7.6)

where Δ_i is 0 or 1, and $\Delta_i \cdot x = x$ if $\Delta_i = 1$ and $\Delta_i \cdot x = $ '-' if $\Delta_i = 0$. σ is the weighted parsimony cost for aligning a set of symbols of the extended alphabet. This cost can be calculated by an upward pass through the tree, using the weighted parsimony algorithm (p. 174), where $S(a,b)$ is now defined not when a,b are pairs of residues, but also when one or both is the gap symbol '-'.

Sankoff & Cedergren's procedure is therefore the following: When we reach $(i_1,i_2,\ldots,i_N)$ in the induction, each of the terms $\alpha_{i_1-\Delta_1,i_2-\Delta_2,...,i_N-\Delta_N}$ in (7.6), for all $2^N - 1$ combinations of $\Delta_1,\ldots,\Delta_N$, will previously have been computed, and the calculation of $\sigma(\Delta_1 \cdot x_{i_1}^1, \Delta_2 \cdot x_{i_2}^2, \ldots, \Delta_N \cdot x_{i_N}^N)$ can be achieved by an upward pass of the tree, requiring of the order of N steps (one step for each edge). The entire computation therefore requires of the order of $N(2n)^N$ steps, where n is the length of the sequences. Unfortunately, this is too large for more than half a dozen or so sequences of normal length (of the order of 100 residues).

Hein's affine cost algorithm

Hein's algorithm [Hein 1989a] uses an affine gap cost which is more realistic than the simple substitution treatment of gaps. It is also much faster than Sankoff & Cedergren's algorithm in most realistic situations, fast enough in fact to allow a search over tree topologies for modest-sized sets of sequences. It is the only current practical algorithm able to align sequences and explore alternative phylogenies effectively. The price paid for these very considerable gains is that the algorithm makes a simplifying assumption in the choice of ancestral sequences which does not always lead to the overall most parsimonious choices.

Suppose that we are given a tree. Recall that the algorithm for traditional parsimony ascends the tree, assigning a list of possible residues to each node. These residues are just those that minimise the number of substitutions along the edges to the two daughter nodes. In this case it is possible to find the minimal number of substitutions for the whole tree by minimising at each node. This same procedure is used in Hein's algorithm: in the upward pass through the tree, the only sequences that are considered at a node are those that have the minimum cost, given the sequences at the two daughter nodes. We shall see later that, unlike traditional parsimony, this procedure is not guaranteed to find the minimum cost for the whole tree. But first, let us see how the minimum cost sequences at each node are determined.

The aim is to find sequences z at a given node aligned to both of the sequences

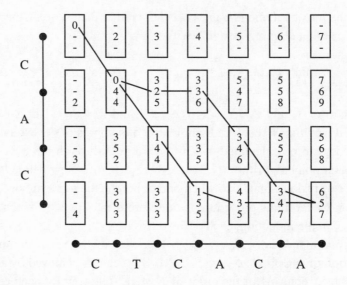

Figure 7.13 *The dynamic programming matrix for two sequences, showing cells with V^M, V^X and V^Y written in this order, from top to bottom. Optimal paths are shown as lines between the cells. Here $d = 2$, $e = 1$, and there is a mismatch cost of 1. Note that we include costs arising from a gap in one sequence followed by a gap in the other (eg the entry $V^Y = 4$ in the second cell from the left in the second row down, which arises from $V^X = 2$ in the cell above), even though such matches are non-optimal.*

x and y at the daughter nodes and satisfying

$$S(x,z) + S(z,y) = S(x,y), \qquad (7.7)$$

where S here denotes the total cost for a given alignment of two sequences. Assuming a mismatch cost of one, with zero cost otherwise, (7.7) can be satisfied at any site if z shares a residue at each site with either x or y (or with both x and y when they have the same residue). Hein's algorithm can also be extended to general weighted parsimony (see Exercise 7.13).

We have not yet shown that sequences z satisfying (7.7) can be found, because we have to deal with gaps. To do this, we use the dynamic programming method for affine gaps described in Chapter 2. Let $V^M(i,j)$, $V^X(i,j)$, $V^Y(i,j)$ denote the minimum costs for alignments up to position i in sequence x, j in sequence y, in the cases where (1) the ith residue in x is aligned to the jth in y, (2) the ith residue in x is aligned to a gap in y, and (3) the jth residue in y is aligned to a gap in x. These correspond to Viterbi costs up to the match state $M(i,j)$, and insert states X and Y, respectively. We write the three numbers V^M, V^X and V^Y in the (i,j)th cell in the dynamic programming matrix (Figure 7.13). Let the affine gap cost for a gap of length k be $d + (k-1)e$, where $e \leq d$. Then the recursion is

$$
\begin{aligned}
V^M(i,j) &= \min\{V^M(i-1,j-1), V^X(i-1,j-1), V^Y(i-1,j-1)\} \\
&\quad + S(x_i, y_j), \\
V^X(i,j) &= \min\{V^M(i-1,j)+d, V^X(i-1,j)+e\}, \\
V^Y(i,j) &= \min\{V^M(i,j-1)+d, V^Y(i,j-1)+e\}. \tag{7.8}
\end{aligned}
$$

Here we assume that the mismatch cost is less than $2e$, which ensures that an optimal alignment will never have a gap in one sequence immediately followed by a gap in the other, e.g. $\begin{smallmatrix}\text{TTAC--}\\\text{TT--GG}\end{smallmatrix}$, but will prefer to match residues, e.g. $\begin{smallmatrix}\text{TTAC}\\\text{TTGG}\end{smallmatrix}$ in this example.

Let us mark all the transitions that occur on paths that give the minimal cost (e.g. those marked in Figure 7.13). Any path that we piece together using these transitions will give an optimal alignment of x and y. Suppose now that x and y are the sequences at the two daughter nodes of a node n. Any path using our marked transitions also serves to define eligible ancestral sequences at n, as follows. If a transition corresponds to a match of two residues in x and y, we choose one of these residues for the ancestral sequence. If a transition corresponds to a match between a gap and a residue, we choose either a gap or a residue in the ancestral sequence.

This will yield a sequence z aligned to x and y. From the way z was constructed, it is clear that if, at some site, both x and y have a residue, then z shares a residue with either x or y (or possibly both of them). Thus equal contributions will be made to the two sides of (7.7). The same will be true when either x or y has a gap provided we take some care with gap-opening costs. In fact, our recipe for making ancestral sequences needs the following extra rule: If a block of consecutive gaps in one sequence occurs on a path, and these are aligned to a set of residues in the other sequence, then the ancestral sequence must either skip this entire set of residues or include them all.

For instance, given the two sequences CAC and CTCACA (see Figure 7.13), the sequence CTC can be derived by following the lower path in the matrix, corresponding to the alignment $\begin{smallmatrix}\text{CAC---}\\\text{CTCACA}\end{smallmatrix}$, choosing a T in the second position, and skipping the block of three gaps. It is a possible ancestral sequence because it can be aligned to CAC by $\begin{smallmatrix}\text{CTC}\\\text{CAC}\end{smallmatrix}$ with a cost of one for the A,T mismatch, and to CTCACA by $\begin{smallmatrix}\text{CTC---}\\\text{CTCACA}\end{smallmatrix}$ with a cost of $d + 2e$. The sum of these two costs, $d + 2e + 1$, is the cost of the original alignment $\begin{smallmatrix}\text{CAC---}\\\text{CTCACA}\end{smallmatrix}$. Similarly CACACA is another eligible ancestral sequence, derived by choosing the residues from the block that are matched to the gaps. What is not allowed is to use only some of these residues. For instance, CACAC is not an ancestral sequence. In fact, optimal alignments to the daughter sequences are $\begin{smallmatrix}\text{CACAC}\\\text{CAC--}\end{smallmatrix}$, with cost $d + e$, and $\begin{smallmatrix}\text{CACAC-}\\\text{CTCACA}\end{smallmatrix}$, with cpst $d + 1$, and both include gap-opening terms. The sum of both costs, $2d + e + 1$, exceeds $d + 2e + 1$, the cost of the original alignment, since we are assuming $d > e$.

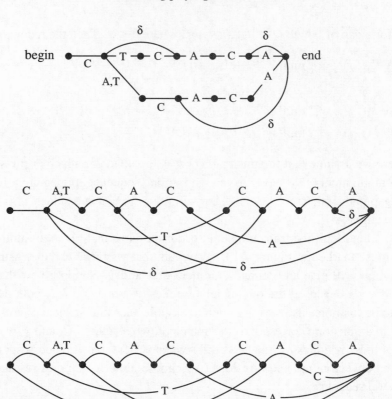

Figure 7.14 *The sequence graph derived from the paths through the dynamic programming matrix in Figure 7.13. Top: the graph, with its dummy edges (marked by a δ). Middle: the same graph, with its nodes arranged in a line. Bottom: the dummy edges have been replaced by an edge that goes back to the preceding vertex and emits the residues attached to that edge.*

We now formalise the idea of following paths through the dynamic programming matrix by deriving a graph from this matrix. Whenever any of the three entries of a cell of the dynamic programming matrix is used by an optimal path, we represent it by a vertex in the graph. It is important to note that different entries in the same cell need different vertices. The situation in Figure 7.13 shows why: The two optimal paths cross in the penultimate cell, one using the M state with cost 3, and the other using the X state with cost 4. If we switched paths in this cell, we would lose track of whether we were opening or extending a gap (see Altschul & Erickson [1986]).

The directed edges of the graph are the transitions that occur in an optimal alignment. We assign residues to all these edges: if a transition in the matrix ends in the match of two residues, then both residues are attached to the edge; if there is a match of a residue and a gap, only that one residue is attached to the edge. Finally, we add 'dummy edges' corresponding to consecutive blocks of two or more gaps. These run from the vertex at the start of the block to that at the end of block and are assigned no residues.

We now consider paths through this graph, running from the initial point to the point matching the last residues in each sequence. The rule is that any path emits the symbols on the edges it uses, choosing one of the symbols if more than one is available. It's easy to see that any path through the graph emits an eligible sequence. The graph will be referred to as a *sequence graph*.

This construction applies to the case where each of the two daughters is a leaf of the tree and so has a single sequence associated to it. What happens as we ascend the tree, and the daughters can have many eligible sequences assigned to them? Hein's ingenious idea is to carry out exactly the same construction, but with graphs rather than sequences as the objects to be matched in the dynamic programming matrix.

To achieve this, we first stretch out each graph so its vertices lie in a line (middle diagram in Figure 7.14); this can always be done so that all the edges point in the same direction. Suppose we have two graphs, G_1 and G_2. Again, we keep track of the values of V^M, V^X and V^Y in each cell. However, instead of considering transitions from the preceding residue in the sequences, we now define 'preceding' by the incoming edges in the graph. When these include a dummy edge, we can skip back to the vertex at its start, and the preceding non-dummy edge then defines a preceding vertex. Note that, because dummy edges span the whole of a block of gaps, and because the condition that a mismatch cost is less than $2e$ excludes a block in one sequence following a block in the other, there cannot be a chain of consecutive dummy edges. Thus a combinatorial explosion in the number of preceding nodes cannot occur.

The procedure for dummy edges can be carried out by first modifying each sequence graph, removing all the dummy edges (marked δ in Figure 7.14) and replacing each of them by an edge going one step back beyond the start of the dummy edge and carrying the symbols associated to that preceding edge. The middle and bottom graphs of Figure 7.14 show how this replacement works. Now the transitions in the dynamic programming matrix are easily described: they are just those obtained by following edges in the modified G_1 or G_2 (see Figure 7.15).

Having defined the values of V in each cell in the matrix, the optimal paths are defined as before by backtracking, and the next sequence graph G_3 is constructed. There is one new feature. The edges in a sequence graph can have several symbols associated to them, these being the sets R_k of the traditional parsimony algorithm. The same procedure as for traditional parsimony governs the combining of sets of symbols: If V^M is defined from an edge in G_1 and an edge in G_2, and if there is a shared symbol (i.e. if there was no mismatch cost in the path using both edges), then only the shared symbols are attached to the derived edge

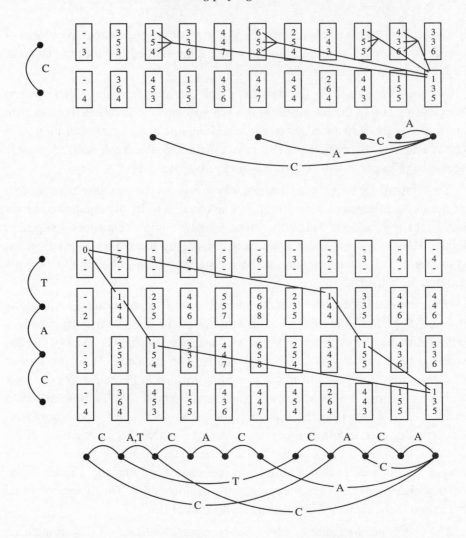

Figure 7.15 *The dynamic programming matrix for a sequence graph, the bottom graph in Figure 7.14, against the sequence* TAC. *The sequence graph in this matrix generates possible ancestral sequences for the top node of the tree in Figure 7.16. The values of V^M, V^X and V^Y in a cell are determined by taking the minimum over all 'preceding' vertices, these being the vertices that can be reached by an edge going back from the current vertex. This is illustrated in the figure above for the computation of a value of V^M in a cell.*

in G_3. If there is no shared symbol, the derived edge acquires all the symbols from both edges.

We can now carry out the recursion (7.8) on the matrix. The optimal paths through the matrix define another graph, and so we continue, ascending the tree until the root is reached. We can then descend the tree, reconstructing the daughter sequences corresponding to a given ancestral sequence. To do this, we follow

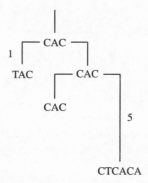

Figure 7.16 *Possible ancestral sequences for the leaf sequences* TAC, CAC, CTCACA, *given the tree shown in the figure.*

the sequence of edges in each daughter graph as the ancestral sequence is traced, choosing symbols in the daughters that are compatible with those in the ancestral sequence. If a delete, or a succession of deletes, skips successive nodes in one of the daughter graphs, the skipped edges must be filled in, with arbitrary choices of symbols. For instance, in tracing the ancestral sequence CAC through the lower path in Figure 7.13, the first three symbols CTC... of the daughter sequence CTCACA are generated, and the last three symbols ...ACA must be added, even though they are skipped in the ancestral path by using a dummy edge (the δ edge of length 3 shown at the top of Figure 7.14). Possible ancestral sequences for a tree are shown in Figure 7.16; the sequence graphs for this tree are those shown in Figure 7.13 and Figure 7.15.

We have now described how sequences are aligned on a given tree. It is also necessary to search through trees, for which Hein [1989b] proposes his own efficient search algorithm. The entire procedure will be manageable if the sequence graphs do not become too complicated. If we have to include most of the transitions in the dynamical programming matrix, the computation will increase in complexity like Sankoff & Cedergren's. The assumption is that most alignments will have a few main routes giving the minimal cost. This will be true if the sequences are similar enough, for then there will be long stretches of unambiguous matches which define a single path through the matrix.

Exercise

7.13   Hein's algorithm can be extended to general weights $S(a,b)$ by attaching a set of minimal costs $S_k(a)$ (as in the weighted parsimony algorithm) to each edge in a sequence graph instead of the set R_k. Show that (7.7) can be satisfied by having z share a residue with x or y provided that $S(a,a) = 0$ for all a. Evaluate the minimal costs (assuming a nucleic acid alphabet) for the sequence graphs shown in Figure 7.13, Figure 7.14 and Figure 7.15.

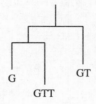

Figure 7.17 *A case where Hein's rule of for choosing optimal ancestral sequences fails to produce the optimal overall assignment to ancestors.*

Limitations of Hein's model

Now let us return to Hein's procedure of taking the minimal cost sequences at each node in the upward pass. To see how this can fail to give an overall optimum for the tree, suppose the cost for a gap of length k is $13 + 3(k - 1)$ and the mismatch cost is 4 (values used by Hein in an example alignment of 5S RNAs). The eligible ancestral sequences for G and GTT are just G and GTT themselves; each requires a consecutive pair of gaps to one of its daughters, with cost $13 + 3 = 16$. The sequence GT is not an eligible ancestral sequence, since it requires single gaps in both alignments, with a total cost of $2 \times 13 = 26$, i.e. $\frac{G-}{GT}$ and $\frac{GT-}{GTT}$. But suppose we have a tree in which the root branches to the ancestor of G and GTT, and also to a third leaf with sequence GT (see Figure 7.17). Then the total tree cost for the ineligible ancestor GT is smaller because two gaps of size one are required in the tree in that case, as opposed to gaps of sizes one and two when either eligible ancestor is used.

Note that in this example the tree topology is not the optimal one for the indicated sequences; the second and third sequences should have been grouped rather than the first and second. In fact, the kind of misbehaviour seen in this example will tend not to occur when the tree topology is correct, and for that reason will be less damaging.

7.7   Further reading

Parsimony was first formulated by Edwards & Cavalli-Sforza [1963; 1964] in the case of continuous parameters, where it amounts to finding a minimum-length tree joining points in Euclidean space. Counting algorithms for parsimony based on sequence data or other discrete variables were introduced by Camin & Sokal [1965], Eck & Dayhoff [1966], Fitch [1971] and others. The combination of the simplicity of these algorithms and the richness of sequence data has combined to make these methods very popular.

Parsimony is sometimes alleged to be a direct philosophical descendant of Occam's razor, and to be free of specific evolutionary assumptions, e.g. 'The use of parsimony in phylogenetic systematics is no different from its use in any other branch of biology or any other science, [and] does not invoke any particular evolutionary mechanism' [Brooks & McLennan 1991, p. 65]. For an instructive debunking of this notion, see Edwards [1996]. See also Section 8.6 for an interpretation of parsimony that relates it to probabilistic models described in this book.

Phylogenetic distance methods were also first described by Edwards & Cavalli-Sforza in their papers mentioned above. They proposed a least-squares match between the observed distances and the summed lengths of a tree. Neighbour-joining can also be given a least-squares interpretation: in this case, the observed distances are compared with those in a simplified tree [Saitou & Nei 1987]. There are many other distance methods besides those discussed here; to mention only one other, that of Fitch & Margoliash [1967a] combines clustering with the distance definition of (7.3).

There are a number of mathematical results on trees with additive lengths. In addition to Studier & Keppler's theorem given in Section 7.8, Buneman [1971] has shown that, when a set of distances satisfies the four-point condition (p. 172), a tree and a set of edge lengths can be found that generate these distances as the sum of edge lengths.

Horizontal transfer of genetic material gives an interesting twist to phylogeny. At first sight, it prevents the use of simple tree structures, since a recombinational event would seem to create a link between a sequence and its two ancestors and thereby give rise a loop. However, the fragments on either side of a recombination point each have a single parent, and the genome may be described as a concatenation of segments, each with its own tree [Hein 1993]. Recombinational events are likely to be particularly important in viruses and prokaryotes, where horizontal transfer is frequent. (The recombination that occurs in diploid 'family trees' is of course even more frequent, but it requires a different kind of model, being dominated by crossing-over events with little evolution of sequence.)

A generalisation of trees has been proposed by Bandelt & Dress [1992]. They showed how to build networks from distance data that branch like a tree where the evidence for a topology is strong and that generate a mesh covering regions of ambiguity.

Useful general references for phylogeny are Waterman [1995], Swofford & Olsen [1996]; for recent reviews see Saitou [1996] and Felsenstein [1996].

7.8 Appendix: proof of neighbour-joining theorem

For completeness, and because it is a pleasing mathematical result, we include the proof by Studier & Keppler [1988] that leaves with minimal neighbour-joining distance are neighbours. This ensures that a tree with additive lengths will be

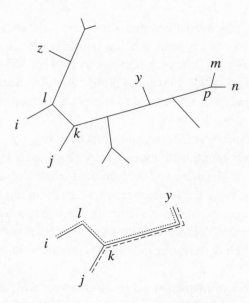

Figure 7.18 *Above: If the leaves i, j are not neighbours, there are at least two nodes on the path joining them, here shown as k and l. The branch from k that does not go to i or j is called L_k and is shown here with a pair of neighbours, m and n, on it. The branch from l is called L_l, and has a branch to a leaf z. Below: The path from i to y (dots) and from j to y (dashes) are shown, giving a simple visual proof that $d_{iy} + d_{jy} = d_{ij} + 2d_{ky}$.*

correctly reconstructed by neighbour-joining. A recent paper extends this result to show that neighbour-joining also correctly reconstructs trees where additivity only holds approximately [Atteson 1997].

Theorem: For a tree with additive lengths, D_{ij} minimal implies i, j are neighbouring leaves.

Proof: Suppose the smallest D is D_{ij}, and suppose furthermore that i and j are not neighbouring leaves. We seek a contradiction.

Since i and j are not neighbours, there must be at least two nodes on the path connecting them (see Figure 7.18). Call these nodes k, l, and let L_k be the set of leaves which derive from the third branch from k, i.e. not the edge towards i or j, and let L_l be the equivalent set for l. Let m and n be a pair of neighbouring leaves in L_k with joining node p (if no such pair exists, an alternative argument is available; see Exercise 7.14). Let d_{uv} denote the summed edge lengths of the path connecting any two nodes u, v. By additivity, this is the correct distance d_{uv} when they are both leaves. For any y in L_k, it is clear that $d_{iy} + d_{jy} = d_{ij} + 2d_{ky}$ (see lower figure in Figure 7.18). Similarly $d_{my} + d_{ny} = d_{mn} + 2d_{py}$. Thus

$$d_{iy} + d_{jy} - d_{my} - d_{ny} = d_{ij} + 2d_{ky} - 2d_{py} - d_{mn}. \tag{7.9}$$

Likewise, for z in L_l, we find

$$d_{iz} + d_{jz} - d_{mz} - d_{nz} = d_{ij} - d_{mn} - 2d_{pk} - d_{lk}. \qquad (7.10)$$

From the definition of D_{ij},

$$D_{ij} - D_{mn} = d_{ij} - d_{mn} - \frac{1}{N-2}\left(\sum_{\text{all leaves } u} d_{iu} + d_{ju} - d_{mu} - d_{nu}\right),$$

and it is easy to check from (7.9) and (7.10) that the coefficients of d_{ij} and d_{mn}, summed over all leaves u in the tree, are both $(N-2)$ (see Exercise 7.15). Thus the term $d_{ij} - d_{mn}$ cancels, and we can write

$$D_{ij} - D_{mn} = \frac{1}{N-2}\left(\sum_{y \text{ in } L_k}(2d_{py} - 2d_{ky}) + \sum_{z \text{ in } L_l}(2d_{pk} + d_{lk})\right) + C$$

where C is the sum of all the extra positive terms coming from other branches on the path between i and j besides k and l. Letting $|L_l|$ and $|L_k|$ denote the numbers of nodes in L_l, L_k, respectively, and the using the fact that $d_{py} - d_{ky} > -d_{pk}$,

$$D_{ij} - D_{mn} > 2d_{pk}(|L_l| - |L_k|)/(N-2).$$

We must have $D_{mn} > D_{ij}$, since D_{ij} is the minimum, so $|L_l| < |L_k|$. But the argument can be applied with the two nodes l and k reversed, so we must also have $|L_k| < |L_l|$. Hence the assumption is false, and i, j are neighbouring leaves.

Exercises

7.14   If the branch from k has only a single leaf m (so it is not possible to find a pair of neighbours in L_k), show that the presence of other nodes besides k on the path from i to j implies $D_{ij} > D_{jm}$, contradicting the assumption that D_{ij} is the minimum.

7.15   Show that the term $2d_{py}$ is absent in (7.9) when $y = m$ or $y = n$. Show that this means that the term $2d_{py}$ can be included in the sum $\sum(2d_{py} - 2d_{ky})$ in (7.11) for all y in L_k, including $y = m$ and $y = n$, provided we subtract $2d_{pm} + 2d_{pn}$ from the sum. Show that the term d_{mn} then cancels, and also check the case where $y = i$ and $y = j$.

8

Probabilistic approaches to phylogeny

8.1 Introduction

Our goal in this chapter is to formulate probabilistic models for phylogeny and show how trees can be inferred from sets of sequences, either by maximum likelihood or by sampling methods. We also review the phylogenetic methods of the previous chapter, and show that they often have probabilistic interpretations, though they are not usually presented this way.

Overview of the probabilistic approach to phylogeny

The basic aim of probability-based phylogeny is to rank trees either according to their likelihood $P(\text{data}|\text{tree})$, or, if we are taking a more Bayesian view, according to their posterior probability $P(\text{tree}|\text{data})$. There may be subsidiary aims, such as finding the likelihood or posterior probability of some particular taxonomic feature, such as a grouping of a set of organisms on a single branch. To achieve any of these aims, we must be able to define and compute $P(x^{\bullet}|T,t_{\bullet})$, the probability of a set of data given a tree. Here the data are a set of n sequences x^j for $j = 1 \ldots n$, which we write compactly as $x^{\bullet}$. T is a tree with n leaves with sequence j at leaf j, and the $t_{\bullet}$ are the edge lengths of the tree. To define $P(x^{\bullet}|T,t_{\bullet})$ we need a model of evolution, i.e. of the mutation and selection events that change sequences along the edges of a tree.

Let us assume that we can define a probability $P(x|y,t)$ for an ancestral sequence y to evolve to a sequence x along an edge of length t. The probability of T with a specific set of ancestors assigned to its nodes can then be calculated by multiplying all the evolutionary probabilities, one for each edge of the tree. For instance, for the tree shown in Figure 8.1 the probability would be

$$P(x^1,\ldots,x^5|T,t_{\bullet}) = P(x^1|x^4,t_1)P(x^2|x^4,t_2)P(x^3|x^5,t_3)P(x^4|x^5,t_4)P(x^5),$$

where $P(x^5)$ denotes the probability of x^5 occurring at the root of the tree. In general (apart from laboratory evolution experiments, like those of Hillis *et al.* [1992]) the ancestral sequences will be unknown, and to obtain the probability $P(x^1,\ldots,x^3|T,t_{\bullet})$ of the known sequences for the given tree we need to sum over

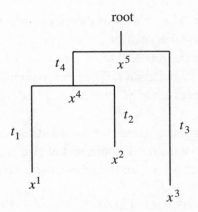

Figure 8.1 *An example of a tree with three sequences.*

all the possible ancestors x^4, x^5. This is similar to summing over all the different paths in a HMM to obtain the probability of the observed data (see Chapter 3).

Given this model, we can seek the maximum likelihood tree, namely the tree with topology T and edge lengths $t_{\bullet}$ that maximises $P(x^{\bullet}|T, t_{\bullet})$. Finding this maximum requires: (1) a search over tree topologies, with the order of assignment of sequences at the leaves specified; (2) for each topology, a search over all possible lengths of edges $t_{\bullet}$.

As we have seen (p. 164), there are $(2n-3)!!$ rooted binary trees with n leaves, and this number grows very large for more than half a dozen sequences. An efficient search procedure (e.g. p. 176 and Felsenstein [1981a]) is therefore required to carry out (1). Part (2), maximising the likelihood of edge lengths, can be achieved by a variety of optimisation techniques (Section 8.3).

An alternative strategy is to search stochastically over trees by sampling from the posterior distribution $P(T, t_{\bullet}|x^{\bullet})$ (Section 8.4). This has only been explored recently, but the method is very promising.

8.2 Probabilistic models of evolution

We have not yet specified the form of $P(x|y, t)$, the probability of a sequence x arising from an ancestral sequence y over an edge of length t. For this we need a model of evolution. We know that, in the course of evolution, residues are substituted by others, that deletion and insertion of groups of residues occur, and that there are more complex constraints imposed by structures of nucleic acids and proteins. Later we shall consider models for deletions and insertions, but to begin with we make some radical simplifying assumptions: that every site of the given data sequences can be treated as independent and that deletions and insertions do not occur. Our sequences therefore form an ungapped alignment, with independent evolution at each site.

Let $P(b|a,t)$ denote the probability of a residue a having being substituted by a residue b over an edge length t. Then our assumption implies that for two aligned, gapless sequences x and y, $P(x|y,t) = \prod_u P(x_u|y_u,t)$, where u indexes sites in the alignment.

Let us look now at possible forms for the substitution probabilities $P(b|a,t)$, for residues a and b. Given a residue alphabet of size K, we can write these as a $K \times K$ matrix that depends on t, and which we denote by $S(t)$:

$$S(t) = \begin{pmatrix} P(A_1|A_1,t) & P(A_2|A_1,t) & \dots & P(A_K|A_1,t) \\ P(A_1|A_2,t) & P(A_2|A_2,t) & \dots & P(A_K|A_2,t) \\ \dots & \dots & \dots & \dots \\ P(A_1|A_K,t) & P(A_2|A_K,t) & \dots & P(A_K|A_K,t) \end{pmatrix}.$$

For several important families of substitution matrices, the family is *multiplicative*, in the sense that

$$S(t)S(s) = S(t+s) \tag{8.1}$$

for all values of the lengths s and t. This is equivalent to saying that the substitution probabilities satisfy

$$\sum_b P(a|b,t)P(b|c,s) = P(a|c,s+t)$$

for all a, c, s and t. If we adopt a viewpoint in which t is regarded as a 'time' variable,[1] then multiplicativity is a consequence of the substitution process being Markovian and stationary, the latter meaning that the probability of substituting a at time t by b at time s depends only on the time interval $(s - t)$ (Exercise 8.2).

For nucleotide sequences, one model is that of Jukes & Cantor [1969]. This assumes that the matrix, R, of *rates* of substitution takes the form

$$\begin{array}{cccc} & A & C & G & T \end{array}$$
$$\begin{array}{c} A \\ C \\ G \\ T \end{array} \begin{pmatrix} -3\alpha & \alpha & \alpha & \alpha \\ \alpha & -3\alpha & \alpha & \alpha \\ \alpha & \alpha & -3\alpha & \alpha \\ \alpha & \alpha & \alpha & -3\alpha \end{pmatrix}, \tag{8.2}$$

which means that all nucleotides undergo transitions at the same rate α. The substitution matrix for a short time $S(\varepsilon)$ is approximately given by $S(\varepsilon) \simeq (I + R\varepsilon)$, where I is the identity matrix with ones down the diagonal and zeros elsewhere.

---

[1] For instance, t might be proportional to mutation rate $\times$ evolutionary time.

Thus

$$I + R\varepsilon = \begin{pmatrix} 1-3\alpha\varepsilon & \alpha\varepsilon & \alpha\varepsilon & \alpha\varepsilon \\ \alpha\varepsilon & 1-3\alpha\varepsilon & \alpha\varepsilon & \alpha\varepsilon \\ \alpha\varepsilon & \alpha\varepsilon & 1-3\alpha\varepsilon & \alpha\varepsilon \\ \alpha\varepsilon & \alpha\varepsilon & \alpha\varepsilon & 1-3\alpha\varepsilon \end{pmatrix}.$$

By multiplicativity, $S(t+\varepsilon) = S(t)S(\varepsilon) \simeq S(t)(I + R\varepsilon)$. We can write this as $(S(t+\varepsilon) - S(t))/\varepsilon \simeq S(t)R$, and in the limit of small ε we get $S'(t) = S(t)R$. From this we can derive a substitution matrix for time t. The symmetry of the rate matrix suggests that we try giving $S(t)$ the following form:

$$S(t) = \begin{pmatrix} r_t & s_t & s_t & s_t \\ s_t & r_t & s_t & s_t \\ s_t & s_t & r_t & s_t \\ s_t & s_t & s_t & r_t \end{pmatrix}. \tag{8.3}$$

Substituting this into $S'(t) = S(t)R$, we get the equations

$$\begin{aligned} \dot{r} &= -3\alpha r + 3\alpha s, \\ \dot{s} &= -\alpha s + \alpha r, \end{aligned}$$

and we easily check that these are satisfied by

$$\begin{aligned} r_t &= \tfrac{1}{4}\left(1 + 3e^{-4\alpha t}\right), \\ s_t &= \tfrac{1}{4}\left(1 - e^{-4\alpha t}\right). \end{aligned} \tag{8.4}$$

The matrix (8.3) with these values of r_t and s_t constitutes the *Jukes–Cantor model*. Note that, when $t = \infty$, $r_t = s_t = \frac{1}{4}$. This means that the nucleotide equilibrium frequencies implied by the model are $q_A = q_C = q_G = q_T = \frac{1}{4}$.

The Jukes–Cantor model does not capture some important features of nucleotide substitution. For instance, transitions, namely purine to purine or pyrimidine to pyrimidine substitutions, are more common than transversions, which change the type of nucleotide.[2] To account for this, Kimura [1980] proposed a model with the rate matrix

$$\begin{pmatrix} -2\beta-\alpha & \beta & \alpha & \beta \\ \beta & -2\beta-\alpha & \beta & \alpha \\ \alpha & \beta & -2\beta-\alpha & \beta \\ \beta & \alpha & \beta & -2\beta-\alpha \end{pmatrix}. \tag{8.5}$$

This can be integrated, by the same procedure we used with the Jukes–Cantor

[2] Thus the transitions are A $\leftrightarrow$ G, C $\leftrightarrow$ T, and the transversions are A $\leftrightarrow$ T, G $\leftrightarrow$ T, A $\leftrightarrow$ C, and C $\leftrightarrow$ G

model, to give the general time-dependent form

$$S(t) = \begin{pmatrix} r_t & s_t & u_t & s_t \\ s_t & r_t & s_t & u_t \\ u_t & s_t & r_t & s_t \\ s_t & u_t & s_t & r_t \end{pmatrix}, \tag{8.6}$$

where

$$
\begin{aligned}
s_t &= \tfrac{1}{4}\left(1 - e^{-4\beta t}\right), \\
u_t &= \tfrac{1}{4}\left(1 + e^{-4\beta t} - 2e^{-2(\alpha+\beta)t}\right), \\
r_t &= 1 - 2s_t - u_t.
\end{aligned}
$$

This model, though widely used, is still far from realistic, as its equilibrium frequencies are equal, $q_A = q_C = q_G = q_T = \tfrac{1}{4}$, whereas many organisms show strong bias in their AT to GC ratio. For a model which allows for this as well as inequality of transitions and transversions, see Hasegawa, Kishino & Yano [1985].

Turning to protein sequences, we saw in Section 2.8 that the PAM matrices of conditional probabilities for integers n are defined by $S(n) = S(1)^n$, i.e. by raising the PAM1 matrix to the nth power. We can extend this to all values of t (i.e. not just integers, but all positive real numbers), and obtain a matrix formally very similar to those of the Jukes–Cantor and Kimura DNA models.

Recall that $S(1)$ was defined by normalising the rows of the symmetric matrix A (p. 42) and then rescaling (to give a 1 PAM matrix). The same result is obtained if these operations are interchanged, rescaling first, then normalising. Since the matrix obtained by rescaling A is symmetric, it can be diagonalised [Mathews & Walker 1970]. Rescaling does not change the diagonal form of the matrix, so we can write $S(1) = U D(\lambda_i) U^{-1}$, where U is a coordinate transformation and $D(\lambda_i)$ is the diagonal matrix with the eigenvalues $\lambda_1 \ldots \lambda_{20}$ down the diagonal. These eigenvalues lie in the range 0 to 1, so can be written $\lambda_i = \exp(-\mu_i)$. Now, the powers of $S(1)$ take a simple form in the diagonal matrix coordinate system; for instance, $S(2) = S(1)S(1) = U D(\lambda_i) U^{-1} U D(\lambda_i) U^{-1} = U D(\lambda_i^2) U^{-1}$, and generally $S(t) = U D(\lambda_i^t) U^{-1}$. Thus we can write

$$S(t) = U \begin{pmatrix} e^{-\mu_1 t} & 0 & \ldots & 0 \\ 0 & e^{-\mu_2 t} & \ldots & 0 \\ \ldots & \ldots & \ldots & \ldots \\ 0 & 0 & \ldots & e^{-\mu_{20} t} \end{pmatrix} U^{-1}.$$

This shows that each entry of $S(t)$ can be expressed as a sum of exponentials: If A_i denotes the ith amino acid, then $P(A_j | A_i, t) = \sum_k u_{ik} \exp(-\mu_k t) v_{kj}$, where u_{ik} and v_{kj} are the entries in U and U^{-1}, respectively.

This resembles the rate matrices for the DNA models, and we can easily see why. If the Jukes–Cantor rate matrix is diagonalised, so $R = U D(\lambda_i) U^{-1}$ for a suitable coordinate transform U, the equation $S'(t) = S(t)R$ becomes $T'(t) =$

$T(t)D$, where $S(t) = UT(t)U^{-1}$. But the equation $T'(t) = T(t)D$ is easily solved, with the initial condition that $S(0)$ is diagonal; in fact $T(t)$ itself must be diagonal with the terms $\exp(\lambda_i t)$ down its diagonal. Since the eigenvalues are 0 and -4α, it is easy to see that we obtain the Jukes–Cantor matrix entries, (8.4), in a way analogous to the above derivation of the PAM matrices.

Putting $t = \infty$, the PAM matrices become

$$
\begin{pmatrix}
q_{A_1} & q_{A_2} & \cdots & q_{A_{20}} \\
q_{A_1} & q_{A_2} & \cdots & q_{A_{20}} \\
\cdots & \cdots & \cdots & \cdots \\
q_{A_1} & q_{A_2} & \cdots & q_{A_{20}}
\end{pmatrix},
$$

where the q_{A_i} are the equilibrium frequencies for amino acids, close to the amino acid frequencies in the database from which Dayhoff, Schwartz & Orcutt [1978] originally constructed their matrices.

Exercise

8.1 Show that the Jukes–Cantor and Kimura substitution matrices are multiplicative.

8.2 Let $P(a(t_2)|b(t_1))$ denote the probability of a residue b, present at time t_1, having been substituted by an a by time t_2. Stationarity means that we can write this as $P(a|b, t_2 - t_1)$. The Markov property means that $P(a(t_2)|b(t_1)) = P(a(t_2)|b(t_1), c(t_0))$ if $t_0 < t_1$, i.e. that the probability of the substitution of b by a is not influenced by the residue being c at the earlier time t_0. Show that

$$
\sum_{b(t)} P(a(s+t)|b(t), c(0)) P(b(t)|c(0)) = P(a(s+t)|c(0)),
$$

and deduce that multiplicativity, (8.1), holds.

8.3 Calculating the likelihood for ungapped alignments

We show here how the likelihood of a tree can be computed using the preceding model. We begin with the case of two sequences, and then proceed to the general case of n sequences.

The case of two sequences

Suppose we have two sequences x^1 and x^2. In this case there is only one tree, namely the one with two branches and a root node which represents the hypothetical common ancestor of x^1 and x^2 (Figure 8.2). Thus we only have to investigate how the likelihood varies with the lengths t_1 and t_2.

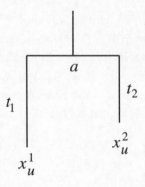

Figure 8.2 *A simple tree.*

Consider a site u. The residues at the leaves 1 and 2 are then x_u^1, x_u^2, respectively. We assign a residue a to the root, using a different notation from the leaves to emphasise the fact that it is a variable, and not specified by the dataset x^1, x^2:

$$P(x_u^1, x_u^2, a | T, t_1, t_2) = q_a P(x_u^1 | a, t_1) P(x_u^2 | a, t_2).$$

This is the probablity of drawing a from the root distribution (which we assume to be the equilibrium distribution of the substitution matrix family) and of making substitutions of a by x_u^1 and x_u^2. Note that we include the cases where either or both of x_u^1, x_u^2 is the same as a. Since in general we do not know what the root residue was, we must sum over all possible as to get the probability of x_u^1, x_u^2. Formally

$$P(x_u^1, x_u^2 | T, t_1, t_2) = \sum_a q_a P(x_u^1 | a, t_1) P(x_u^2 | a, t_2). \tag{8.7}$$

If there are N sites, we can write the full likelihood as

$$P(x^1, x^2 | T, t_1, t_2) = \prod_{u=1}^{N} P(x_u^1, x_u^2 | T, t_1, t_2). \tag{8.8}$$

Example: The likelihood of two nucleotide sequences

Suppose we have two nucleotide sequences, and for simplicity, that only two of the nucleotides, C and G, are present. For instance, the sequences might be

CCGGCCGCGCG
CGGGCCGGCCG

What is the likelihood $P(x^1, x^2 | T, t_1, t_2)$ of these sequences, assuming the Jukes–Cantor model?

Using the substitution probabilities (8.4), (8.7) gives the probability of C occurring at both leaves of the tree T as:

$$P(\text{C}, \text{C} | T, t_1, t_2) = q_\text{C} r_{t_1} r_{t_2} + q_\text{G} s_{t_1} s_{t_2} + q_\text{A} s_{t_1} s_{t_2} + q_\text{T} s_{t_1} s_{t_2} = \tfrac{1}{4} \left(r_{t_1} r_{t_2} + 3 s_{t_1} s_{t_2} \right).$$

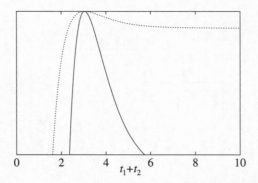

Figure 8.3 *The log likelihood $P(x^1, x^2 | T, t_1, t_2)$ given by (8.9), with $n_1 = 100, n_2 = 250$, and with $n_1 = 1000, n_2 = 2500$. The latter curve is sharper, as there are more data to define the maximum likelihood peak. The curves have been shifted so their peaks superimpose.*

By symmetry $P(\mathrm{G}, \mathrm{G} | T, t_1, t_2) = P(\mathrm{C}, \mathrm{C} | T, t_1, t_2)$. Similarly,

$$P(\mathrm{C}, \mathrm{G} | T, t_1, t_2) = P(\mathrm{G}, \mathrm{C} | T, t_1, t_2) = \tfrac{1}{4} \left(r_{t_1} s_{t_2} + s_{t_1} r_{t_2} + 2 s_{t_1} s_{t_2} \right).$$

Substituting the values r and s gives

$$
\begin{aligned}
P(\mathrm{C}, \mathrm{C} | T, t_1, t_2) &= \tfrac{1}{16} \left(1 + 3 e^{-4\alpha(t_1 + t_2)} \right), \\
P(\mathrm{C}, \mathrm{G} | T, t_1, t_2) &= \tfrac{1}{16} \left(1 - e^{-4\alpha(t_1 + t_2)} \right).
\end{aligned}
$$

Now suppose there are n_1 sites where the residues in the two sequences are identical and n_2 sites where a substitution occurs. Then (8.8) gives

$$P(x^1, x^2 | T, t_1, t_2) = \frac{1}{16^{n_1 + n_2}} \left(1 + 3 e^{-4\alpha(t_1 + t_2)} \right)^{n_1} \left(1 - e^{-4\alpha(t_1 + t_2)} \right)^{n_2}. \qquad (8.9)$$

Note that the likelihood depends only on the sum of t_1 and t_2. This is because the Jukes–Cantor substitution process is time-symmetrical, so there is no information available to specify the position of the root. The likelihood remains unchanged if the root slides while the sum $t_1 + t_2$ remains constant. This indeterminacy of the root will be discussed more fully on p. 202. Figure 8.3 shows an example of how the likelihood (plotted as the log likelihood) varies with $t_1 + t_2$. $\qquad \square$

The likelihood for an arbitrary number of sequences

We can now extend these calculations to the case of n sequences. Suppose we have a tree T with edge lengths $t_\bullet$. Let $\alpha(i)$ denote the immediate ancestral node to i, i.e. the node at the top of the edge above i. Let $x_u^1 \ldots x_u^n$ denote as usual the residues at the uth site of the n sequences $x^1, \ldots, x^n$. The probability $P(x_u^1 \ldots x_u^n | T, t_\bullet)$ of generating these residues at the n leaves of T is given by

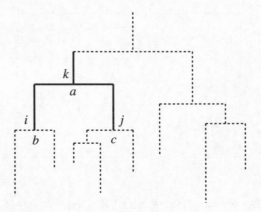

Figure 8.4 *Labelling at a branch in a tree.*

multiplying the probabilities of substitutions at all edges of the tree. Thus

$$P(x_u^1 \ldots x_u^n | T, t_\bullet) =$$

$$\sum_{a^{n+1}, a^{n+2}, \ldots a^{2n-1}} q_{a^{2n-1}} \prod_{i=n+1}^{2n-2} P(a^i | a^{\alpha(i)}, t_i) \prod_{i=1}^{n} P(x_u^i | a^{\alpha(i)}, t_i) \qquad (8.10)$$

where the sum is over all possible assignments of residues a^k to non-leaf nodes k (these nodes being numbered $n+1$ to $2n-1$).

This probability can be computed by working up the tree from the leaves, using post-order traversal [Felsenstein 1981a]. Let $P(L_k | a)$ denote the probability of all the leaves below node k given that the residue at k is a. Then we compute $P(L_k | a)$ from the probabilities $P(L_i | b)$ and $P(L_j | c)$ for all b and c, where i and j are the daughter nodes of k (Figure 8.4):

Algorithm: Felsenstein's algorithm for likelihood

Initialisation:

   Set $k = 2n - 1$.

Recursion: Compute $P(L_k | a)$ for all a as follows:

   If k is leaf node:

      Set $P(L_k | a) = 1$ if $a = x_u^k$, $P(L_k | a) = 0$ if $a \neq x_u^k$.

   If k is not a leaf node:

      Compute $P(L_i | a)$, $P(L_j | a)$ for all a at the daughter nodes i, j,

      and set $P(L_k | a) = \sum_{b,c} P(b | a, t_i) P(L_i | b) P(c | a, t_j) P(L_j | c)$.

Termination:

   Likelihood at site $u = P(x_u^\bullet | T, t_\bullet) = \sum_a P(L_{2n-1} | a) q_a$.   ◁

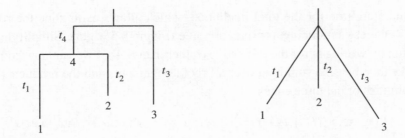

Figure 8.5 *A tree with three leaves (left), which can be simplified to a tri-furcating tree (right) for the purpose of computing the likelihood.*

Note the resemblance to the weighted parsimony algorithm (p. 174). We discuss this further on p. 224.

The concluding step in computing the likelihood is to use the assumption of independence at sites to write:

$$P(x^\bullet|T,t_\bullet) = \prod_{u=1}^{N} P(x_u^\bullet|T,t_\bullet). \tag{8.11}$$

Example: A tree with three nucleotide sequences

We extend the example on p. 198 now to a tree with three leaves (Figure 8.5, left). The data are three nucleotide sequences composed only of Cs and Gs, for instance:

CCGGCCGCGCG

CGGGCCGGCCG

GCCGCCGGGCC

We compute the likelihood according to the Jukes–Cantor model. As before, we consider the sites with different assignments of residues separately. Consider the case where C occurs at all leaves. We have

$$
\begin{aligned}
P(C,C,C|T,t_1,t_2,t_3) &= q_C r_{t_3}\left(r_{t_4}r_{t_1}r_{t_2} + 3s_{t_4}s_{t_1}s_{t_2}\right) \\
&\quad + (q_A + q_G + q_T)s_{t_3}\left(r_{t_4}s_{t_1}s_{t_2} + 2s_{t_4}s_{t_1}s_{t_2} + s_{t_4}r_{t_1}r_{t_2}\right) \\
&= \tfrac{1}{4}r_{t_1}r_{t_2}(r_{t_3}r_{t_4} + 3s_{t_3}s_{t_4}) \\
&\quad + \tfrac{3}{4}s_{t_1}s_{t_2}(2s_{t_3}s_{t_4} + s_{t_3}r_{t_4} + s_{t_4}r_{t_3}) \\
&= \tfrac{1}{4}\left(r_{t_1}r_{t_2}r_{t_3+t_4} + 3s_{t_1}s_{t_2}s_{t_3+t_4}\right),
\end{aligned}
$$

where the first equation simply computes the terms in (8.10), beginning with the equilibrium probabilities at the root, the second regroups them, and the third follows from the multiplicativity of the Jukes–Cantor matrices (Exercise 8.3). Once again, we find that the lengths of the edges adjoining the root, here t_3 and t_4, appear only as their sum. This holds true for all leaf values, not just 'C, C, C';

it is therefore true for the total likelihood, which allows us to slide the root to node 4, thereby producing a trifurcating tree (Figure 8.5, right). Simplifying the notation by writing t_3 for the third edge (rather than $t_3 + t_4$) we can now compute the likelihood easily, summing over all root assignments and the products of the probabilities of the three edges:

$$P(x_u^1, x_u^2, x_u^3 | T, t_1, t_2, t_3) = \sum_a q_a P(x_u^1 | a, t_1) P(x_u^2 | a, t_2) P(x_u^3 | a, t_3).$$

There are four possible types of terms, where all residues are the same, or where one differs from the other two; for instance:

$$P(\mathsf{C}, \mathsf{C}, \mathsf{C} | T, t_1, t_2, t_3) = \tfrac{1}{4} \left(r_{t_1} r_{t_2} r_{t_3} + s_{t_1} s_{t_2} s_{t_3} \right),$$
$$P(\mathsf{C}, \mathsf{C}, \mathsf{G} | T, t_1, t_2, t_3) = \tfrac{1}{4} \left(r_{t_1} r_{t_2} s_{t_3} + s_{t_1} s_{t_2} r_{t_3} + 2 s_{t_1} s_{t_2} s_{t_3} \right).$$

If there are n_1 sites with the same residue, n_2 of type CCG or GGC, n_3 of type CGC or GCG, and n_4 of type GCC or CGG, then by symmetry

$$
\begin{aligned}
P(x^1, x^2 | T, t_1, t_2) &= 4^{-3(n_1 + n_2 + n_3 + n_4)} a(t_1, t_2, t_3)^{n_1} b(t_1, t_2, t_3)^{n_2} \\
&\quad \times b(t_1, t_3, t_2)^{n_3} b(t_3, t_2, t_1)^{n_4}
\end{aligned}
\tag{8.12}
$$

where $a(t_1, t_2, t_3)$ and $b(t_1, t_2, t_3)$ are sums of exponentials (see Exercise 8.4). For an illustration of this likelihood function, see Figure 8.6. □

Exercises

8.3   Show that $r_{t_3} r_{t_4} + 3 s_{t_3} s_{t_4}$ and $2 s_{t_3} s_{t_4} + s_{t_3} r_{t_4} + s_{t_4} r_{t_3}$ are terms arising from the product of the Jukes–Cantor matrices for times t_3 and t_4, and deduce that they can be written as $r_{t_3 + t_4}$ and $s_{t_3 + t_4}$, respectively.

8.4   Show that $a(t_1, t_2, t_3)$ and $b(t_1, t_2, t_3)$ are given by

$$a(t_1, t_2, t_3) = 1 + 3\mathrm{e}^{-4\alpha(t_1 + t_2)} + 3\mathrm{e}^{-4\alpha(t_1 + t_3)} + 3\mathrm{e}^{-4\alpha(t_2 + t_3)} + 6\mathrm{e}^{-4\alpha(t_1 + t_2 + t_3)}$$

and

$$b(t_1, t_2, t_3) = 1 + 3\mathrm{e}^{-4\alpha(t_1 + t_2)} - \mathrm{e}^{-4\alpha(t_1 + t_3)} - \mathrm{e}^{-4\alpha(t_2 + t_3)} - 2\mathrm{e}^{-4\alpha(t_1 + t_2 + t_3)}$$

Reversibility and independence of root position

With the parsimony method, we only need to search over unrooted tree topologies. It is much less obvious that the likelihood is independent of the position of the root, but under certain reasonable assumptions this is true. In fact, two assumptions suffice. One is that the substitution matrix family is multiplicative

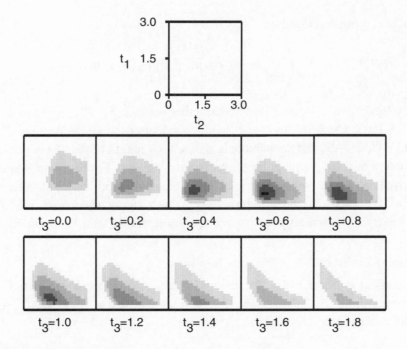

Figure 8.6 *The likelihood function given by (8.12), for $n_1 = 10$, $n_2 = 20$, $n_3 = 15$, $n_4 = 17$. White and the five grey levels indicate likelihood values in the ranges separated by 0, 0.001, 0.01, 0.07, 0.3, 0.9, 1. Each square shows the likelihood for a particular value of t_3 (indicated below the square), the two axes of the square representing t_1 and t_2, whose ranges are indicated in the square at the top of the figure.*

(8.1), which, as we have seen, holds for the Jukes–Cantor and PAM matrices, amongst others. The other is that *reversibility* should hold. This means that

$$P(b|a,t)q_a = P(a|b,t)q_b \qquad (8.13)$$

for all a, b and t. It is clear from their symmetry that reversibility holds for the Jukes–Cantor and Kimura matrices. Reversibility for the PAM matrices follows from the fact that information about the direction of evolutionary time is discarded when the counts are being collected: a substitution from an ancestral residue a to a descendant residue b is treated as equivalent to a substitution in the reverse direction (see Section 2.8 and the example on the PAM family below).

To show that multiplicativity and reversibility imply that all positions of the root give the same likelihood, suppose the two nodes below the root node $2n - 1$ are i and j. From the definition of $P(L_{2n-1}|\cdot)$ in terms of $P(L_i|\cdot)$ and $P(L_j|\cdot)$ in Felsenstein's algorithm, we write the likelihood of the sequences $x^\bullet$ at site u as

$$P(x_u^\bullet|T,t_\bullet) = \sum_a q_a P(L_{2n-1}|a) = \sum_{b,c,a} q_a P(b|a,t_i)P(c|a,t_j)P(L_i|b)P(L_j|c),$$

and hence, using reversibility,

$$P(x_u^{\bullet}|T,t_{\bullet}) = \sum_{b,c}\left(\sum_a P(c|a,t_j)P(a|b,t_i)\right)q_b P(L_i|b)P(L_j|c).$$

By multiplicativity, we can rewrite the inner sum as $\sum_a P(c|a,t_j)P(a|b,t_i) = P(c|b,t_i+t_j)$, which means that P is independent of the assignments a of symbols to node $2n-1$, and depends only on the total length of the two edges below the root. Thus the root can be moved freely between i and j, and hence can be moved anywhere within the tree. This is the 'pulley principle' of Felsenstein [1981a]. It implies that the search for the best tree only needs to be carried out on unrooted trees when multiplicative and reversible matrix families are being used.

Example: Reversibility of the PAM family

As remarked in Section 2.8, the counts matrix A used to construct the PAMs is symmetric, i.e. $A_{ab} = A_{ba}$ for all a, b, and since $p_a = \sum_b A_{ab}/\sum_{cd} A_{cd}$, it follows that the normalised matrix B satisfies

$$p_a B_{ab} = A_{ab}/\sum_{cd} A_{cd} = A_{ba}/\sum_{cd} A_{cd} = p_b B_{ba}.$$

Since $S(1)$ is obtained by scaling B, this implies reversibility for $t = 1$. To show reversibility of $S(n)$ for all n, suppose we have proved it for all k less than n. Then, applying reversibility for $n-1$ and 1:

$$p_a P_n(b|a) \;\; = \sum_c p_a P_{n-1}(c|a)P_1(b|c) \;\; = \sum_c P_{n-1}(a|c)p_c P_1(b|c)$$

$$= \sum_c P_{n-1}(a|c)P_1(c|b)p_b \;\; = P_n(a|b)p_b.$$

□

Example: A non-reversible matrix family

What might a non-reversible matrix look like? Suppose that for two residues A and B the substitution A $\to$ B occurs more often than B $\to$ A. In order for the frequency of the residues to remain constant, there must be balancing substitutions between other residues. The simplest case is where there are three residues in the alphabet, with a cyclic substitution pattern, giving the instantaneous substitution matrix

$$\begin{pmatrix} -\alpha & \alpha & 0 \\ 0 & -\alpha & \alpha \\ \alpha & 0 & -\alpha \end{pmatrix}. \tag{8.14}$$

This leads to a t-dependent family

$$S(t) = \begin{pmatrix} r_t & s_t & u_t \\ u_t & r_t & s_t \\ s_t & u_t & r_t \end{pmatrix},$$

with

$$r_t = \tfrac{1}{3}\left(1 + 2e^{-3\alpha t/2}\cos(\sqrt{3}\alpha t/2)\right),$$
$$s_t = \tfrac{1}{3}\left(1 - e^{-3\alpha t/2}\cos(\sqrt{3}\alpha t/2) + \sqrt{3}e^{-3\alpha t/2}\sin(\sqrt{3}\alpha t/2)\right),$$
$$u_t = 1 - r_t - s_t.$$

$\square$

Exercises

8.5   Show that the above family is multiplicative, has positive entries, and substitution rates at $t = 0$ given by (8.14). Find its limiting distribution and show that reversibility, i.e. (8.13), fails for all $t > 0$.

8.6   We have shown that reversibility allows the root to be moved to any position in the tree. What happens when the root is moved to one of the leaf nodes?

8.4   Using the likelihood for inference

We have now reached the heart of probabilistic phylogeny. Having formulated an evolutionary model and defined an algorithm for computing the likelihood from this model, we need to put this machinery to work and infer phylogenetic properties of sets of data. We now give an overview of probabilistic inference methods, beginning with the most venerable and widely used of them, maximum likelihood. We can also use probabilistic methods to assess the quality of the probabilistic model and any variants we devise, but we defer discussing that till we have seen some examples of more elaborate models (see Section 8.5).

Maximising the likelihood

One candidate for the 'best' tree is the tree that maximises the likelihood. Recall that the strategy is to search over trees, and for each topology T to find the lengths $t_\bullet$ that maximise the likelihood $P(x_u^\bullet | T, t_\bullet)$. The topology and the assignment of edge lengths that give the overall maximum of this likelihood is the desired tree.

Given a small number of sequences, say two to five, it is easy to enumerate all trees. For each tree, we can write down the likelihood explicitly as a function of the edge lengths, and maximise it by a suitable numerical technique. This in essence is what Kishino, Miyata & Hasegawa [1990] do, using Newton's method of optimisation [Press *et al.* 1992]. Their method is intended for maximum likelihood phylogeny of protein sequences, and they use PAM matrices.

For a larger number of sequences, the likelihood can be computed by Felsenstein's algorithm (p. 200). Felsenstein [1981a] also gave an EM algorithm for finding the optimal edge lengths in this case. Alternatively, we can use a standard optimiser, such as conjugate gradients, [Press *et al.* 1992]. This requires the derivatives of the likelihood with respect to the edge lengths, but this is straightforward to compute: we replace $P(y^k|y^{\alpha(k)}, t_k)$ wherever it occurs in (8.10) by its derivative $\partial P(y^k|y^{\alpha(k)}, t_k)/\partial t_k$.

Even with the best optimiser, maximising the likelihood is computationally demanding, and it is more so for protein sequences because the core computation uses a 20×20 substitution matrix rather than a 4×4 one. To tackle large datasets calls for another strategy; one approach is to use sampling methods.

Exercise

8.7   The maximum likelihood edge lengths can be calculated in certain simple cases. Show that, for our example of two nucleotide sequences (p. 198), the ML solution is given by

$$t_1 + t_2 = -\tfrac{3}{4} \ln \frac{3n_1 - n_2}{3n_1 + 3n_2}.$$

Sampling from the posterior distribution

As we have seen, maximum likelihood is computationally taxing. Furthermore, it is not clear that it is ultimately the best strategy. If we knew the prior $P(T, t_\bullet)$, we could use Bayes' rule to compute the posterior probability $P(T, t_\bullet \mid x^\bullet)$ by

$$P(T, t_\bullet \mid x^\bullet) = \frac{P(x^\bullet|T, t_\bullet)P(T, t_\bullet)}{P(x^\bullet)}.$$

The posterior provides the information we really seek, namely how probable each phylogenetic model is, given the data.

Several authors have used Bayesian methods on small sets of data, where all tree topologies can easily be enumerated (for example Rannala & Yang [1996]). Recently Mau, Newton & Larget [1996] have shown how quite large sequence sets can be handled by sampling from the posterior distribution on the space of trees and edge lengths using the Metropolis algorithm.

To sample from the space of trees is to pick trees randomly with probabilities given by some distribution, in this case their posterior distribution. If we have a large number of samples, then the frequency with which some property of trees is present in the sample converges, in the limit of a large number of samples, to the posterior probability of that property according to the model. For instance, if a particular tree topology is present in some fraction f of the samples, then f is an estimate for the posterior probability of this topology. We could also determine the probability that a given group is monophyletic, or that one branch point occurs between two others, say, by counting the fraction of cases in which the relevant condition holds. Such questions cannot easily be tackled by likelihood methods, since they require integration over variables, and likelihoods are not probability distributions (see p. 311).

The particular sampling method used by Mau *et al.*, the Metropolis algorithm, is a sampling procedure that generates a sequence of trees, each from the previous one. It assumes that a mechanism is available to generate one tree randomly given another, by sampling from a *proposal* distribution. Let $P_1 = P(T, t_\bullet | x^\bullet)$ be the posterior probability of the current tree, and $P_2 = P(\tilde{T}, \tilde{t}_\bullet | x^\bullet)$ that of a proposed new tree. The Metropolis rule is that the new tree is accepted as the next item in the sequence if $P_2 \geq P_1$; if $P_2 < P_1$, the new tree is accepted with probability P_2/P_1. Otherwise the original tree constitutes the next sample (and this repetition of $T, t_\bullet$, if it occurs, is an important part of the process, since we are going to be counting the number of samples with particular properties).[3]

This procedure is guaranteed to sample correctly from the posterior distribution provided that the proposal distribution is symmetrical, in the sense that the probability of proposing $\tilde{T}, \tilde{t}_\bullet$ from $T, t_\bullet$ is the same as that of proposing $T, t_\bullet$ from $\tilde{T}, \tilde{t}_\bullet$ (see Section 11.4).

Exercise

8.8   Consider a simplified phylogenetic space consisting of two trees T and $\tilde{T}$ with probabilities $P(T)$ and $P(\tilde{T})$. If the proposal procedure always selects the other tree, i.e. the one that is not the current tree, show that the Metropolis algorithm produces a sequence where the frequencies of T and $\tilde{T}$ converge to their probabilities.

A proposal distribution for phylogenetic trees

The choice of the proposal distribution is all-important in making the Metropolis algorithm work well. If the proposed tree is merely randomly selected from the space of all trees, its posterior probability will generally be small, and there will be many wasted repetitions. On the other hand, if the proposed tree is too close to the current tree, many steps will be needed to explore the space of phylogenies adequately. The art lies in finding a way of proposing trees that are promising variants of the current one.

Mau *et al.* suggested a proposal mechanism with two components, one an

---

[3] Note also that the Metropolis rule only uses the ratio of P_1 and P_2, which is fortunate, because the denominator in Bayes' rule can only be obtained by integrating over the space of trees, and is generally an unknown factor.

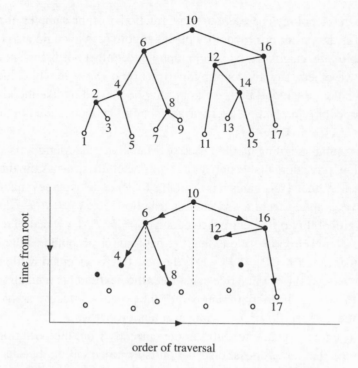

Figure 8.7 *Above: an example of a tree with its nodes numbered in the order of the traversal profile. Below: Reconstruction of the tree from the traversal profile.*

adjustment of lengths of the edges that can also bring about switches in topology in an interesting way, and the other a reordering of the assignment of sequences to leaves. The first uses a representation of a tree that they call a *traversal profile*. This is a diagram completely equivalent to the original tree, but allowing more convenient manipulations of the topology. In the traversal profile,[4] a node is placed at a height corresponding to the sum of the edge lengths from the root to that node. The nodes are regularly spaced horizontally, in the order in which they are encountered during a traversal of the tree. This traversal is defined as follows: Beginning at the leftmost leaf, we traverse the tree depth first from left to right, assigning numbers incrementally, and numbering nodes when we first come to them. This ensures that, for any node with number k, all its left children have numbers lower than k, and all its right children have numbers higher than k. The top diagram in Figure 8.7 shows an example of a tree in which the nodes have been numbered by their order in the traversal profile.

Given a traversal profile, we can reconstruct the tree by a procedure illustrated in Figure 8.7. The root is taken to be the highest node (node 10 in the figure).

---

[4] In Mau, Newton & Larget [1996], the nodes are connected by lines of constant slope, rather than being equally spaced horizontally, as we have chosen to represent them.

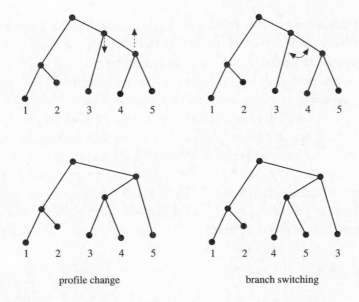

profile change   branch switching

Figure 8.8 *The two parts of the proposal mechanism are changes in the height of the nodes in the profile (left), and reordering of the leaves by switching branches (right). The former can produce changes in the topology, as shown here. The latter does not do this; it just rearranges the existing topology. However, the change of order of the leaves allows new topologies to be reached through further steps of the first type.*

Edges are then drawn to the highest nodes to the left and the right of the root (nodes 6 and 16 in the figure). Suppose now we have reached node k. The daughter nodes of k must lie within the horizontal stretch bounded by any nodes that are higher than k; within this stretch edges are drawn, as before, to the highest nodes to the left and right. Thus, in the figure, the region where the right daughter of node 6 can lie is delimited by the vertical dotted lines. The process stops when a leaf is reached (the leaves have been marked as hollow circles in the figure).

One part of the proposal procedure of Mau *et al.* consists of taking the traversal profile for the current tree and shifting the positions of nodes up and down by an amount chosen from a uniform distribution within certain bounds. Whenever the relative heights of nodes are switched a new topology is produced (Figure 8.8). However, this will never allow leaves which are not adjacent in the traversal order to be neighbours (but see Exercise 8.10). They therefore give an additional proposal mechanism which achieves this. It reorders the leaves by randomly switching the direction of the branches at each node. This produces no change in the posterior probability (so is always accepted), but will lead into a new region of tree space. Adjusting the heights of the traversal profile does of course produce changes in the posterior, but these changes vary continuously with the size of the adjustments, even when there is a change in topology (Exercise 8.9). The proposal mechanism behaves better in this respect than branch-swopping, which is an intuitively obvious way of modifying trees but has the disadvantage that it is likely to make large changes in the posterior probability.

To define the posterior, a prior has to be chosen over trees. As there is little reliable information available about the distribution of trees, Mau *et al.* assumed a flat prior, assigning equal probabilities to all sets of edge lengths $t_1, \ldots, t_{2n-1}$ for n sequences, for any tree topology. (Note that this does not imply that all tree topologies have equal prior probability; see Exercise 8.11.) To ensure a normalisable probability distribution, they imposed an upper bound on the total edge length from root to any leaf.[5] They found that they could reproducibly identify the most probable topologies for datasets of up to 32 sequences. Their method seems to work best when there is a molecular clock, i.e. when the leaves in the traversal profile are all at the same height.

Exercises

8.9　　Consider the profile change shown in the two left-hand figures in Figure 8.8. Suppose the two nodes with arrows in the top figure are at heights h_1 and h_2, and their heights are switched to h_2 and h_1. Show that the resulting change in the likelihood tends to zero as $h_1 - h_2$ tends to zero.

8.10　　Show that the two leaves at the extreme ends of the traversal profile can become neighbours, but no other non-adjacent pair can.

8.11　　The flat prior on edge lengths assigns a prior to any topology that is obtained by integrating over all possible edge lengths for that topology. This integral will be defined if, following Mau *et al.*, we impose a bound on the total edge length from root to any leaf; call this bound B. Consider the case where there is a molecular clock, and show that the tree with four leaves and topology $((01)(23))$ has integrated prior probability $B^3/3$; show that this integral is $B^3/6$ for the topology $((0(12))3)$. (Hint: Define times from the three branch nodes in each tree to the present, and integrate over these three variables.) This shows that different topologies can have different priors. Show, however, that if one defines a *labelled history* to be a specific ordering for the times of branch nodes relative to present time (assuming a molecular clock), then all labelled histories for four leaves have the same prior probability. Extend this to n leaves.

---

[5]　The limit on edge lengths in the prior might seem an artificial constraint, but it actually has little effect, because trees with extremely long edges generally have low likelihoods. This is because the substitution probabilities over a long edge tend to the equilibrium frequencies, q_a; all correlations with other sequences are therefore neglected.

Other phylogenetic uses of sampling

Sampling methods have not only been used for species or gene phylogeny, but also for inferring the history of populations from a set of present-day individuals. Kuhner, Yamato & Felsenstein [1995] used a sampling method to pick plausible trees, T, relating the individuals of the set. Now, the prior on trees depends on the size of the population, θ. Intuitively, this is because the larger a population is, the further back we expect to go to find the common ancestor of any two individuals. Thus, for each tree T, we can use the prior $P(T|\theta)$ as a likelihood for estimating θ. The sampling process then allows us to accumulate likelihood data in proportion to $P(\text{data}|T)$, giving, in the limit of many samples, the desired likelihood function $\int P(\text{data}|T)P(T|\theta)dT = P(\text{data}|\theta)$.

The proposal mechanism used by Kuhner, Yamato & Felsenstein [1995] is closely related to that of Mau, Newton & Larget [1996]. Instead of adjusting the heights of all nodes in a traversal profile, Kuhner *et al.* adjust the relative heights of *two* nodes, and allow their children to be relabelled. This is therefore a local version of the two components in the method of Mau *et al.*. It would be interesting to know which mechanism samples more effectively.

The prior $P(T|\theta)$ might seem difficult to calculate, since it involves summing over all trees in the population that could provide possible phylogenies for the set of present-day individuals. However, there is a remarkably simple way of evaluating $P(T|\theta)$, based on the idea of running time backwards and allowing branches to *coalesce* [Kingman 1982a; 1982b; Hudson 1990]. For a fixed, large, population size, the probability density of a coalescence in time turns out to be $2/\theta$. Imagine a horizontal line rising through the tree T, starting from the level of the leaves. Each time a coalescence occurs, the number of edges will fall by one. Suppose the time between the coalescence from k to $(k-1)$ edges is τ_k. Then the probability of a coalescence in the interval dt between two of the $k(k-1)/2$ pairs of edges is $k(k-1)dt/\theta$, so $(2/\theta)\exp(-\tau_k k(k-1)/\theta)$ is the probability of coalescence occurring at the end of the period τ_k and not before. Taking the product over all intervals τ_k gives the total probability of the tree

$$P(T|\theta) = \left(\frac{2}{\theta}\right)^{n-1} \exp\left(-\sum_{k=2}^{k=n} \frac{k(k-1)\tau_k}{\theta}\right).$$

Closely related to the coalescent is a prior obtained by a simple evolutionary model, where we think of a tree as being formed by a series of splitting events. If there is a constant probability density, λ say, of a split occurring in a growing edge, the splitting is said to follow a *Yule process*. The resulting prior on trees has a simple form, being proportional to $\exp(-\lambda\sum t_i)$, for edge lengths t_i (see Exercise 8.12). This is different from the coalescent prior because it assumes that all the descendants of the root sequence are present at the leaves, without omissions or extinctions whereas the coalescent prior treats the species or genes as being picked randomly from a large pool. Which prior is more appropriate depends on whether a taxonomist is looking at a small closely related family or a wide-ranging selection.

Exercises

8.12   Under a Yule process, the probability density for no split occurring dur-
        ing the interval 0 to t is given by the limit of $(1 - \lambda \delta t)^{t/\delta t}$ as $\delta t \to 0$,
        and is therefore $\exp(-\lambda t)$. Deduce that the Yule prior for a tree with n
        leaves is proportional to $\exp(-\lambda \sum t_i)$, where the t_i are all edge lengths.
        Following the same reasoning as in Exercise 8.11, show that the priors
        for all labelled histories on four leaves are equal under the Yule prior.
        Extend this to the case of n leaves.

8.13   Assuming a molecular clock, calculate the expected lengths of all the
        branches of rooted trees with two, three or four leaves under the Yule
        prior with splitting rate λ and the coalescent prior with population size
        θ. (Hint: Consider the case of three leaves. Let the two short edges have
        length s and the long edge length t. The total edge length of the tree is
        then $2t + s$, so the tree probability for the Yule process is proportional
        to $\exp(-\lambda(2t + s))$. Check that the coalescent probability for the tree is
        proportional to $\exp(-(2t + 4s)/\theta)$. Now compute the means of s and t for
        these distributions in the standard manner, integrating over all $0 \le t \le \infty$
        and $0 \le s \le t$.)

The bootstrap revisited

The bootstrap, p. 179, can be applied to maximum likelihood, just as to other tree building methods. Artificial data are generated by drawing columns randomly with replacement from the true dataset, and then the maximum likelihood tree is found for the artificial dataset. The frequency of occurrence of some feature over many replications of this procedure measures the confidence we have in inferring it by maximum likelihood.

One therefore obtains information similar to that obtained by sampling from the posterior, and in fact the two methods are related: For some phylogenetic models, the bootstrap confidence for a feature approximates the posterior probability of that feature, assuming a flat prior over trees. To gain some intuition for why this is so, consider the simple case of maximum likelihood estimation of the probability of getting a head in a coin toss, on the basis of a set of data.

Example: Bootstrapping the results of a coin toss experiment

A coin is tossed N times, giving m heads (H) and n tails (T). The posterior distribution for the probability p of a head, assuming a flat prior, is given by the

Dirichlet distribution

$$P(p|mH,nT) = p^m(1-p)^n \frac{(N+1)!}{m!n!}. \qquad (8.15)$$

A bootstrap trial starts by drawing a set of N coin tosses from the data, with probability m/N of H, n/N of T. If there are k heads in this set, the maximum likelihood probability estimated from the set is $p^{ML} = k/N$. Thus

$$P\left(p^{ML} = k/N\right) = \left(\frac{m}{N}\right)^k \left(\frac{n}{N}\right)^{N-k} \binom{N}{k}. \qquad (8.16)$$

For large N, we can approximate this by the distribution

$$P\left(p^{ML} = p\right) = \left(\frac{m}{N}\right)^{Np} \left(\frac{n}{N}\right)^{N-Np} (N+1)\binom{N}{Np} \qquad (8.17)$$

where the factor $(N+1)$ appears because we have replaced $(N+1)$ terms of the binomial expansion with a density on $[0,1]$. For large N we can approximate (8.15) by a normal distribution (p. 300):

$$P(p|mH,nT) \simeq \frac{(N+1)}{\sqrt{2\pi Np(1-p)}} \exp\left(-\frac{(m-Np)^2}{2Np(1-p)}\right), \qquad (8.18)$$

and similarly (8.17) becomes

$$P\left(p^{ML} = p\right) \simeq \frac{(N+1)}{\sqrt{2\pi mn/N}} \exp\left(-\frac{(m-Np)^2}{2mn/N}\right). \qquad (8.19)$$

It is a straightforward exercise to show that, if N is large enough, i.e. if there is plenty of data, these two distributions approximate one another closely (Exercise 8.14). □

This result can easily be extended to multinomial distributions. In phylogenetic examples, the probability of drawing particular leaf assignments to make the bootstrap dataset is governed by a multinomial distribution. We consider next a case where this distribution collapses to a binomial, and the coin tossing example above can be directly applied.

Example: The bootstrap for a simple tree

Suppose we have two nucleotide sequences which we wish to model using a tree with two leaves and the Jukes–Cantor substitution matrices. We can place the root at one of the leaves, and set t to be the length of the edge connecting the two leaves. Let n_s denote the number of sites in the original data set for which the two leaves take the same nucleotide, and let n_d be the number of sites where they differ; if there are N sites altogether, $n_d + n_s = N$. Extending the calculation in (8.9), and assuming a flat prior on t, we can write the posterior probability as

$$P(t|\text{data}) = (1+3e^{-\alpha t})^{n_s}(1-e^{-\alpha t})^{n_d}/Z, \qquad (8.20)$$

where Z is a normalising factor from Bayes' theorem. Suppose n_{XY} denotes the number of leaf assignments of type XY in the original dataset, and m_{XY} the corresponding number in a bootstrap dataset. The probability of drawing the bootstrap dataset is given by the multinomial distribution

$$P(m_\bullet|n_\bullet) = \left(\frac{n_{AA}}{N}\right)^{m_{AA}} \left(\frac{n_{AC}}{N}\right)^{m_{AC}} \cdots \frac{N!}{m_{AA}! m_{AC}! \cdots}. \tag{8.21}$$

Now the maximum of the likelihood for the bootstrap dataset is given by an obvious extension of Exercise 8.7 as

$$\exp(-\alpha t^{ML}) = \frac{3m_s - m_d}{3N}, \tag{8.22}$$

where m_s, m_d are the number of equal and differing leaves in the bootstrap set. Thus t^{ML} depends only on m_s and m_d, and not on the individual counts m_{XY}, and all $P(m_\bullet|n_\bullet)$s given by (8.21) that have the same m_s and m_d can be summed in determining the frequency with which the bootstrap value t^{ML} will occur. Summing over these terms gives

$$P\left(e^{-\alpha t^{ML}} = \frac{3m_s - m_d}{N}\right) = \left(\frac{n_s}{N}\right)^{m_s} \left(\frac{n_d}{N}\right)^{m_d} \binom{N}{m_s}. \tag{8.23}$$

Comparing (8.20) with (8.15) and (8.16) with (8.23), and noting that (8.22) implies $m_s = N(1 + 3\exp(-\alpha t^{ML}))/4$ and $m_d = 3N(1 - \exp(-\alpha t^{ML}))/4$, we see that the posterior for this tree approximates the bootstrap, just as for coin tossing. □

Thus the bootstrap distribution can, for certain phylogenetic models, and with enough data (large enough N), give a good approximation to the posterior. However, the labour involved in evaluating a large number of maximum likelihood trees makes this use of the bootstrap an unattractive alternative to sampling. The bootstrap is probably more useful for non-probabilitistic tree building methods. However, the relationship between the bootstrap and the posterior does give some insight. In particular, it helps to counter objections raised by Hillis & Bull [1993]. These authors generated sample datasets from a tree and found the distribution of the frequency with which a given tree topology was reconstructed by parsimony; for each sample dataset they also obtained the bootstrap frequency of that topology. They found the distribution of bootstrap frequencies was far wider than that of the original samples. This is unsurprising, however, since sampling followed by bootstrapping adds variance from two steps. In fact, the bootstrap distribution in their simulations has the correct variance *given the data* [Efron, Halloran & Holmes 1996], i.e. when viewed as a posterior distribution.

Exercise

8.14   Show that, for sufficiently large N, $P(p|mH,nT)$ given by (8.18) and $P(p^{ML} = p)$ given by (8.19) are either both very small or else take nearly equal values.

8.5   Towards more realistic evolutionary models

The evolutionary models used so far have made some fairly drastic simplifying assumptions (p. 193). The restriction to ungapped alignments discards useful phylogenetic information given by the pattern of deletions and insertions. It is also clearly incorrect to model each site in a sequence with the same substitution matrix, as assumed in (8.11), since there are different constraints at different sites, imposed by structure of proteins, base pairing of RNA, and so on. To focus on a single basic property of sites, it has long been known that substitutions occur much more rapidly at some sites than others [Fitch & Margoliash 1967b]. We describe first some attempts to model this behaviour by allowing variable rates of evolution, and then turn to ways of treating gapped alignments.

Allowing different rates at different sites

The basic strategy of maximum likelihood is to pick a tree T and a set of lengths $t_\bullet$ and compute the likelihood over all sites, using

$$P(x^\bullet|T,t_\bullet) = \prod_{u=1}^{N} P(x_u^\bullet|T,t_\bullet).$$

Yang [1993] suggested introducing a site-dependent variable, r_u, that scales all the $t_\bullet$ at the site u. If we knew the values of r_u at each site, we could write the likelihood as

$$P(x^\bullet|T,t_\bullet,r_u) = \prod_{u=1}^{N} P(x_u^\bullet|T,r_u t_\bullet).$$

Since we generally do not know the values of r_u, our best strategy is to assume a prior for them, and integrate over all values of each r. Yang [1993] used a gamma distribution $g(r,\alpha,\alpha)$ as his prior; this has mean 1 and variance $1/\alpha$, and therefore allows a range from a tight distribution (α large) to a broad one (α small). The likelihood is

$$P(x^\bullet|T,t_\bullet,\alpha) = \prod_{u=1}^{N} \int_0^\infty P(x_u^\bullet|T,rt_\bullet)g(r,\alpha,\alpha)dr. \tag{8.24}$$

For each fixed T, this likelihood is maximised with respect to $t_\bullet$ and α. Using a set of globins, Yang derived an optimal tree for four mammals, and showed that the log likelihood decreased significantly when variable times were allowed, compared to the fixed times at all sites used in the original methods of Felsenstein [1981a].

The integrals in (8.24) can be evaluated analytically (they are just gamma function integrals), but the number of terms in the resulting expression grows exponentially with the number of sequences, so optimisation can be computationally slow. Yang [1994] therefore suggested an approximate method which replaces the integral by a discrete sum. The interval $(0, \infty)$, is subdivided into m intervals each containing equal areas of the gamma distribution $g(r, \alpha, \alpha)$. Let r_k denote the mean of the gamma distribution in the kth interval. Then we define

$$P(x^{\bullet}|T, t_{\bullet}, \alpha) = \prod_{u=1}^{N} \sum_{k=1}^{k=m} P(x_u^{\bullet}|T, r_k t_{\bullet})/m. \qquad (8.25)$$

Yang found $m = 3, 4$ sufficed to give a good approximation to the continuous version of the model. Since only m times as much computation is required as for non-varying sites, this is a more tractable algorithm.

Here, as in the continuous model, α is estimated from the given data by maximising the likelihood. This may be acceptable if the data are plentiful, but for smaller amounts of data this approach suffers from the problems encountered with profile HMMs when estimating probabilities from counts. It may therefore be better to infer a value of α from a large data set of trusted phylogenies.

Felsenstein & Churchill [1996] proposed an algorithm similar to Yang's, but in hidden Markov model format. Each position in their model corresponds to a site in the alignment. At each position, there is a number of states, each corresponding to a different rate of evolution. There are transitions between all possible rate-states at adjacent positions. If state i is picked at site u, it contributes a term $P(x_u^{\bullet}|T, r_i t_{\bullet})$ to the likelihood, r_i being the rate for state i. A version of the forward algorithm can be applied to give the summed probability for all possible assignments of rates. The difference between this and the forward algorithm for HMMs, p. 58, is (1) that a path through the model is a set of choices of rates rather than an alignment of a sequence to the model, and (2) that the probabilities are not emission probabilities from a state that sum to 1, but likelihoods for the whole set of sequences at a site. Formally, however, the two algorithms are the same, if we set $e_l(x_i)$ and a_{kl} in the HMM forward algorithm to be the likelihood at site i for rate l, and the transition probability from rate k to rate l, respectively.

The total likelihood from Felsenstein & Churchill's hidden Markov model would be identical to that given by Yang's discrete model, (8.25), except that the hidden Markov model includes the transition probabilities a_{kl} between its states. These transition probabilities can capture any tendency for particular patterns of rates to occur in successive sites. In proteins, substitutions occur more freely at surface sites in a protein, but the pattern of rates will depend on the type of secondary structure. In loops, we might expect a run of exposed sites, which could be modelled by making transitions between states for fast rates more probable. However, beta sheets will show an alternating pattern of buried and exposed residues, and helices will show a rough triplet pattern. A more elaborate model architecture would be needed to represent these structural features.

Evolutionary models with gaps

We turn now to the problem of allowing gaps in the alignment of the leaf sequences. Gaps can be crudely introduced by treating '−' as an extra character in the alphabet of K residues and replacing the $K \times K$-sized substitution matrix for residues with a $(K+1) \times (K+1)$-sized matrix that includes the gap character. This has the usual fault of making gaps at adjacent sites independent, and therefore not allowing for the tendency for gaps to occur in blocks.

A better model can be made by introducing delete or insert states: Allison, Wallace & Yee [1992b] have shown how this could be carried out in principle to give affine-type gaps in phylogenies. Their approach, via minimum description lengths, is closely related to maximum likelihood. Unfortunately, treating affine gap penalties this way seems computationally intractable at present.

Another approach, through a model of fragment substitution [Thorne, Kishino & Felsenstein 1992], has the attraction of a degree of biological plausibility, but has only been applied so far to the case of two sequences.

We describe now a type of model that does allow affine-type gap penalties to be handled in a computationally reasonable way. A *tree HMM* [Mitchison & Durbin 1995] uses a profile HMM architecture, and treats paths through the model as the objects that undergo evolutionary change [Mitchison 1998].

We assume that sequences can be aligned to a hidden Markov model with an architecture simpler than that of the profile HMM of Krogh *et al.* [1994]; it has only match and delete states, which we denote by M_k and D_k, where k is the position in the model. Suppose a sequence y is the ancestor of a sequence x; we can think of these two sequences as lying at two nodes connected by an edge in a tree with length t. Suppose both sequences have been aligned to our model, so each follows a prescribed path through it, emitting specified residues at the match states. Consider the segment of the model shown in Figure 8.9. Both sequences use the match state M_k, and x emits the residue x_i at M_k, and y emits y_j there. The probability of the substitution $y_j \rightarrow x_i$ is taken to be $P(x_i|y_j,t)$, just as in standard maximum likelihood phylogeny.

Next, consider the possibility that y uses different *transitions* from x, so it does not pass through the same state sequence as x. In Figure 8.9, at position k, x goes from match to delete states, $M_k \rightarrow D_{k+1}$, whereas y goes to the next match state, $M_k \rightarrow M_{k+1}$. We assign a probability to this substitution of transitions analogous to the substitution of emissions from y_j to x_i. Writing 'MM' for a match-to-

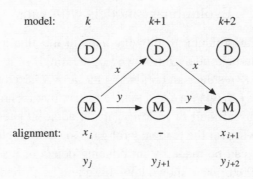

Figure 8.9 *A segment of the HMM, showing the paths followed by x and y.*

match transition, and 'MD' for a match-to-delete, we denote this probability by $P(\mathrm{MD}|\mathrm{MM}, t)$.

At position $k + 1$ in Figure 8.9, x makes a transition $\mathrm{D}_{k+1} \rightarrow \mathrm{M}_{k+2}$, which we abbreviate to 'DM', and y makes an MM transition $\mathrm{M}_{k+1} \rightarrow \mathrm{M}_{k+2}$. The assumption we make here is that x behaves independently of y wherever the two sequences start from different states. Thus if a deletion occurs in x relative to y, the choices it makes of DD or DM during its course are what determine the length of the deletion, and these are assumed to be under the control of a mutational process that operates independently of the sequence y. We assume that the probabilities of transitions in the independent path used by x are given by priors; so the transition $\mathrm{D}_{k+1} \rightarrow \mathrm{M}_{k+2}$ for x has probability q_{DM}.

We can represent both substitutions and priors for transitions in a 4×4 matrix, corresponding to the four transitions that can occur in this particular HMM architecture, namely MM, MD, DM and DD. However, this is not a standard substitution matrix because the probabilities in a row do not sum to one. Instead, it breaks up into four 2×2 blocks, determined by the state (match or delete) that the ancestral and descendant sequences begin their transition from:

$$\left(\begin{array}{cc} \begin{pmatrix} P(\mathrm{MM}|\mathrm{MM}, t) & P(\mathrm{MD}|\mathrm{MM}, t) \\ P(\mathrm{MM}|\mathrm{MD}, t) & P(\mathrm{MD}|\mathrm{MD}, t) \end{pmatrix} & \begin{pmatrix} q_{\mathrm{DM}} & q_{\mathrm{DD}} \\ q_{\mathrm{DM}} & q_{\mathrm{DD}} \end{pmatrix} \\ \begin{pmatrix} q_{\mathrm{MM}} & q_{\mathrm{MD}} \\ q_{\mathrm{MM}} & q_{\mathrm{MD}} \end{pmatrix} & \begin{pmatrix} P(\mathrm{DM}|\mathrm{DM}, t) & P(\mathrm{DD}|\mathrm{DM}, t) \\ P(\mathrm{DM}|\mathrm{DD}, t) & P(\mathrm{DD}|\mathrm{DD}, t) \end{pmatrix} \end{array} \right).$$

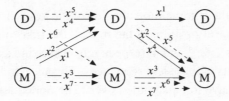

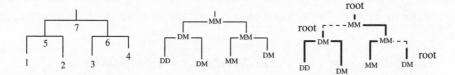

Figure 8.10 *A segment of a tree HMM for a tree with four leaves. The paths taken by the leaf sequences are shown as solid arrows. One possible assignment of ancestral paths is shown by dotted arrows. The tree is shown below with the numbering of its nodes (left), with the transitions that occur in the central position of the model (centre), and with root indicated wherever the prior is used in place of substitution probabilities (right), i.e. wherever a parent and child sequence begin from different states.*

Let us see how this works in the case of the tree HMM shown in Figure 8.9. At position k we have terms $q_{y_j} P(x_i|y_j,t)$ from emissions, q_{y_j} coming from the root prior for y. Transitions contribute $q_{MM} P(MD|MM,t)$, where q_{MM} is the root prior for the transition MM. If it seems confusing to include priors both in the above substitution matrix and in the expression for the tree probability, note that they have the same origin: where an aspect of sequence behaviour is not explained by an ancestor, we fall back on priors. Thus at position $k+1$, transitions give $q_{MM} q_{DM}$, and the two prior terms arise because both sequences are here behaving independently of their ancestors (though of course we know that x's ancestor is y whereas the latter's ancestor is unspecified).

Suppose now that we have an arbitrary tree T with edge lengths $t_\bullet$ and sequences $x^\bullet$ at its leaves that are all aligned to the HMM. By analogy with the probabilistic model for ungapped alignments (8.10), we define $P(x^\bullet|T,t_\bullet)$ by multiplying the substitution probabilities for all edges in T, including terms for priors at the root. However, with the tree HMM there are two types of substitution probabilities to be multiplied, those for emissions and those for transitions. To obtain the total probability for the ungapped model, we sum over all possible assignments of residues to ancestral nodes at each position. With the tree HMM, we likewise sum over all possible assignments of the relevant variables, which in this case are both emissions and transitions. If we define a path not only by the transitions it uses, but also by the symbols it emits, then this amounts to saying that we sum over all paths used by ancestral sequences.

As an illustration of the computation of some of the terms in the likelihood, see Figure 8.10. At the central position in the model, the transitions DD, DM, MM and DM occur at leaves 1, 2, 3 and 4, respectively. These are assumed to be given by the known alignments of these leaf sequences. The ancestral transitions are not known, however, and have to be summed over. The dotted arrows show one set of ancestral paths; there will be many other possible combinations. At the central position in the model, the given set of ancestral paths implies the transition tree shown below (centre). We can compute its probability from our transition matrix as

$$\text{Probability of tree} \quad = \quad q_{\text{MM}} P(\text{MM}|\text{MM}, t_6) P(\text{MM}|\text{MM}t_3)$$
$$\times q_{\text{DM}} \times q_{\text{DM}} P(\text{DD}|\text{DM}, t_1) P(\text{DM}|\text{DM}.t_2).$$

The three terms separated by product signs can be regarded as the probabilities of subtrees created by breaks along edges. These breaks occur where there is a change in the state that a transition starts from.

All these terms can be summed by a dynamic programming algorithm; it is much slower than the forward algorithm for profile HMMs, however, because of the need to keep track simultaneously of the preceding state used by a path and the states used by its ancestral path. This imposes a computational burden that grows exponentially with the number of sequences, and the algorithm is there- fore only suitable for small numbers of sequences. There is, however, a good approximation to this likelihood [Mitchison 1998] that be computed with a cost comparable to that of the original algorithm of Felsenstein [1981a].

Evaluating different probabilistic models

One problem with developing ever more complex models is that it may not be clear how much is gained by the added model structure. If a model M_2 is more complex than another model M_1, the maximum of M_2's likelihood may be larger than that of M_1 (indeed this must generally be true when M_2 is an elaboration of M_1, containing M_1 as a special case). However, M_2 may be a poorer model in the sense that the likelihood is non-negligible only for a very narrow range of parameter values. Instead of comparing the maxima of likelihoods, therefore, a better approach is to compare the probabilities $P(D|M_1)$ and $P(D|M_2)$ obtained by integrating over all the parameters of each model. More precisely, if M_1 has parameters $\theta_\bullet$, with prior probabilities $P(\theta_\bullet)$, we have

$$P(D|M_1) = \int P(D|M_1, \theta) P(\theta_\bullet) d\theta_1 \ldots d\theta_n$$

and similarly for $P(D|M_2)$. The probability $P(D|M)$ is sometimes called the *evidence* for the model M given the data [MacKay 1992]. If the non-negligible contributions to $P(D|M)$ come from a small region of parameter space, with prior

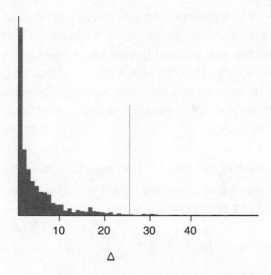

Figure 8.11 *The example of two substitution matrices given in the text. The histogram shows the distribution of the log likelihood differences Δ_i for simulated data. The value of Δ for the original data is shown as the thin vertical bar.*

probability P_r, this sets a bound $P(D|M) < \max_\theta P(D|M,\theta)P_r$, and $P(D|M)$ will be small. The natural way to compare the two models M_1 and M_2, taking into account their prior probabilities $P(M_1)$ and $P(M_2)$ is to compute the posterior probability of M_1 given by

$$P(M_1|D) = \frac{P(D|M_1)P(M_1)}{P(D|M_1)P(M_1) + P(D|M_2)P(M_2)}. \tag{8.26}$$

An alternative method for assessing models was proposed by Goldman [1993], following Cox [1962]. Let $\hat{L}_1(D)$, $\hat{L}_2(D)$ denote the maximum likelihoods of the data D for the models M_1 and M_2, respectively, each maximum being evaluated independently of the other model (the maxima may occur at different values of the parameters they share). Let

$$\Delta = \log(\hat{L}_2(D)) - \log(\hat{L}_1(D)).$$

For the reasons mentioned above, the value of Δ is not in itself a good indicator of any superiority in M_2. But if we now simulate datasets D_i from M_1, using the values of the parameters of M_1 that gave the maximum likelihood for D, we can ask whether the distribution of values of Δ_i for the simulated sets D_i shows the original value Δ to be typical (e.g. to lie within the 95% bounds of the distribution), or to exceed almost all of the the Δ_i. If the latter is the case, the more complex model M_2 has captured some aspect of the data that M_1 cannot mimic, and M_1 can be rejected.

This method is sometimes called the *parametric bootstrap*, and is a more powerful test than the plain bootstrap defined earlier (p. 179), and therefore more appropriate as a significance test for probabilistic models. Goldman shows how the parametric bootstrap can be used to compare a phylogenetic model M_1 of current interest with a very parameter-rich model, M_2, that assigns probabilities to all possible sets of residues at a site. The following example shows how the method works in a much simpler situation, and also illustrates Bayesian model comparison on the same data.

Example: Comparison of two substitution matrix models

Suppose there are two types of residue A and B, and that the two models for substitution are M_1, with one parameter p

$$\left(\begin{array}{cc} 1-p & p \\ p & 1-p \end{array} \right),$$ (8.27)

and M_2 with the two parameters p_1, p_2

$$\left(\begin{array}{cc} 1-p_1 & p_1 \\ p_2 & 1-p_2 \end{array} \right).$$ (8.28)

We first create a basic set of data, D, by sampling from M_2 with parameters $p_1 = 0.5$, $p_2 = 0.4$. We suppose that we have a total of $N = 500$ As and Bs, randomly chosen with equal probability, and we derive residues from them using the conditional probabilities given in matrix (8.28). We denote by n_{AA} and n_{AB} the number of As and Bs, respectively, derived from an A, and by n_{BB} and n_{BA} the number of Bs and As derived from a B. The values of p_1 and p_2 are quite close, so we expect the data to fit not too badly to both M_1 and M_2; the question is whether either of our tests recognises a potentially better fit to M_2.

Given this dataset D, we determine the maximum likelihood value of p for M_1, and simulate 1000 sets of data D_i from M_1. For each set, we compute Δ_i and accumulate the distribution, shown in Figure 8.11 as the histogram. The thin vertical bar marks the value of Δ for the basic dataset D. This gives us an estimate of $P(\Delta_i < \Delta)$ which in this case is 0.985. This tells us that D lies outside the 95% bounds of the distribution, so M_1 is rejected, and we deduce that the two-parameter model M_2 is appropriate.

If we repeated this whole experiment, beginning with sampling M_2 to obtain a new dataset D, we will get a distribution of values for $P(\Delta_i < \Delta)$. It is this distribution which we now compare with the distribution of Bayesian probabilities. To obtain these, we assume a flat prior for all parameters, so

$$P(D|M_1) = \int P(D|M_1, p) dp$$

$$= \beta \int p^{n_{AB}+n_{BA}} (1-p)^{n_{AA}+n_{BB}} dp,$$

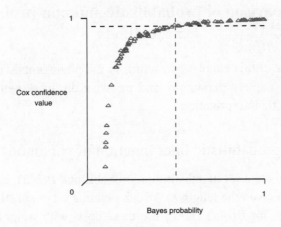

Figure 8.12 *Comparison of the models with one and two parameters. For 100 datasets D of size 500, the Bayesian probability $P(M_2|D)$ is plotted against the Cox confidence value $P(\Delta_i < \Delta)$ estimated from the histogram shown in Figure 8.11. The 95% confidence limit is shown as a dashed horizontal line, and the value $P(M_2|D) = 0.5$ is shown by a dashed vertical line.*

where β is a binomial factor shared by both $P(D|M_1)$ and $P(D|M_2)$. The corresponding expression for $P(D|M_2)$ is

$$P(D|M_2) = \beta \int p_1^{n_{AB}}(1-p_1)^{n_{AA}} p_2^{n_{BA}}(1-p_2)^{n_{BB}} dp_1 dp_2.$$

These integrals can be expressed in terms of factorials; see (11.6):

$$P(D|M_1) = \frac{\beta(n_{AB}+n_{BA})!(n_{AA}+n_{BB})!}{(N+1)!},$$

$$P(D|M_2) = \frac{\beta n_{AB}!n_{BA}!n_{AA}!n_{BB}!}{(n_{AB}+n_{AA})!(n_{BA}+n_{BB})!},$$

from which $P(M_2|D)$ can be computed using (8.26), assuming equal prior probabilities for $P(M_1)$ and $P(M_2)$. Figure 8.12 shows the distribution of values of $P(M_2|D)$ obtained with 100 datasets D, plotted against the estimated value of $P(\Delta_i < \Delta)$. For most of the points where the latter probability exceeds 0.95, so M_1 is rejected with 95% significance, we also find $P(M_2|D) > 0.5$, indicating a preference for M_2. These tests, very different in their character, therefore show some agreement on these particular data. However, when N, the number of data points in D, is increased, the Bayesian method often prefers M_1 when it is rejected by the parametric bootstrap, and the reverse tendency is seen with small numbers of data points. The relationship between the two methods deserves to be explored further, particularly in view of the increasing use of likelihood ratio methods [Huelsenbeck & Rannala 1997]. $\square$

8.6   Comparison of probabilistic and non-probabilistic methods

For the remainder of this chapter, we return to the phylogenetic methods of the previous chapter, namely parsimony and pairwise distance methods, and give them a probabilisitic interpretation.

A probabilistic interpretation of parsimony

Suppose we are given a set of substitution probabilities $P(b|a)$, in which we neglect the dependence on the length t. We can obtain a set of substitution costs by setting $S(a,b) = -\log P(b|a)$. If we use these costs with weighted parsimony, then, as Felsenstein [1981b] pointed out, the minimal cost at site u for the whole tree T obtained by the weighted parsimony algorithm (p. 174) can be regarded as an approximation to the likelihood. In fact, it is the Viterbi approximation to the full probability $P(x_u^1, \ldots, x_u^n | T)$ given by (8.10). Just as the full probability sums over all paths in HMMs, whereas the Viterbi method finds the most probable path, so the probability given by (8.10) sums over all assignments of residues to ancestral nodes whereas parsimony, by minimising the sum of the negative probabilities $-\log P(b|a)$, finds a set of ancestral assignments that maximise the probability. The correspondence is not complete, because the equivalent of the probabilistic model's root distribution is not usually included in parsimony. However, if we assume this distribution is flat, then it contributes a constant term which can be neglected in computing the parsimony optimum of the tree.

Not all sets of costs $S(a,b)$ can be realised as probabilities in this way. However, the costs of traditional parsimony, i.e. 1 for any substitution and 0 for identical residues, can readily be interpreted as log probabilities. In fact, any substitution matrix with probabilities α down the diagonal and β elsewhere, with $\beta < \alpha$, will do. For then parsimony using $S(a,a) = -\log(\alpha)$ and $S(a,b) = -\log(\beta)$, for $a \neq b$, will be equivalent to traditional parsimony (see Exercise 8.15).

Parsimony is an attractive method because of its speed. In fact, the main computational gain of parsimony is that it does not require the optimisation of edge lengths that maximum likelihood uses. If we interpret parsimony as the Viterbi approximation to maximum likelihood, then it achieves this simplification by discarding the time parameter t in $P(a|b,t)$. This can have unfortunate consequences, as the following example shows.

Example: Comparison of parsimony and ML

A simple method of testing the performance of tree-building algorithms is to generate trees probabilistically, by sampling, and then see how often a given algorithm reconstructs them correctly. The sampling process works by picking a residue a at the root, with probability q_a, then accepting a substitution to b along the edge down to node i with probability $P(b|a, t_i)$, and so on, working down the tree. This generates an assignment of residues at the leaves; sequences of length N are generated by N independent repetitions of this procedure. For an unrooted tree, any node can be picked as a root and the procedure carried out. Provided the generating model is reversible, the choice of node for root is irrelevant.

If the same probabilistic model is used to reconstruct the tree, then because of its consistency, maximum likelihood should tend to reconstruct the tree correctly in the limit of a large amount of data. The interesting question is how well other algorithms perform at the task.

The tree with four leaves shown in Figure 8.13 has been the workhorse of many such simulation studies. Of particular interest is the case where two sister leaves have short edges, and the other two long edges. This case was first studied by Felsenstein [1978a] and Cavender [1978], who showed that parsimony gave a wrong answer even with large amounts of data. Following Felsenstein, we assume for simplicity that the alphabet has two characters, {A, B}, with the substitution matrix[6]

$$
\begin{pmatrix}
1-p & p \\
p & 1-p
\end{pmatrix}.
\tag{8.29}
$$

We take $p = 0.3$ for leaves 1 and 3, $p = 0.1$ for leaves 2 and 4, and $p = 0.09$ for the edge connecting the leaves. This tree is drawn in Figure 8.13.

There are three possible unrooted trees on four leaves (p. 164); we call the original tree T_1 and the other two possibilities T_2 and T_3. The tables below show the result of 1000 test runs with various sequence lengths, N, reconstructing sampled trees with maximum likelihood or parsimony. The columns show the number of times that each T_i was preferred.

Reconstruction of trees by maximum likelihood:

N	T_1	T_2	T_3
20	419	339	242
100	638	204	158
500	904	61	35
2000	997	3	0

---

[6] This can be made into a multiplicative matrix family by putting $p = \frac{1}{2}(1 - \exp(-\alpha t))$, but we do not use this here.

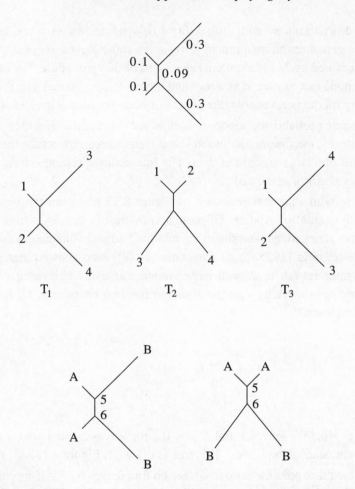

Figure 8.13 *Top: An unrooted tree with very unequal edge lengths. Middle row: The original tree T_1, with the two alternative unrooted trees (T_2 and T_3). Bottom row: A particular assignment of residues to the numbered leaves, shown for topologies T_1 and T_2.*

Reconstruction of trees by parsimony:

N	T_1	T_2	T_3
20	396	378	224
100	405	515	79
500	404	594	2
2000	353	646	0

Note that as N increases, T_1 is increasingly preferred by maximum likelihood, as would be expected. This is not true for parsimony, where a marked bias in favour of T_2 increases with N. To see why parsimony fails, consider the assignment A, A, B, B to leaves $1, 2, 3$ and 4 respectively (left figure of the bottom row in Figure 8.13); this will occur quite often with the given edge lengths because substitutions are likely to occur on the long edges to leaves 3 and 4, whereas leaves 1 and 2 are close. This assignment has a parsimony cost of two mismatches in tree T_1, but needs only one mismatch in tree T_2 (right figure of bottom row) if a substitution occurs along the 'bridge' between nodes 5 and 6. Maximum likelihood is not caught out in this way. When the edges have the correct lengths, substitution between nodes 5 and 6 is improbable because the edge is short. So the most probable explanation for the assignment needs two substitutions in T_2 as in T_1. This shows very clearly the drawbacks of the time-independence implicit in parsimony.

The tree in this example may be regarded as somewhat pathological, since the lengths differ considerably between terminal edges, and the tree strongly contravenes a molecular clock assumption. However, there are examples of trees with five leaves that do satisfy a molecular clock, and yet are incorrectly reconstructed by parsimony [Hendy & Penny 1989]. □

Exercise

8.15   Show that finding the most parsimonious tree using the costs $S(a,a) = -\log(\alpha)$, $S(a,b) = -\log(\beta)$, for $a \neq b$, is equivalent to traditional parsimony with a mismatch cost of 1.

Maximum likelihood and pairwise distance methods

We return now to pairwise distance methods, and explore a link between them and probabilistic modelling.

Suppose we are given a tree T with edge lengths $t_\bullet$, and we sample sequences of length N at the leaves, as described on p. 224, using a multiplicative, reversible substitution matrix. Pick two leaves i and j. It is easy to see that the sampled sequences we get at these leaves are also samples from the 'stripped-down' tree which is left when all edges are removed except those on the path connecting i and j (see the leftmost diagram in Figure 8.14). This follows because only the sampling steps made along the edges from the root down to i and j are relevant to the choice of residues at i and j. Furthermore, the parts of the tree above the top node of the stripped-down tree (node 8 in Figure 8.14) are irrelevant because the distribution at the top node is the same as that at the root, by reversibility.

Using multiplicativity, we can sum all the edge lengths down each of the paths from the top node to i or j. For instance, given the tree shown in Figure 8.14, with $i = 1$, $j = 3$, multiplicativity implies

$$P(a^1|a^8, t_1 + t_6) = \sum_{a^6} P(a^1|a^6, t_1) P(a^6|a^8, t_6)$$

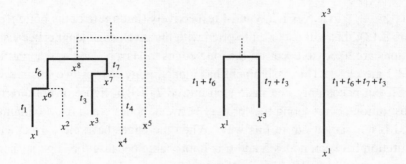

Figure 8.14 *The edges along the shortest path connecting the two leaves* 1 *and* 3 *are shown in bold.*

where we are using a^k to denote a residue at node k. This implies that, given some choice of a^8, samples made along an edge of length $t_1 + t_6$ will pick residues at leaf 1 with the same probabilities as samples made successively at node 6 and then at leaf 1 (see central diagram in Figure 8.14).

Reversibility implies that we can go further and 'straighten out' the stripped-down tree by reversing one of its legs. For instance, given the central tree in Figure 8.14 and a root distribution q, the probabilities of residues a^1 and a^3 are the same as if a^3 were picked with probability q, and a^1 then picked by sampling from the tree with one edge of length $t_1 + t_6 + t_7 + t_3$ (see the right-hand diagram in Figure 8.14). This follows because

$$\sum_{a^8} P(a^1|a^8, t_1 + t_6) P(a^3|a^8, t_7 + t_3) q_{a^8} = P(a^1|a^8, t_1 + t_6) P(a^8|a^3, t_3 + t_7) q_{a^3}$$

$$= P(a^1|a^3, t_1 + t_6 + t_3 + t_7) q_{a^3}.$$

For the general tree, suppose the edge lengths linking i to j are $t_{k_1}, t_{k_2}, \ldots, t_{k_r}$. Then our sampling argument shows

$$P(x_u^i, x_u^j | T, t_\bullet) = q_{x_u^j} P(x_u^i | x_u^j, t_{k_1} + t_{k_2} + \ldots + t_{k_r}).$$

Define the *maximum likelihood distance*[Felsenstein 1996] by

$$d_{ij}^{ML} = \underset{t}{\mathrm{argmax}} \left\{ \prod_u q_{x_u^j} P(x_u^i | x_u^j, t) \right\},$$

with the product taken over all sites u. Since the term $q_{x_u^j}$ is independent of t, we can write this as

$$d_{ij}^{ML} = \underset{t}{\mathrm{argmax}} \left\{ \prod_u P(x_u^i | x_u^j, t) \right\}. \qquad (8.30)$$

Then, when N is large, the consistency of maximum likelihood (p. 311) implies

$$d_{ij}^{ML} \simeq t_{k_1} + t_{k_2} + \ldots + t_{k_r}. \qquad (8.31)$$

If the probabilistic model is correct, therefore, maximum likelihood distances between the leaf sequences should be very close to additive, given a large amount of data. Now we know that neighbour-joining correctly reconstructs an additive tree, so it follows that neighbour-joining will also correctly reconstruct any tree, if we use maximum likelihood distances derived from a multiplicative, reversible model, and if there are plenty of data (and, of course, if the underlying probabilistic model is correct). The example below shows that neighbour-joining does indeed do as well as maximum likelihood for the tree in Figure 8.13 that parsimony failed at so conspicuously.

Neighbour-joining is in general far faster than any probabilistic approach, avoiding as it does the need to search through the space of trees, so it is tempting to think that we could discard probabilistic methods altogether. However, this neglects the power of such methods to assess the reliability of trees, and also to evaluate the plausibility of the model itself, using the posterior probability of the model. Neighbour-joining, or other distance methods, should therefore be thought of not as a replacement for probabilistic methods, but as a means of generating plausible trees, given such a model. The tree it provides might, for instance, provide a good starting point for a sampling procedure.

Example: Reconstruction of a tree by neighbour-joining

As an example of the successful performance of neighbour-joining, data from the tree in Figure 8.13 were simulated as described on p. 224, using the substitution probabilities from the matrix (8.29). Maximum likelihood distances were derived using this same matrix, and then neighbour-joining was used to construct a tree. The number of times this procedure yielded each of the possible three unrooted trees is shown below:

Reconstruction of trees by neighbour-joining:

N	T_1	T_2	T_3
20	477	301	222
100	635	231	134
500	896	85	19
2000	995	5	0

Clearly neighbour-joining generates the correct tree, T_1, with high reliability, given plenty of data. There is, in fact, little reason to favour maximum likelihood over neighbour-joining in this particular test. $\qquad \square$

We conclude this section by looking briefly at some particular cases of maximum likelihood distances. For DNA, the Jukes–Cantor model leads to a simple distance formula, for Exercise 8.7 implies that $d^{ML} = -\frac{1}{4\alpha}\log_e(1 - \frac{4f}{3})$, where f is the fraction of sites where nucleotides differ. The *Jukes–Cantor distance* is usually expressed not in time units, but in terms of the expected number of substitutions over the length d^{ML}. From the rate matrix (8.2), we see this number is $3\alpha d^{ML} = -\frac{3}{4}\ln(1 - \frac{4f}{3})$.

The Kimura matrix, (8.6), also leads to a compact expression for distance. Kimura [1980] defines Q to be the fraction of transversions, P the fraction of transitions, in an alignment of two sequences. He then sets $s_t = Q/2$ and $u_t = P$, in the notation of (8.6), from which it follows, after a little manipulation, that $\alpha t = -\frac{1}{2}\log(1 - 2P - Q) + \frac{1}{4}\log(1 - 2Q)$, and $\beta t = -\frac{1}{4}\log(1 - 2Q)$. From (8.5), the expected total number K of substitutions over an edge of length t is $(2\beta + \alpha)t$, so

$$K = (2\beta + \alpha)t = -\tfrac{1}{2}\log(1 - 2P - Q) - \tfrac{1}{4}\log(1 - 2Q).$$

K is the *Kimura distance*. The way it is derived can be interpreted as follows: Write the log of the likelihood in (8.30) as

$$\sum_u \log P(x_u^i | x_u^j, t) = N((1 - P - Q)\log r_t + Q/2 \log s_t + P \log u_t + Q/2 \log s_t),$$

where N is the total number of aligned sites. This is the relative entropy of the probabilities r_t, s_t, u_t, s_t occurring in a row of the Kimura matrix (8.6) with respect to the frequencies of the corresponding substitution types, $1 - P - Q$, $Q/2$, P, $Q/2$. We know (Figure 11.5) that the relative entropy is maximised when these sets of probabilities are equal, which implies Kimura's equations $s_t = Q/2$ and $u_t = P$.

Now, the maximum relative entropy cannot be achieved in general if we maximise over t alone. There may not be a value of t which satisfies both of the preceding equations simultaneously. However, if we maximise over both t and the ratio α/β while keeping $\alpha + \beta$ constant, then the number of unknowns is matched to the number of equations, and Kimura's equations can be satisfied. When the amount of data is large, estimating α/β from the data this way may be a sound procedure, but when comparing two sequences that are not very long, we might prefer to include a prior for α/β. For instance, we might use a gamma function, and define $\tilde{K} = \text{argmax}_t \max_{\alpha/\beta}\{g(\alpha/\beta, a, b)\prod_u P(x_u^i | x_u^j, t, \alpha, \beta)\}$, where a and b are suitable constants, and $P(x_u^i | x_u^j, t, \alpha, \beta)$ denotes the substitution probability from the Kimura matrix.

Finally, turning to protein sequences, the PAM matrix $S(t)$ can be used to define the $P(x_u^i | x_u^j, t)$ in (8.30). The maximising value of t cannot be expressed analytically, but can be easily found by gradient ascent, or some more efficient optimising technique.

Exercise

8.16   Obtain the Jukes–Cantor distance from the maximum relative entropy principle (Figure 11.5).

A probabilistic interpretation of Sankoff & Cedergren

If the scores in Sankoff & Cedergren's algorithm are interpreted as log probabilities, and if their procedure is carried out with a '+' in place of a 'max', then the resulting algorithm will compute the full likelihood, as pointed out by Allison, Wallace & Yee [1992a]. The tree score $S(\Delta_1 \cdot x_{i_1}^1, \Delta_2 \cdot x_{i_2}^2, \ldots, \Delta_N \cdot x_{i_N}^N)$ will become the sum over all assignments at ancestral nodes, and the recursion (7.6) will take the sum over the preceding αs and therefore sum over all possible alignments. Like Sankoff & Cedergren's original algorithm, this computation is not practical for most problems.

Interpreting Hein's algorithm probabilistically

As remarked above (p. 224), parsimony can be regarded as the Viterbi approximation to the full probability if the scores are interpreted as $\log P(x|y)$, where the $P(x|y)$ are substitution probabilities that don't depend on time. If derived this way, scores will generally take different values for different residue substitutions. This means that there will usually only be one optimal alignment of two sequences, and hence that Hein's sequence graphs will consist of only one path. There will, however, generally be a great many paths that are only slightly suboptimal. Parsimony therefore gives a poor approximation to the full probability in this case.

If we attempt to remedy the situation by using '+' instead of 'max', then we have to include all paths through the dynamic programming matrix in the sequence graph. At the first node above the leaves, this graph has size N^2, at the next-highest node it will have size N^3 or N^4, and so on. It is clear that we lose all the advantage gained over the comprehensive but slow Sankoff–Cedergren approach.

As a compromise, we could try to select near-optimal paths in the hope of approximating the full probability while keeping the sequence graphs down to manageable size. Such a strategy might produce a good alignment/phylogeny algorithm, but would probably need clever heuristics for selecting the paths.

8.7   Further reading

Maximum likelihood was first applied to phylogeny by Edwards & Cavalli-Sforza [1963; 1964], who examined the case of continuous variables, such as the size of skeletal features of a species, or the frequency of genes in a population. They described the evolution of these variables by a random walk combined with a Yule process allowing bifurcations [Edwards 1970]. Thompson [1975] devised computational methods for implementing this, and applied them to some examples of interest.

An important paper by Felsenstein [1981a] showed how to carry maximum likelihood methods over to the case of discrete characters, such as the residues in a sequence. In this paper, Felsenstein introduced the basic algorithm for computing the likelihood of trees of any size (p. 200), gave an effective procedure for maximising this likelihood with respect to edge lengths (p. 205), and showed how reversibility could be used to reduce the problem to unrooted trees (p. 202). This laid the foundations for the likelihood methods most commonly used in molecular phylogeny nowadays.

In this chapter and the previous one, we have treated DNA and protein sequences as essentially similar types of data, apart from alphabet size. But of course their biological roles are very different, and this makes them suitable for different purposes. For instance, the rapid changes in the third position in codons allows us to explore recent evolutionary events, whereas the more conserved regions of proteins may carry information about early speciation events in the Earth's history [Doolittle *et al.* 1996]. In many cases we should treat the DNA and protein levels simultaneously. Goldman & Yang [1994] have shown how this can be done by using a Markov model whose states are codons, and whose transition probabilities reflect both DNA substitution patterns and (when there is a change in the residue coded for) amino acid properties.

The future of phylogeny seems very promising. The spectacular advance of genome science means that vast amounts of sequence data will become available, and it is likely that new types of sequence information will be used for phylogeny. Already, it is clear that the presence of various repeat families can be a useful phylogenetic marker [Shimamura *et al.* 1997], as can chromosomal inversions and other genomic rearrangements [Hannenhalli *et al.* 1995]. For once, the forest of data may enable us to see the trees more clearly.

9
Transformational grammars

Until now, we have treated biological sequences as one-dimensional strings of independent, uncorrelated symbols. This assumption is computationally convenient but not structurally realistic. The three-dimensional folding of proteins and nucleic acids involves extensive physical interactions between residues that are not adjacent in primary sequence. Can probabilistic models of proteins and nucleic acid sequences be developed that allow for longer range interactions? Can we compute efficiently with such models? In this chapter, we will step back from models of particular sequence problems and address these more theoretical issues. We will see how many of the methods described in previous chapters fit into a more general view of modelling sequences.

A general theory for modelling strings of symbols has been developed by computational linguists [Chomsky 1956; 1959]. This theory is known as the *Chomsky hierarchy of transformational grammars*. In the Chomsky hierarchy, most of the models we have used so far in this book are the lowest of four types of model of increasing complexity and descriptive power. Transformational grammars were developed in an attempt to understand the structure of natural languages. They became important in theoretical computer science [Hopcroft & Ullman 1979; Gersting 1993] because computer languages, unlike natural languages, can be precisely specified as formal grammars. Recently, transformational grammars have been applied to sequence analysis problems in molecular biology [Searls 1992; Dong & Searls 1994; Rosenblueth *et al.* 1996].

An example of the application of grammar theory to higher-order structure in biological sequence analysis is the use of *stochastic context-free grammars* (SCFGs) in RNA secondary structure analysis [Eddy & Durbin 1994; Sakakibara *et al.* 1994; Grate 1995; Lefebvre 1995; 1996]. Although many sequence alignment methods in computational molecular biology are implicitly *stochastic regular grammars*, they have a long history of their own and can live in happy ignorance of the Chomsky hierarchy. In contrast, the application of SCFGs to probabilistic modelling of RNA secondary structure is a more recent development, and the jargon of RNA SCFGs remains very close to its roots in computational linguistics. We need to understand the basics of computational linguistics to understand RNA SCFGs. The main purpose of this chapter is to set the stage for applying SCFG-based probabilistic modelling to RNA secondary structure problems. We start with an overview of transformational grammars in their non-probabilistic form. We then introduce stochastic grammars as a formalised system for full probabilistic modelling of sequences with long-range correlations and constraints. We conclude by giving generalised alignment algorithms for stochastic context-free grammars, of which the RNA models of the next chapter are a subset.

9.1   Transformational grammars

Though nonsensical, 'colourless green ideas sleep furiously' is a grammatically correct English sentence. Most English speakers (except those who have read Chomsky) have never before seen this sentence or even any of its combinations of adjacent words. Nonetheless, they will recognise it, parse its grammar correctly, and speak it with the correct intonation of an English sentence.

Chomsky was interested in how a brain or a computer program could algorithmically determine whether a novel sentence was grammatical or not. He constructed finite formal machines called 'grammars' which recursively enumerate an infinite number of sentences that belong to a language. For the question 'does the language contain this sentence?' grammar theory substitutes 'can the grammar generate this sentence?' The first question is intractable (the set of possible sentences is infinite) but the second question can be practically answered for many useful forms of grammars. How well this works depends on how well the grammar models the constraints on the language; i.e. how many grammatical sentences there are that the grammar fails to generate, and how many ungrammatical sentences the grammar generates erroneously.

Transformational grammars are sometimes called *generative grammars*. One speaks in terms of generating sequence even if the primary use of the model is for recognising, scoring, and/or parsing strings. In Chapter 3, we described hidden Markov models as generative probabilistic models that 'emit' sequences. Whether a given sequence belonged to a family or not was inferred by calculating the probability that the sequence would be generated by a hidden Markov model of the family. When a hidden Markov modeller speaks of generating sequences, biologists sometimes find this concept confusing. Obviously biological evolution generated the sequences, not an HMM. The terms 'generation' and 'emission' are part of a convenient formalism that is largely due to Chomsky.

Definition of a transformational grammar

A transformational grammar consists of a number of *symbols* and a number of *rewriting rules* $\alpha \rightarrow \beta$ (also called *productions*) where α and β are both strings of symbols. There are two kinds of symbols: abstract *nonterminal* symbols and *terminal* symbols that actually appear in an observed string. The left-hand side α contains at least one nonterminal, which in general is transformed into a new string of terminals and/or nonterminals on the right-hand side of the production. If we were modelling sentences, the terminals might be words; if we were modelling protein sequences, the terminals might be amino acid symbols. We will use lower-case letters to represent terminals and upper-case letters to represent nonterminals.

The easiest way to see how a transformational grammar works is by example. We will use a two-letter terminal alphabet $\{a, b\}$ and a single nonterminal S. A special blank terminal symbol ϵ is used to end the process. Here is a transformational grammar that generates any string of as and bs:

$$S \rightarrow aS, \qquad S \rightarrow bS, \qquad S \rightarrow \epsilon.$$

To generate a string of as and bs, we carry out a series of transformations according to the grammar's rules starting from some initial string. By convention, we usually start from a special start nonterminal S (which in this case is our only nonterminal). An applicable production is chosen which has the string S on its left-hand side, and S is replaced by the string on the right-hand side of that production. The process of choosing a substring and rewriting it in place according to one of the allowed rewriting rules continues until the string consists entirely of terminals and no further rewritings are possible. The succession of strings that result from this process is called a *derivation* from the grammar. An example derivation of our simple example grammar is:

$$S \Rightarrow aS \Rightarrow abS \Rightarrow abbS \Rightarrow abb.$$

For convenience, we will usually specify multiple possible productions using an abbreviated representation like $S \rightarrow aS \mid bS \mid \epsilon$, where the symbol $\mid$ indicates 'or'. In this example, we would have three choices of what to transform S into.

When transformational grammars are used for a sequence analysis problem, we often have a particular sequence in mind. The question is whether the sequence 'matches' (could be generated by) the grammar. We work backwards to determine whether a derivation exists for the string. If a derivation exists, then the string is a valid member of the language modelled by the grammar. Finding a valid derivation for a given sequence is called *parsing*, and in this context, a derivation is called a *parse* of the sequence. We can think of a parse as an *alignment* of the grammar and the sequence. Just as a Viterbi alignment of a sequence to an HMM is an assignment of sequence positions to HMM states, so a parse of a sequence with a grammar is essentially an assignment of sequence positions to grammar nonterminals.

The Chomsky hierarchy

Chomsky [1959] described four sorts of restrictions on a grammar's rewriting rules. The resulting four classes of grammar fall into a hierarchy known as the Chomsky hierarchy of transformational grammars. In the following examples, we use W to represent any nonterminal, a to represent any terminal, α and γ to represent any string of nonterminals and/or terminals including the null string, and β to represent any string of nonterminals and/or terminals not including the null string.

regular grammars Only production rules of the form $W \rightarrow aW$ or $W \rightarrow a$ are allowed.

context-free grammars Any production rule of the form $W \rightarrow \beta$ is allowed. The left-hand side of the production rule must consist of just one nonterminal but the right-hand side can be any string.

context-sensitive grammars Productions are of the form $\alpha_1 W \alpha_2 \rightarrow \alpha_1 \beta \alpha_2$. The allowed transformations of nonterminal W are dependent on its context α_1 and α_2. It is provably equivalent to require that the right-hand side contains at least as many symbols as the left-hand side; context-sensitive grammar productions never shrink [Chomsky 1959]. This allows context-sensitive productions of the form $AB \rightarrow BA$, for instance.

unrestricted (phrase structure) grammars Any production rule of the form $\alpha_1 W \alpha_2 \rightarrow \gamma$ is allowed.

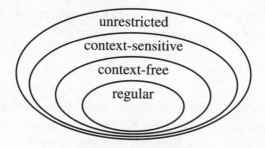

Figure 9.1 *The Chomsky hierarchy of transformational grammars, nested according to the increasing restrictions placed on the production rules in the grammar. In terms of allowed productions, regular grammars are the simplest and most restricted grammars, and therefore the easiest to parse. However, the regular grammars also have the least power to describe 'structural' constraints on strings.*

Automata

In computer science, each grammar has a corresponding abstract computational device called an *automaton*. Grammars are described as generative models, while automata are usually described as parsers that accept or reject a given sequence. We will find automata useful here for two limited purposes. First, automata are often intuitively more easy to describe and understand than their equivalent grammars. In particular, finite state automata have a nice graphical representation that is easier to understand than a laborious enumeration of a regular grammar's rewriting rules. Secondly, automata give a more concrete idea of how we might recognise a sequence using a formal grammar.

Grammar	Parsing automaton
regular grammars	finite state automaton
context-free grammars	push-down automaton
context-sensitive grammars	linear bounded automaton
unrestricted grammars	Turing machine

Table 9.1. *Parser abstractions associated with the hierarchy of grammars.*

9.2 Regular grammars

All the production rules in a regular grammar are of the form $W \rightarrow aW$ or $W \rightarrow a$, where W and a represent any nonterminal or terminal in the grammar, respectively. We will also sometimes allow an additional production of $W \rightarrow \epsilon$ for terminating derivations, where ϵ is the null string.[1] Essentially, regular grammars generate sequence from left to right. Regular grammars cannot efficiently describe long-range correlations between the terminal symbols. They are 'primary sequence' models.[2]

Example: An odd regular grammar

The first grammar in this chapter was a regular grammar that generated any string consisting of as and bs: a rather boring language. Regular grammars are capable of more interesting and sometimes surprising behaviour. Here's an example of a regular grammar that generates only strings of as and bs that have an odd number

---

[1] The rule $W \rightarrow \epsilon$ is a 'shrinking' production. The right side is shorter than the left. Technically, this makes it an unrestricted grammar rule. However, it can be proved that a regular grammar can always be expanded to absorb the ϵ. For instance, the nearly regular grammar $S \rightarrow aS \mid bS \mid \epsilon$ is the same as the regular grammar $S \rightarrow aS \mid bS \mid a \mid b$. ϵ productions are not a serious problem for either regular grammar or context-free grammar parsing algorithms, but they do present some technical difficulties in proofs.

[2] We may also have right-to-left grammars with productions only of the form $W \rightarrow Wx$ or $W \rightarrow x$. These are also regular grammars. Allowing *both* $W \rightarrow Wx$ and $W \rightarrow xW$ productions in the same grammar gives a context-free grammar.

of *a*s [Searls 1992]:

$$\text{start from } S,$$
$$S \rightarrow aT \mid bS,$$
$$T \rightarrow aS \mid bT \mid \epsilon.$$

Whenever a string contains an odd number of *a*s, the derivation is in nonterminal *T*; when it has an even number of *a*s, it is in nonterminal *S*. Since it can only terminate from nonterminal *T*, it only generates strings with odd numbers of *a*s.

□

Finite state automata

The parsing automaton corresponding to a regular grammar is a *finite state automaton*. We saw finite state automata used in Chapter 2 as a general model of pairwise alignment algorithms. We now consider them more generally. A finite state automaton is a device which reads one symbol at a time from an input string. The symbol may be accepted, in which case the automaton enters a new state; or the symbol may not be accepted, in which case the automaton halts and rejects the string. If the automaton reaches a final 'accepting' state, the input string has been successfully recognised and parsed by the automaton.

A finite state automaton is a model composed of a number of *states*, and the states are interconnected by *state transitions*. The states and state transitions correspond to the nonterminals and productions of the equivalent regular grammar. Finite state automata are often drawn in abstract form with circles representing states and arrows for transitions.

Example: FMR-1 triplet repeat region

The human FMR-1 gene sequence contains a triplet repeat region in which the sequence CGG is repeated a number of times. The number of triplets is highly variable between individuals, and increased copy number is associated with fragile X syndrome, a genetic disease that causes mental retardation and other symptoms in one out of 2000 children. The finite state automaton shown in Figure 9.2 compactly models the CGG repeat region of FMR-1 by allowing a cyclic transition back into a new CGG.

To check if a sequence matches this description of the FMR-1 CGG repeat, the sequence is fed to the automaton one symbol at a time. If the first symbol is a G, the automaton enters state 1; otherwise it quits and rejects the sequence. If the automaton is in state 1 and it reads a C, it successfully moves to state 2, and so on, until the automaton successfully recognises the sequence by reaching the end state E with no symbols left to examine.

The finite state automaton will match any string from the 'language' that contains the strings GCG CTG, GCG CGG CTG, GCG CGG CGG CTG, GCG CGG CGG

(a) Human FMR-1 mRNA sequence, fragment

```
 . . . GCG CGG CGG CGG CGG CGG CGG CGG CGG
CGG CGG AGG CGG CGG CGG CGG CGG CGG CGG
CGG CGG AGG CGG CGG CGG CGG CGG CGG CGG
CGG CGG CTG . . .
```

(b)

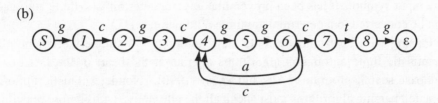

Figure 9.2 *(a) The sequence of the FMR-1 triplet repeat region, from* GEN-BANK *HSFMR1A, accession X69962. Two variant* AGG *triplets in the repeat are underlined. (b) A finite state automaton that recognises FMR-1 triplet repeat regions with any number of triplets. Note the presence of a transition that accepts the variant* AGG *triplets.*

CGG CTG, *ad infinitum* for any number of copies of CGG. A regular grammar that is equivalent to this finite state automaton is:

$$
\begin{aligned}
S &\rightarrow gW_1 & W_5 &\rightarrow gW_6 \\
W_1 &\rightarrow cW_2 & W_6 &\rightarrow cW_7 \mid aW_4 \mid cW_4 \\
W_2 &\rightarrow gW_3 & W_7 &\rightarrow tW_8 \\
W_3 &\rightarrow cW_4 & W_8 &\rightarrow g \\
W_4 &\rightarrow gW_6
\end{aligned}
$$

$\square$

Moore vs. Mealy machines

In the FMR-1 automaton of Figure 9.2, terminal symbols are associated with the transitions in the automaton. Finite automata that accept on transitions are called *Mealy machines*. In contrast, in the hidden Markov models of Chapter 3, we associated terminal symbols with states, and separated symbol emission events from state transition events. Finite automata which accept on states are called *Moore machines*. The two types of machines are interconvertible. For example, we could label state 1 in the FMR-1 automaton with a G, and have the state, rather than the transition into the state, accept the G. The grammar production corresponding to state 1 in the FMR-1 automaton is $S \rightarrow gW_1$ in the Mealy machine, but could be written as $S \rightarrow \hat{W}_1$, $\hat{W}_1 \rightarrow gW_1$ in a Moore machine, where $\hat{W}_1$ is an added intermediate nonterminal. (Since the two forms are equivalent, we need not be too concerned that the rule $S \rightarrow \hat{W}_1$ in the Moore machine is not a strictly conforming regular grammar rule.)

Deterministic vs. nondeterministic automata

The FMR-1 automaton is an example of a *nondeterministic* finite automaton. When the automaton is in state 6 and the next input symbol is a C, the automaton can accept the C by moving *either* to state 4 or state 7. In a *deterministic* finite automaton, no more than one accepting transition is possible for any state and any input symbol. It has been proven that any nondeterministic finite automaton can be converted to a deterministic finite automaton.

Parsing with deterministic finite state automata is extremely efficient. Deterministic finite automaton algorithms operate at the heart of the fast BLAST database search programs [Altschul *et al.* 1990]. Nondeterministic finite automaton parsing algorithms must check all the alternative paths before rejecting a sequence, but can still be made efficient. The UNIX text pattern-matching utilities in programs such as GREP SED, AWK, and VI implement highly efficient nondeterministic finite automata; UNIX 'regular expressions' are equivalent to regular grammars.

Exercises

9.1   Convert the FMR-1 automaton in Figure 9.2 to a Moore machine in which each state accepts a particular symbol, instead of each transition accepting a particular symbol.

9.2   Convert the FMR-1 automaton to a deterministic automaton.

PROSITE patterns

An excellent example of a biological application of regular grammars is the PROSITE database compiled by Amos Bairoch and his colleagues in Geneva [Bairoch, Bucher & Hofmann 1997]. A PROSITE entry includes a sequence pattern for a highly conserved signature motif shared by all or almost all of the members of a protein family. Unlike methods which assign scores to alignments, PROSITE patterns either match a sequence or don't; they are regular grammars that are matched to sequences using finite state automata.

A PROSITE pattern consists of a string of pattern elements separated by dashes and terminated by a period. In a pattern element, a letter indicates the single-letter code for one of the amino acids; square brackets indicate that any one of the enclosed residues can occur; curly brackets indicate that anything *but* one of the enclosed residues can occur; and an x indicates that any residue can occur at this position. Lengths or ranges of lengths are given in parentheses, such as -x(4)- to match a spacer of four residues of any type and -x(2,4)- to match a spacer of two, three, or four residues of any type. Figure 9.3 shows an example of one of the 1029 PROSITE patterns in the February 1995 release of the PROSITE database.

(a)

```
RU1A_HUMAN    S R S L K M R G Q A F V I F K E V S S A T
SXLF_DROME    K L T G R P R G V A F V R Y N K R E E A Q
ROC_HUMAN     V G C S V H K G F A F V Q Y V N E R N A R
ELAV_DROME    G N D T Q T K G V G F I R F D K R E E A T
```
RNP-1 motif

(b)

```
[RK]-G-{EDRKHPCG}-[AGSCI]-[FY]-[LIVA]-x-[FYM].
```

Figure 9.3 *(a) Part of a multiple sequence alignment showing the highly conserved 'RNP-1' sequence motif of a major family of RNA binding proteins. (b) The RNP-1* PROSITE *pattern PS00030.*

Any PROSITE pattern is a regular grammar, and can be matched with a nondeterministic finite automaton. The syntax of PROSITE patterns is close to standard regular expression syntax. Some popular PROSITE pattern searching implementations use UNIX GREP implementations as their search engine by first converting the PROSITE pattern to a UNIX regular expression, which GREP then builds an automaton for.

Example: A PROSITE pattern in regular grammar form

A regular grammar that corresponds to the PROSITE RNP-1 pattern in Figure 9.3 is as follows. We use a starting nonterminal S and eight nonterminals $W_1, \ldots, W_8$ corresponding to the eight positions of the conserved motif. For brevity, some of the productions are written with brackets as in the PROSITE description: for instance, $[ac]W$ means $aW \mid cW$.

$$
\begin{aligned}
S &\rightarrow rW_1 \mid kW_1 \\
W_1 &\rightarrow gW_2 \\
W_2 &\rightarrow [afilmnqstvwy]W_3 \\
W_3 &\rightarrow [agsci]W_4 \\
W_4 &\rightarrow fW_5 \mid yW_5 \\
W_5 &\rightarrow lW_6 \mid iW_6 \mid vW_6 \mid aW_6 \\
W_6 &\rightarrow [acdefghiklmnpqrstvwy]W_7 \\
W_7 &\rightarrow f \mid y \mid m
\end{aligned}
$$

□

Exercise

9.3   The PROSITE pattern for a C2H2 zinc finger, an important DNA binding protein motif, is C-x(2,4)-C-x(3)-[LIVMFYWC]-x(8)-H-x(3,5)-H. Draw a finite automaton that accepts this pattern.

What a regular grammar can't do

Two classic examples [Chomsky 1956] of languages *L* that regular grammars cannot describe arise when:

(i) *L* contains all the strings of the form *aa*, *bb*, *abba*, *baab*, *abaaba*, etc. that read the same forwards as backwards (a palindrome language).

(ii) *L* contains all the strings of the form *aa*, *abab*, *aabaab* that consist of two identical halves (a copy language).

Regular grammars can generate palindromic strings as part of their language. The point is that a regular grammar cannot efficiently generate *only* palindromes, and hence cannot distinguish a correct palindrome from a non-palindrome. Describing more and more specific constraints on the grammatical strings in a language requires grammars more complex than regular grammars.

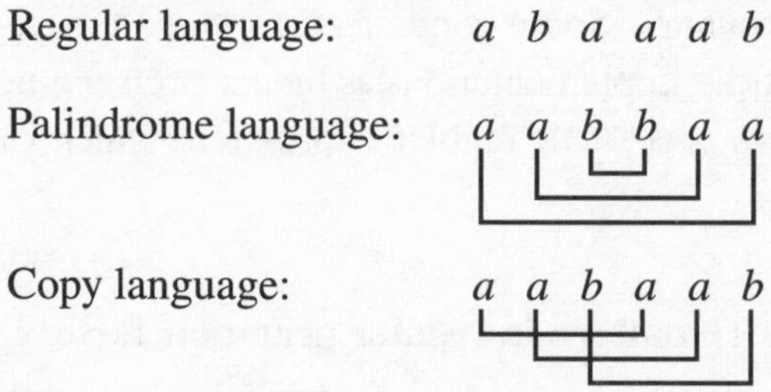

Regular language: *a b a a a b*

Palindrome language: *a a b b a a*

Copy language: *a a b a a b*

Figure 9.4 *Unlike regular languages, palindrome and copy languages have correlations between distant positions. Lines indicate correlated positions in strings from the palindrome and the copy language.*

As shown in Figure 9.4, the interactions in palindrome languages are *nested*, i.e. the lines of the interactions do not cross; in the copy languages, crossing interactions can occur. This distinction is important in determining the type of grammar that generates each language.

9.3   Context-free grammars

The palindrome languages are dealt with by the next level in Chomsky's hierarchy, the *context-free grammars* (CFGs). Obviously the problem of parsing 'Doc, note. I dissent. A fast never prevents a fatness. I diet on cod.'[3] arises rarely in computational biology. The reason to look carefully at the context-free grammars is that RNA secondary structure is a kind of palindrome language, as illustrated in the example below. RNA secondary structure presents a problem in which the

---

[3] A palindrome credited to Peter Hilton, a member of the British cryptography team that cracked the German Enigma code in World War II.

sequence may not matter as long as strong base pair correlations are maintained between certain nested pairs of positions.

The context-free grammars permit additional rules that allow the grammar to create nested, long-distance pairwise correlations between terminal symbols. The left side of a production rule must still be a single nonterminal, but the right side of a production rule can be any combination of terminals and nonterminals. The right side can therefore generate a correlated base pair from a single nonterminal, unlike regular grammar productions which must generate a symbol pair independently from two different nonterminals. An example of a CFG that can generate a palindrome language would be:

$$S \rightarrow aSa \mid bSb \mid aa \mid bb.$$

A derivation of the palindrome 'aabaabaa' from this CFG is:

$$S \Rightarrow aSa \Rightarrow aaSaa \Rightarrow aabSbaa \Rightarrow aabaabaa.$$

Whereas regular grammars generate strings from left to right, context-free grammars can generate strings from outside in. Only nested correlations can be captured because of this outside-in generation. The crossing correlations of the copy language (Figure 9.4) violate this nesting constraint, so copy languages are not context-free languages.

Example: A context-free grammar for an RNA stem loop

In the picture below, *seq1* and *seq2* can fold into the same RNA secondary structure despite having different sequences because they share the same pattern of base pairs (A-U and C-G). *Seq3*, though identical in sequence to the first half of *seq2* and the second half of *seq1*, cannot fold into a similar structure. The consensus RNA secondary structure imposes a set of nested pairwise constraints like a palindrome language, except that the correlated pairs are complementary instead of identical.

A CFG that models RNA stem loops with three base pairs and a GCAA or GAAA loop like *seq1* and *seq2* would be:

$$
\begin{aligned}
S &\rightarrow aW_1u \mid cW_1g \mid gW_1c \mid uW_1a, \\
W_1 &\rightarrow aW_2u \mid cW_2g \mid gW_2c \mid uW_2a, \\
W_2 &\rightarrow aW_3u \mid cW_3g \mid gW_3c \mid uW_3a, \\
W_3 &\rightarrow gaaa \mid gcaa.
\end{aligned}
$$

$\square$

Exercises

9.4   Write derivations for *seq1* and *seq2* using the context-free grammar in the example above.

9.5   Write a regular grammar that generates *seq1* and *seq2* but not *seq3* in the example above.

9.6   Consider the complete language generated by the CFG in the example above. Describe a regular grammar that generates exactly the same language. Does describing this sequence family with a regular grammar seem like a good idea?

Parse trees

An alignment of a context-free grammar to a sequence (i.e. a parse) has an elegant representation called a *parse tree*. The root of the tree is the start nonterminal S. Leaves are the terminal symbols in the sequence. Internal nodes are nonterminals. The children of an internal node are the productions of that nonterminal, in left-to-right order.

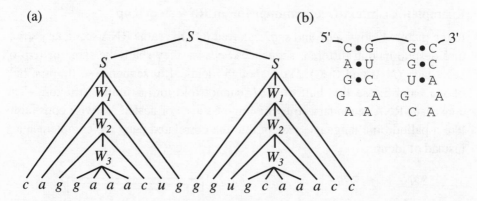

Figure 9.5 *(a) A parse tree for* CAG GAA ACU GGG UGC AAA CC *and the stem-loop grammar, extended with a production rule* $S \rightarrow SS$ *to make a more interesting tree. (b) The RNA secondary structure for the same sequence, which corresponds closely to the parse tree representation.*

A *subtree* is a fragment of a parse tree rooted at an internal node. Any sub-tree derives a contiguous segment of the observed sequence. This property is important. It allows algorithms to build optimal parse trees for a sequence by recursively building larger and larger optimal parse subtrees for larger and larger subsequences. An example of a parse tree for a CFG and a small RNA is shown in Figure 9.5.

Example: Parse tree for a PROSITE pattern

Regular grammars are a subset of the context-free grammars. Therefore, alignments of regular grammars to sequences can also be represented as parse trees. Figure 9.6 shows a parse tree for the regular grammar of the RNP-1 PROSITE pattern in Figure 9.3. The correspondence between alignments and parse trees should be clear.   □

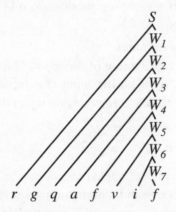

Figure 9.6 *Parse tree for the RNP-1 motif* RGQAFVIF *using the regular grammar from page 241. Regular grammars are linear special cases of the context-free grammars, and hence the parse tree for a regular grammar is essentially just a standard linear alignment of the grammar nonterminals onto sequence terminals.*

Push-down automata

The parsing automaton for CFGs is called a *push-down automaton*. Whereas finite state automata required no memory except for keeping track of the current state, a push-down automaton keeps a limited memory of symbols in the form of a push-down stack.[4]

A push-down automaton parses a sequence from left to right according to the following algorithm. The automaton's stack is initialised by pushing the start

---

[4] A push-down stack is an array or list which is accessed last in, first out. Elements are 'pushed' onto and 'popped' off of the 'top' of the stack, like a stack of plates.

nonterminal onto it. The following steps are then iterated until no input symbols remain. If the stack is empty when no input symbols remain then the sequence has been successfully parsed.

Algorithm: Parsing with a push-down automaton

Pop a symbol off the stack.
If the popped symbol is a nonterminal:
  - Peek ahead in the input from the current position and choose a valid production for the nonterminal. For a deterministic push-down automaton there is at most one possible choice. For a nondeterministic automaton, all possible choices need to be evaluated individually. If there is no valid production, terminate and reject the sequence.
  - Push the right side of the chosen production rule onto the stack, rightmost symbols first.

If the popped symbol is a terminal:
  - Compare it to the current symbol of the input. If it matches, move the automaton to the right on the input (the input symbol is accepted). If it does not match, terminate and reject the sequence.   ◁

Push-down automata are not efficient recognisers for nondeterministic context-free grammars. All series of valid automaton moves must be tried exhaustively until either the input string is successfully accepted or no more series of moves remain to be tried. Although it is possible to use this brute-force algorithm to recognise strings with many not-too-complex nondeterministic CFGs, there is potentially a combinatorial explosion of different derivations that need to be tested. Later in the chapter, we will describe the more sophisticated, polynomial time Cocke–Younger–Kasami (CYK) parsing algorithm for context-free grammars.

Example: Parsing an RNA stem loop with a push-down automaton

Consider parsing the sequence GCC GCA AGG C using the context-free grammar of a three base pair RNA stem loop from page 244. Below are shown the series of operations that occur on the automaton's stack while parsing the sequence. The position of the automaton on the input (left column) is shown by a box. The symbols in the push-down stack are shown (middle column) with the top of the stack to the left. Based on the current position in the input and the current stack, the next automaton operations are described (right column). For brevity, nonterminals are denoted by their numbers, so that 1 is used for W_1, etc.

Input string	Stack	Automaton operation on stack and input
[G]CCGCAAGGC	S	Pop S. Peek at input; produce $S \to g1c$.
[G]CCGCAAGGC	$g1c$	Pop g. Accept g; move right on input.
G[C]CGCAAGGC	$1c$	Pop 1. Peek at input; produce $1 \to c2g$.
G[C]CGCAAGGC	$c2gc$	Pop c. Accept c; move right on input.
GC[C]GCAAGGC	$2gc$	Pop 2. Peek at input; produce $2 \to c3g$.
GC[C]GCAAGGC	$c3ggc$	Pop c. Accept c; move right on input.
GCC[G]CAAGGC	$3ggc$	Pop 3. Peek at input; produce $3 \to gcaa$.
GCC[G]CAAGGC	$gcaaggc$	Pop g. Accept g; move right on input.
⋮		(several acceptances)
GCCGCAAGG[C]	c	Pop c. Accept c; move right on input.
GCCGCAAGGC[ ]	-	Stack empty. Input string empty. Accept.

□

Exercise

9.7 Modify the push-down automaton parsing algorithm so that it randomly *generates* one of the possible valid sequences in a context-free grammar's language.

9.4 Context-sensitive grammars

Though at first sight the copy language appears no more complex than the palindrome language, copy languages are not context-free languages. In general, copy languages require context-sensitive grammars. A context-sensitive grammar that generates even our simple example of a copy language is complicated. Consider, for example, the copy language consisting of strings like cc, $acca$, $abaccaba$, $bbabccbbab$; i.e. all strings consisting of two copies of a string of as and bs, with a pair of cs between them. A context-sensitive grammar that generates this language is:

initialisation:
$$S \to CW$$
nonterminal generation:
$$W \to A\hat{A}W \mid B\hat{B}W \mid C$$
nonterminal reordering:
$$\hat{A}B \to B\hat{A}$$
$$\hat{A}A \to A\hat{A}$$
$$\hat{B}A \to A\hat{B}$$
$$\hat{B}B \to B\hat{B}$$

terminal generation:
$$CA \to aC$$
$$CB \to bC$$
$$\hat{A}C \to Ca$$
$$\hat{B}C \to Cb$$
termination:
$$CC \to cc$$

We have seven different nonterminals, S, A, $\hat{A}$, B, $\hat{B}$, C, and W. A and $\hat{A}$ are destined to generate an a symbol (and likewise B and $\hat{B}$ are destined to generate a b symbol, and C is destined to generate a c). A and B nonterminals generate the left half of the string, and $\hat{A}$ and $\hat{B}$ generate the right half of the string.

The context-sensitive grammar does not directly generate the crossing pairwise interactions between symbols in a copy language. Instead the W nonterminal generates them as pairs with uncrossed interactions, then the grammar reorders the nonterminals appropriately by examining their local context. The reordering rules swap nonterminals, moving the hat nonterminals rightwards past the non-hat nonterminals. Since any production rule can be used any time its left hand side appears during a derivation, the grammar is carefully constructed so as to not start generating terminal symbols until the nonterminals are properly ordered.

An example derivation of the string *aabccaab* from this grammar would be:

$$S \Rightarrow CW \Rightarrow CA\hat{A}W \Rightarrow CA\hat{A}A\hat{A}W \Rightarrow CA\hat{A}A\hat{A}B\hat{B}W \Rightarrow CA\hat{A}A\hat{A}B\hat{B}C$$
$$\Rightarrow CAA\hat{A}\hat{A}B\hat{B}C \Rightarrow CAA\hat{A}B\hat{A}\hat{B}C \Rightarrow CAAB\hat{A}\hat{A}\hat{B}C \Rightarrow CAAB\hat{A}\hat{A}Cb$$
$$\Rightarrow CAAB\hat{A}Cab \Rightarrow CAABCaab \Rightarrow aCABCaab \Rightarrow aaCBCaab$$
$$\Rightarrow aabCCaab \Rightarrow aabccaab.$$

The parsing automaton for a context-sensitive grammar is a *linear bounded automaton*. A linear bounded automaton is a mechanism for systematically working backwards through all possible derivations of the observed string until either a derivation reaches the starting nonterminal, or all possible derivations have been exhausted without finding a valid one. Because a context-sensitive grammar is restricted so that the left side of a production rule cannot be longer than the right side, there must be a finite number of possible derivations to examine. No intermediate in the derivation can be longer than the observed string itself. Computer science textbooks describe a linear bounded automaton as an abstract 'tape' of linear memory and a read/write head; the term 'bounded' refers to the knowledge that the amount of tape required is guaranteed to be less than or equal to the length of the observed string. Nonetheless, the number of possible derivations is exponentially large. No general polynomial-time algorithms for parsing context-sensitive grammars are known to exist. This is a serious concern in considering any practical context-sensitive grammar applications.

NP problems and 'intractability'

Problems in which there is no *known* polynomial-time algorithm for finding a solution, but a solution can be checked for correctness in polynomial time, are called *nondeterministic polynomial* or *NP* problems. Many problems of interest are NP problems, including context-sensitive grammar parsing. *Proving* that no polynomial-time algorithm exists for solving NP problems is a holy grail in computer science. A subclass of NP problems, including both context-free grammar parsing and the famous travelling salesman problem, are called *NP-complete* problems. A polynomial-time algorithm that solves one NP-complete problem will solve all NP-complete problems.

NP problems are sometimes called 'intractable'. However, it is important to remember that many NP problems can be addressed usefully. Exact brute-force algorithms such as branch-and-bound may be applied if the problem is not too large. Approximate algorithms such as simulated annealing may be applied to even larger problems.

It is also important to recognise that for many modelling problems, tractable special cases may exist for problems which are NP in more general form. An example is the problem of finding tandem repeats in DNA sequences. Although tandem repeats are clearly copy languages and therefore apparently an NP parsing problem, a variety of efficient polynomial-time algorithms exist which locate tandem repeats. A very crude polynomial-time algorithm is to simply enumerate all the different subsequences (i, j) in the sequence; there are about L^2 of these. Each subsequence can then be aligned individually against the complete sequence to test if it is tandemly repeated somewhere using the $O(L^2)$ repeated-matches dynamic programming algorithm in Chapter 2, giving an overall time complexity of $O(L^4)$.

Unrestricted grammars and Turing machines

An unrestricted grammar is a transformational grammar in which the left and right sides of the production rules can be any combination of symbols. The equivalent parsing automaton is a Turing machine. There is *no* general algorithm that is guaranteed to determine whether a string has a valid derivation from an unrestricted grammar in less than infinite time. Intuitively this is because productions can shrink to fewer symbols on the right-hand side. The intermediate strings in working backwards through possible Turing machine derivations can grow longer than the input, and thus the number of possible derivations can grow without bound. In contrast, the number of intermediate strings in a context-sensitive grammar derivation must be finite because the intermediate strings on the linear bounded automaton's tape can only get smaller as the automaton works backwards towards possible solutions. The properties of Turing machines are of great theoretical interest in computer science, but the lack of any parsing algorithm that is guaranteed to halt makes unrestricted grammars unappealing for practical applications, except perhaps for more limited special cases of these grammars. Many problems which could be formulated as unrestricted grammars are instead formulated as optimisation problems and 'parsing' is done by (for instance) simulated annealing in a non-exact way, as discussed above for context-sensitive grammars and NP problems.

9.5   Stochastic grammars

Careful consideration of PROSITE patterns reveals a drawback in using simple finite automata for computational biology. As more sequences are determined and the family grows, it gets increasingly difficult to create a specific pattern. Exceptions to the rules of the pattern may occur at any position. For instance, the RNP-1 motif of another RNA binding protein, the SRP55 protein SR55_DROME which is involved in mRNA splicing in fruit flies, has the sequence NGYGFVEF. The first N fails to match the PROSITE pattern, which requires an R or a K at this position. The pattern has to be modified to allow N. As exceptions accumulate and the pattern is loosened, the specificity of the pattern degrades. As a result, it may have so little information content that it matches unrelated, random sequences. For some diverse protein families, it has proved impossible to produce a discriminative PROSITE pattern. The logical solution is to allow the exceptions, but instead of considering all possibilities equal, give the exceptions less score than a strong match to the consensus. This idea leads to stochastic (probabilistic) regular grammars like sequence 'profiles' (Chapter 5) and hidden Markov models (Chapter 3).

Any of the grammars in the Chomsky hierarchy can be used in a stochastic form as a basis for a probabilistic modelling system for sequences. A stochastic grammar model θ generates different strings x with probabilities $P(x \mid \theta)$, whereas non-stochastic grammars either generate a string x or not.

In a stochastic regular grammar or stochastic context-free grammar, the sum of the probabilities of all the possible productions from any given nonterminal is 1. The resulting stochastic grammar defines a probability distribution over sequences x, i.e. $\sum_x P(x|\theta) = 1$. For example, in the first production rule of our PROSITE example, $S \to r W_1 \mid k W_1$, a stochastic regular grammar might assign probabilities of 0.5 for the productions:

$$S \ \to \ r W_1, \qquad S \ \to \ k W_1.$$
$$(0.5) \qquad\qquad\qquad (0.5)$$

The stochastic regular grammar can then admit exceptions without grossly degrading the recognition of more convincing motifs, by giving the exceptions low but non-zero probabilities. For example, the non-consensus N in the first position of the RNP-1 motif of SR55_DROME might be modelled with production rules like:

$$S \ \to \ r W_1, \qquad S \ \to \ k W_1, \qquad S \ \to \ n W_1.$$
$$(0.45) \qquad\qquad (0.45) \qquad\qquad (0.10)$$

If the production rules allow a probability for all possible symbols (any of the twenty amino acids) and the grammar is designed in such a way that it can generate sequences of any length, then the language specified by a stochastic grammar includes *all* possible strings, not just a subset of them. A stochastic grammar can therefore be used to specify a probability distribution over all of an infinite sequence space.

Stochastic context-sensitive or unrestricted grammars

We will not explore stochastic context-sensitive or stochastic unrestricted grammars in any detail, as we are unaware of any practical applications of these in computational biology. However, we should note here that production rules for the stochastic versions of context-sensitive and unrestricted grammars must be formulated more carefully than the description we have just given of regular grammars and context-free grammars. A nonterminal W may have different production rules in different contexts and the contexts are not necessarily unique. Consider for example the context-sensitive grammar $S \rightarrow aW$, $S \rightarrow bW$, $bW \rightarrow bb$, $W \rightarrow a$, $W \rightarrow b$ with probabilities $p_1, \ldots, p_5$. The language generated by this grammar is $\{aa, ab, ba, bb\}$ with probabilities $\{p_1 p_4, p_1 p_5, p_2 p_4, (p_2 p_3 + p_2 p_5)\}$. It can readily be shown algebraically that simply requiring that the productions for S and W sum to one, i.e. $p_1 + p_2 = 1$ and $p_3 + p_4 + p_5 = 1$, does not give a probability distribution over the language except for the special cases where $p_1 = 0$ or $p_3 = 0$. This problem can be solved by first rearranging the grammar so that the context of a nonterminal uniquely determines a set of possible production rules and no nonterminal ever has a choice between more than one form of left-hand side. Then, setting the probabilities for transforming a nonterminal in a given context to sum to one leads to a stochastic grammar. For example, the above grammar can be changed to $S \rightarrow aW$, $S \rightarrow bW$, $bW \rightarrow bb$, $bW \rightarrow ba$, $aW \rightarrow aa$, and $aW \rightarrow ab$ with probabilities $p_1, \ldots, p_6$, where now the conditions $p_1 + p_2 = 1$, $p_3 + p_4 = 1$, $p_5 + p_6 = 1$ give a proper stochastic grammar.

Hidden Markov models are stochastic regular grammars

Hidden Markov models are equivalent to stochastic regular grammars. The only difference is that the two kinds of model are traditionally represented differently. HMMs are normally described as Moore machines which emit symbols on a state, independent of transitions. Stochastic regular grammar productions correspond to Mealy machines which emit a terminal on transition to a new nonterminal (i.e. productions are of the form $W_1 \rightarrow aW_2$). As we saw previously in this chapter, Moore and Mealy machines are interchangeable. For instance, any HMM state which makes N transitions to new states that each emit one of M symbols can also be modelled by a set of NM stochastic regular grammar productions. Thus, the algorithms for aligning, scoring, and training stochastic regular grammars are the same algorithms we used for hidden Markov models (Chapter 3).

Exercises

9.8   G-U pairs are accepted in base paired RNA stems but occur with lower
        frequency than G-C and A-U Watson–Crick pairs. Make the RNA stem
        loop context-free grammar from page 244 into a stochastic context-free
        grammar, allowing G-U pairs in the stem with half the probability of a
        Watson-Crick pair.

9.9   Extend the push-down automaton algorithm from page 246 to gener-
        ate sequences from a stochastic context-free grammar according to their
        probability. (Note: This gives an efficient algorithm for sampling se-
        quences from any SCFG, including the more complex RNA SCFGs in
        the next chapter.)

9.10   Consider a simple HMM that models two kinds of base composition in
        DNA. The model has two states fully interconnected by four state transi-
        tions. State 1 emits CG-rich sequence with probabilities $(p_a, p_c, p_g, p_t) =$
        $\{0.1, 0.4, 0.4, 0.1\}$ and state 2 emits AT-rich sequence with probabilities
        $(p_a, p_c, p_g, p_t) = \{0.3, 0.2, 0.2, 0.3\}$. (a) Draw this HMM. (b) Set the tran-
        sition probabilities so that the expected length of a run of state 1s is 1000
        bases, and the expected length of a run of state 2s is 100 bases. (c) Give
        the same model in stochastic regular grammar form with terminals, non-
        terminals, and production rules with their associated probabilities.

9.6   Stochastic context-free grammars for sequence modelling

We can now write down stochastic context-free grammars as models of sequences.
However, writing down a stochastic grammar is only the first step in creating a
useful probabilistic modelling system for a sequence analysis problem. As with
HMMs, we must also have algorithms to address the following three problems:

 (i) Calculate an optimal alignment of a sequence to a parameterised stochas-
     tic grammar. (The alignment problem.)
 (ii) Calculate the probability of a sequence given a parameterised stochastic
     grammar. (The scoring problem.)
(iii) Given a set of example sequences/structures, estimate optimal probabil-
     ity parameters for an unparameterised stochastic grammar. (The training
     problem.)

In Chapter 3, we saw solutions to each problem for hidden Markov models
(and hence for stochastic regular grammars). The Viterbi algorithm solves the
alignment problem. The forward pass of the forward–backward algorithm solves
the scoring problem. The forward–backward algorithm is used in Baum–Welch

expectation maximisation to address the training problem. Analogous dynamic programming algorithms also exist for stochastic context-free grammars.

Normal forms for stochastic context-free grammars

CFGs can have an unlimited variety of symbol strings on the right-hand side of their rewriting rules. To express a general CFG parsing algorithm, it is very useful to adopt a restricted 'normal form' for the rewriting rules. One such normal form is *Chomsky normal form*. Chomsky normal form requires that all CFG production rules are of the form $W_v \rightarrow W_y W_z$ or $W_v \rightarrow a$. Any CFG can be recast into Chomsky normal form by expanding a non-conforming rewriting rule into a series of normal form productions from additional nonterminals. A parsing algorithm that applies to CFGs in Chomsky normal form is therefore generally applicable to any CFG. For example, the production rule $S \rightarrow aSa$ from our palindrome CFG on page (243) could be expanded to $S \rightarrow W_1 W_2$, $W_1 \rightarrow a$, $W_2 \rightarrow SW_1$ in Chomsky normal form.

Exercises

9.11 Convert the production rule $W \rightarrow aWbW$ to Chomsky normal form. If the probability of the original production is p, show the probabilities for the productions in your normal form version.

9.12 Convert the production rules $W_3 \rightarrow gaaa \mid gcaa$ from the RNA stem model grammar on page 244 to Chomsky normal form. Assuming that $W_3 \rightarrow gaaa$ has probability p_1 and $W_3 \rightarrow gcaa$ has probability $p_2 = 1 - p_1$, assign probabilities to your normal form productions. Show that your normal form version correctly assigns probabilities p_1 and p_2 for GAAA and GCAA loops, respectively.

The inside algorithm

The inside–outside algorithm for SCFGs in Chomsky normal form [Lari & Young 1990] is the natural counterpart of the forward–backward algorithm for HMMs (Chapter 3). The inside algorithm calculates the probability (score) of a sequence given an SCFG, just as the forward algorithm is used for HMMs. A best path variant of the inside algorithm, the Cocke–Younger–Kasami (CYK) algorithm, finds the maximum probability alignment of the SCFG to the sequence, just as the Viterbi algorithm is used for HMMs. Inside–outside is a recursive dynamic programming algorithm like forward–backward, but the computational complexity of inside–outside is substantially greater.

Let us define some notation. Consider a Chomsky normal form SCFG with M different nonterminals $W = W_1, \ldots, W_M$. The start nonterminal is W_1. Let v, y and z be indices for nonterminals W_v, W_y, and W_z. Production rules are

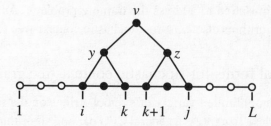

Figure 9.7 *Illustration of the iteration step of the inside calculation of $\alpha(i,j,v)$, the probability of the parse subtree rooted at state v for the subsequence from i to j. This is calculated recursively by summing parse subtrees for states y and z and smaller subsequences i to k and $k+1$ to j, for all y, z, and k, weighted by the transition probability $v \to yz$.*

of the form $W_v \to W_y W_z$ and $W_v \to a$ (where a is a possible symbol in the terminal alphabet). Let the probability parameters for these productions be called $t_v(y,z)$ and $e_v(a)$, respectively (for *transition* and *emission*). The sequence x has L symbols, indexed by $x_1,\ldots,x_L$. Let i, j and k be indices for symbols x_i, x_j and x_k in the sequence x.

The inside algorithm calculates the probability $\alpha(i,j,v)$ of a parse subtree rooted at nonterminal W_v for subsequence $x_i,\ldots,x_j$ for all i, j and v [Lari & Young 1990]. The calculation requires an $L \times L \times M$ three-dimensional dynamic programming matrix. The calculation starts with subsequences of length 1 ($i = j$), then does subsequences of length 2, and works outwards recursively on longer and longer subsequences until a probability of a parse tree has been determined for the complete parse tree rooted at the start nonterminal. A schematic illustration of the recursive nature of the algorithm is given in Figure 9.7. Formally, the inside algorithm is:

Algorithm: Inside

Initialisation: for $i = 1$ to L, $v = 1$ to M:
$$\alpha(i,i,v) = e_v(x_i).$$

Iteration: for $i = 1$ to $L-1$, $j = i+1$ to L, $v = 1$ to M:
$$\alpha(i,j,v) = \sum_{y=1}^{M} \sum_{z=1}^{M} \sum_{k=i}^{j-1} \alpha(i,k,y)\alpha(k+1,j,z)t_v(y,z).$$

Termination:
$$P(x|\theta) = \alpha(1,L,1). \qquad \triangleleft$$

The inside algorithm thus calculates the probability (score) of a sequence with an SCFG. The memory complexity of the inside algorithm is $O(L^2M)$, as is apparent from the three indices for α. The time complexity of the algorithm is $O(L^3M^3)$, as is apparent from the recursive loops over three sequence position indices i, j, k and three grammar nonterminal indices v, y and z.

The outside algorithm

The outside algorithm calculates a probability called $\beta(i,j,v)$ of a complete parse tree rooted at the start nonterminal for the complete sequence x, *excluding* all parse subtrees for the subsequence $x_i, \ldots, x_j$ rooted at nonterminal W_v for all i, j and v [Lari & Young 1990]. Like the inside algorithm, the calculation is done in an $L \times L \times M$ three-dimensional matrix. Calculating outside $\beta(i,j,v)$ probabilities requires the results $\alpha(i,j,v)$ from a previous inside calculation. The outside algorithm starts from the largest excluded subsequence $x_1, \ldots, x_L$ and recursively works its way inward. A schematic illustration of the outside algorithm is given in Figure 9.8. Formally, the algorithm is:

Algorithm: Outside

Initialisation:

$$\beta(1, L, 1) = 1;$$
$$\beta(1, L, v) = 0 \qquad \text{for } v = 2 \text{ to } M.$$

Iteration: for $i = 1$ to L, $j = L$ to i, $v = 1$ to M:

$$\beta(i,j,v) = \sum_{y,z} \sum_{k=1}^{i-1} \alpha(k, i-1, z)\beta(k, j, y)t_y(z, v)$$
$$+ \sum_{y,z} \sum_{k=j+1}^{L} \alpha(j+1, k, z)\beta(i, k, y)t_y(v, z).$$

Termination:

$$P(x|\theta) = \sum_{v=1}^{M} \beta(i, i, v)e_v(x_i) \qquad \text{for any } i. \qquad \triangleleft$$

Parameter re-estimation by expectation maximisation

The inside variables α and the outside variables β can be used to re-estimate the probability parameters of an SCFG by expectation maximisation much as we used the forward and backward variables in HMM training by EM [Lari & Young 1990]. The expected number of times that state v is used in a derivation is

$$c(v) = \frac{1}{P(x|\theta)} \sum_{i=1}^{L} \sum_{j=i}^{L} \alpha(i,j,v)\beta(i,j,v).$$

This can be further expanded to find the expected number of times that W_v is occupied and then production rule $W_v \to W_y W_z$ is used:

$$c(v \to yz) = \frac{1}{P(x|\theta)} \sum_{i=1}^{L-1} \sum_{j=i+1}^{L} \sum_{k=i}^{j-1} \beta(i,j,v)\alpha(i,k,y)\alpha(k+1,j,z)t_v(y,z).$$

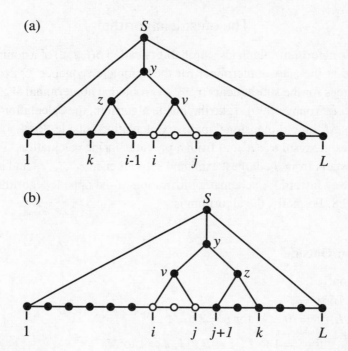

Figure 9.8 *Illustration of the recursive calculation of $\beta(i,j,v)$, the summed probabilities of all parse trees* excluding *subtrees rooted at nonterminal v that generate the subsequence i,j (open circles). Diagram (a) corresponds to the first part of the outside iteration equation for the contributions to $\beta(i,j,v)$ of combining the outside value for nonterminal y and subsequence $1,\ldots,k-1,j+1,\ldots,L$, the inside value for nonterminal z filling in the subsequence $k,\ldots,i-1$, and the transition probability for $y \rightarrow zv$. Diagram (b) corresponds to the second part of the iteration equation, which combines the outside probability for nonterminal y on the excluded subsequence $i,\ldots,k$, the inside probability for state z filling in the subsequence $j+1,\ldots,k$, and the transition probability for $y \rightarrow vz$.*

It then follows that the EM re-estimation equation for the probabilities of the production rules $W_v \rightarrow W_y W_z$ is

$$
\begin{aligned}
\hat{t}_v(y,z) &= \frac{c(v \rightarrow yz)}{c(v)} \\
&= \frac{\sum_{i=1}^{L-1} \sum_{j=i+1}^{L} \sum_{k=i}^{j-1} \beta(i,j,v)\alpha(i,k,y)\alpha(k+1,j,z)t_v(y,z)}{\sum_{i=1}^{L} \sum_{j=i}^{L} \alpha(i,j,v)\beta(i,j,v)}.
\end{aligned}
$$

Similar equations hold for the other production rules $W_v \rightarrow a$, giving

$$
\hat{e}_v(a) = \frac{c(v \rightarrow a)}{c(v)} = \frac{\sum_{i|x_i=a} \beta(i,i,v)e_v(a)}{\sum_{i=1}^{L} \sum_{j=i}^{L} \alpha(i,j,v)\beta(i,j,v)}.
$$

Extension of these re-estimation equations from a single observed sequence x

to the case of multiple independent observed sequences is straightforward. Expected counts are simply summed over all sequences.

The CYK alignment algorithm

The remaining problem is to find an optimal parse tree (alignment) for the sequence. This is solved with the Cocke–Younger–Kasami (CYK) algorithm, a variant of the inside algorithm with max operations replacing the sums.[5] It calculates a variable $\gamma(i,j,v)$ which ultimately leads to $\log P(x,\hat{\pi}|\theta)$, where $\hat{\pi}$ is the most probable parse tree. We also keep a traceback 'variable' $\tau(i,j,v)$ which is a triplet of numbers (y,z,k) that we need for tracing back through the three-dimensional dynamic programming matrix and recovering the optimal alignment. Formally, the matrix fill stage of the algorithm is:

Algorithm: CYK

Initialisation: for $i = 1$ to L, $v = 1$ to M:

$$\gamma(i,i,v) = \log e_v(x_i);$$
$$\tau(i,i,v) = (0,0,0).$$

Iteration: for $i = 1$ to $L-1$, $j = i+1$ to L, $v = 1$ to M:

$$\gamma(i,j,v) = \max_{y,z} \max_{k=i\dots j-1}$$
$$\{\gamma(i,k,y) + \gamma(k+1,j,z) + \log t_v(y,z)\};$$
$$\tau(i,j,v) = \text{argmax}_{(y,z,k),k=i\dots j-1}$$
$$\{\gamma(i,k,y) + \gamma(k+1,j,z) + \log t_v(y,z)\}.$$

Termination:

$$\log P(x,\hat{\pi}|\theta) = \gamma(1,L,1).$$ ◁

This is followed by a traceback to recover the best alignment which is done by pushing and popping triplets (i,j,v) on and off a push-down stack:

Algorithm: CYK traceback

Initialisation:
     Push $(1,L,1)$ on the stack.
Iteration:
     Pop (i,j,v).
     $(y,z,k) = \tau(i,j,v)$.
     If $\tau(i,j,v) = (0,0,0)$ (implying $i = j$), attach x_i as the child of v;
     else:

---

[5] As originally described by Cocke, Younger and Kasami independently, the CYK algorithm is an exact match algorithm for nonstochastic CFGs. Our use of the name 'CYK algorithm' for the SCFG parsing algorithm is thus a bit imprecise, but we are not aware of any other name for the SCFG form of the algorithm in the literature.

Attach y, z to parse tree as children of v.
Push $(k+1, j, z)$.
Push (i, k, y).   ◁

Just as the Viterbi alignment algorithm can be used as an approximation to the EM training algorithm for HMMs, CYK can be used as an approximation of inside–outside training. Instead of calculating expected numbers of counts probabilistically using inside–outside, we calculate optimal CYK alignments for the training sequences and then count the transitions and emissions that occur in those alignments.

Summary of SCFG algorithms

Using inside–outside and CYK algorithms, SCFGs can be used as a full probabilistic modelling system just as we have used HMMs. The following table summarises the properties of SCFG algorithms compared to their HMM counterparts:

Goal	HMM algorithm	SCFG algorithm
optimal alignment	Viterbi	CYK
$P(x\|\theta)$	forward	inside
EM parameter estimation	forward–backward	inside–outside
memory complexity:	$O(LM)$	$O(L^2M)$
time complexity:	$O(LM^2)$	$O(L^3M^3)$

The computational complexity of SCFG algorithms appears intimidating, but much of it results from the generality of the algorithm. More restricted SCFGs have faster algorithms. RNA SCFG algorithms in the next chapter are $O(L^3M)$ in time. This is still bad, but much better than $O(L^3M^3)$.

It is sometimes said that the inside–outside algorithm can only be applied to SCFGs in Chomsky normal form, implying that SCFGs must first be laboriously converted to Chomsky normal form before any parsing can be done. This is true only for a pedantic definition of the inside–outside algorithm. The inside–outside algorithm is given for Chomsky normal form SCFGs solely for purposes of generality and notational convenience (recall that any SCFG, however complicated its productions may be, can be rewritten to Chomsky normal form). Essentially identical algorithms follow for other SCFG 'normal forms' that restrict the right-hand side of productions. We will see natural alternatives to Chomsky normal form for RNA modelling in the next chapter.

9.7 Further reading

Our description of formal language theory in this chapter is not rigorous. Readers interested in more detail should consult texts such as Harrison's [1978] *Introduction to Formal Language Theory* or Hopcroft & Ullman's [1979] *Introduction to Automata Theory, Languages, and Computation*. Both texts give substantial detail about nonstochastic context-free grammars, push-down automata, and fast CFG parsing algorithms, since these are important in the design of computer languages and efficient language compilers. Gene Myers [1995] has also written on the topic of context-free grammar parsing algorithms.

Our description of SCFG algorithms is based on the work of Lari & Young [1990;1991] in the field of speech recognition.

Transformational grammar theory has been applied to formalised descriptions of biological problems other than sequence analysis with varying degrees of usefulness. These problems include modelling of metabolic pathways [Collado-Vides 1989; 1991] and of developmental pathways [Lindenmayer 1968]. Additionally, there are other 'linguistic' approaches in computational sequence analysis which are based on k-tuple ('word') frequencies rather than transformational grammar theory [Brendel, Beckmann & Trifonov 1986; Pesole, Attimonelli & Saccone 1994; Pietrokovski, Hirshon & Trifonov 1990].

10

RNA structure analysis

Many interesting RNAs conserve a secondary structure of base-pairing interactions more than they conserve their sequence. This makes RNA sequence analysis more complicated and difficult than protein or DNA sequence analysis. RNA secondary structure problems are a natural application for probabilistic models based on the stochastic context-free grammars introduced in Chapter 9. In this chapter, we will examine two RNA analysis problems of biological interest.

The first problem is RNA secondary structure prediction for a single sequence. We will outline two well-known dynamic programming algorithms for RNA secondary structure prediction, the Nussinov and the Zuker algorithms. Then we will use RNA secondary structure prediction as an introductory example for the use of SCFGs for RNA analysis, by developing a small SCFG that implements a probabilistic version of the Nussinov algorithm.

The second is a related set of problems, having to do with the analysis of multiple alignments of families of related RNAs. Like Chapter 5, where profile HMMs were used for both multiple alignment and for database searching, we develop RNA structure profiles called 'covariance models' (CMs) for dealing with RNA multiple alignments with secondary structure constraints included. Covariance models are used for both RNA multiple alignment and database searches. Consensus structure prediction from RNA multiple alignments, a process called comparative RNA sequence analysis, is also somewhat automated by RNA covariance model training algorithms.

As you read this chapter, bear in mind that SCFG-based RNA analysis methods are not widely known or used. All of the SCFG methods we describe are in their infancy and have considerable problems with computational complexity. Improved SCFG methods for RNA analysis might be around the corner. Here, we try to give the fundamentals of SCFG-based probabilistic methods for RNA analysis without getting mired in details that may soon change. At the least, RNA SCFGs provide us with a pedagogical counterpoint to profile HMMs. We will see how much of the same probabilistic machinery developed for HMMs also applies to a different and more complex class of model.

10.1 RNA

To many people, RNA is merely the passive intermediary messenger between DNA genes and the protein translation machinery. Messenger RNA is often described as a linear, unstructured sequence, uninteresting but for the protein amino acid sequence that it encodes. However, many non-coding RNAs exist which adopt sophisticated three-dimensional structures, and some even catalyse biochemical reactions. Since the startling discovery of catalytic RNAs in the early 1980s [Cech & Bass 1986], a number of interesting new structural and catalytic RNAs have been discovered. More recently, novel RNAs have been invented using *in vitro* evolution technologies to screen repertoires of random RNA sequences for new catalysts and new specific ligands [Gold *et al.* 1995].

The discovery of RNA catalysis revived a notion now widely known as the 'RNA world' hypothesis for the origin of life [Gilbert 1986; Gesteland & Atkins 1993]. The RNA world hypothesis posits a primordial world before DNA genomes and protein catalysts when RNA genomes were replicated by RNA catalysts. It is sometimes argued that many modern structural and catalytic RNAs are 'molecular fossils' that have been handed down in evolutionary time from an extinct RNA world.

Structural and catalytic RNAs are also important in the molecular biology of modern organisms. The peptidyl transferase activity of ribosomes is thought to be catalysed by ribosomal RNA [Noller, Hoffarth & Zimniak 1992]. RNA splicing (removal of introns from eukaryotic pre-mRNA transcripts) is catalysed by a complex RNA/protein machine (the spliceosome) which contains five major species of small nuclear RNAs [Baserga & Steitz 1993]. The signal recognition particle that is involved in translocating proteins across the plasma membrane is an RNA/protein complex [Larsen & Zwieb 1993]. Proper ribosomal RNA processing and modification require a host of small nucleolar RNAs [Maxwell & Fournier 1995]. In messenger RNA transcripts, RNA structure (particularly in 5′ and 3′ untranslated regions) is used in a variety of ways to effect post-transcriptional genetic regulation. Known post-transcriptional regulatory mechanisms include alternative mRNA splicing control [McKeown 1992], modulation of translational efficiency [Melefors & Hentze 1993] and regulation of mRNA stability [Peltz & Jacobson 1992].

Terminology of RNA secondary structure

RNA is a polymer of four different nucleotide subunits. The four nucleotides are abbreviated A, C, G and U, for adenine, cytosine, guanine and uracil. In DNA, thymine (T) replaces uracil.

G-C and A-U form hydrogen bonded base pairs and are said to be complementary. G-C pairs form three hydrogen bonds and tend to be more stable than

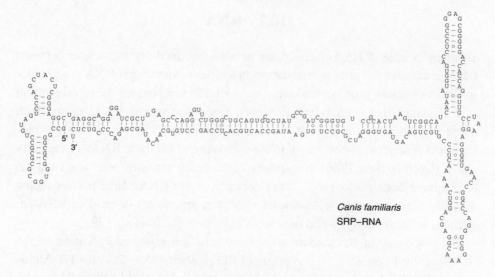

Figure 10.1 *The RNA secondary structure of signal recognition particle (SRP) RNA from the dog,* Canis familiaris.

A–U pairs, which form only two. Base pairs are approximately coplanar and are almost always *stacked* onto other base pairs in an RNA structure. Contiguous stacked base pairs are called *stems*. In three-dimensional space, RNA stems generally form a regular (A-form) double helix. Unlike DNA, RNA is typically produced as a single stranded molecule which then folds intramolecularly to form a number of short base-paired stems. This base-paired structure is called the *secondary structure* of the RNA. RNA secondary structures are typically represented by two-dimensional pictures like the one shown in Figure 10.1.

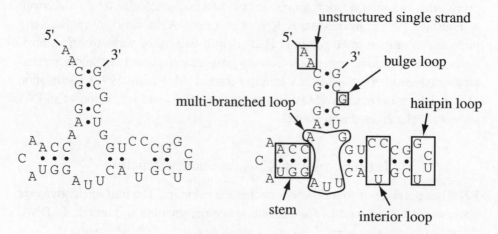

Figure 10.2 *The fundamental elements of RNA secondary structure are indicated for a hypothetical example.*

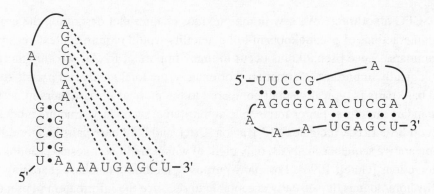

Figure 10.3 *Base pairs between a loop and positions outside the enclosing stem are called a pseudoknot (left). Another representation of the same pseudoknot is shown on the right. In three-dimensional space, the two stems can stack coaxially and mimic a contiguous A-form helix. This particular example is an artificially selected RNA inhibitor of the human immunodeficiency virus reverse transcriptase [Tuerk, MacDougal & Gold 1992].*

The elements of an RNA secondary structure are named as shown in Figure 10.2. Single stranded subsequences bounded by base pairs are called *loops*. A loop at the end of a stem is called a *hairpin loop*. Simple substructures consisting of a simple stem and loop are called *stem loops* or *hairpins* (because the structure resembles a hairpin when drawn). Single stranded bases occurring within a stem are called a *bulge* or *bulge loop* if the single stranded bases are on only one side of the stem, or an *interior loop* if there are single stranded bases interrupting both sides of a stem. Finally, there are *multi-branched loops* from which three or more stems radiate.

In addition to canonical A-U and G-C base pairs, non-canonical pairs also occur in RNA secondary structure. The most common non-canonical pair is the G-U pair, which is almost as thermodynamically favourable as Watson–Crick pairs. Other pairs form as well. Non-canonical pairs distort regular A-form RNA helices. These distortions seem to be a favoured target of proteins specialised for recognising RNA.

Base pairs almost always occur in a nested fashion in RNA secondary structure. Informally, this means that if we draw arcs over an RNA sequence connecting the base pairs, none of the arcs need to cross each other. More formally, a base pair between positions i and j and a base pair between positions i' and j' are nested if and only if $i < i' < j' < j$ or $i' < i < j < j'$. (Recall that this is the condition met by the constraints on palindrome languages in Chapter 9 – this is why context-free grammars apply to RNA secondary structure.) When non-nested base pairs occur, they are called *pseudoknots*. An example of a pseudoknot is given in Figure 10.3.

None of the dynamic programming algorithms that we describe can deal with pseudoknots, including the Zuker and Nussinov RNA folding algorithms as well

as SCFG algorithms. We saw in the previous chapter that describing the crossing interactions of pseudoknots in full generality would require context-sensitive
grammars. Since pseudoknots occur in many important RNAs, we are ignoring
biologically important information. Fortunately, the total *number* of pseudoknotted base pairs is typically small compared to the number of base pairs in nested
secondary structure. For example, one authoritative secondary structure model of
E. coli SSU rRNA indicates 447 Watson–Crick and G-U base pairs supported by
comparative sequence analysis, only eight of which are in non-nested pseudoknot
interactions [Gutell 1993]. For many purposes, including database searching for
RNA homologues, it is usually acceptable to sacrifice the information in pseudoknots in return for efficient dynamic programming algorithms. For other purposes
such as three-dimensional structure prediction, pseudoknots must be considered
and the same sacrifice cannot be made.

RNA sequence evolution is constrained by structure

It is relatively common to find examples of homologous RNAs that have a common secondary structure without sharing significant sequence similarity. Drastic
changes in sequence can often be tolerated as long as compensatory mutations
maintain base-pairing complementarity. It would be advantageous to be able to
search for conserved secondary structure in addition to conserved sequence when
searching databases for homologous RNAs.

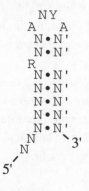

Figure 10.4 *The consensus binding site for R17 phage coat protein.* N, Y
and R *are standard 'degenerate' symbols for multiple possible nucleotides.*
N *indicates {*A,C,G,U*}*, Y *indicates {*C,U*} and* R *indicates {*A,G*}*. N'
indicates a complementary base pairing to N.

The structure shown in Figure 10.4 is the consensus RNA binding site for the
coat protein of the bacterial RNA virus R17 [Witherell, Gott & Uhlenbeck 1991].
R17 coat protein binds this site and represses translation of its replicase as part of
the normal timing of an R17 lytic cycle. Only four primary sequence positions are
specified in the consensus, and two of them are degenerate. If we were interested

in searching a nucleotide sequence for occurrences of consensus R17 coat protein binding sites, it would be useless to use a standard sequence alignment method.

How useless? It is instructive to extract some rules of thumb from Shannon information theory. In information theoretic terms, a consensus base pair conveys as much information as a conserved base. The information (relative entropy) contributed by a completely conserved base ($p_x = 1$) is $\sum_x p_x \log_2 \frac{p_x}{f_x} = 2$ bits (assuming equiprobable initial expected base frequencies, $f_x = \frac{1}{4}$). Similarly, the degenerate R and Y in Figure 10.4 each convey 1 bit of information, and the N is worth 0. The information contributed by a Watson–Crick base pair of any sequence is also 2 bits, since $\sum_x \sum_y p_{xy} \log_2 \frac{p_{xy}}{f_{xy}} = 2$ (again assuming that our initial expectation is equiprobable, $f_{xy} = \frac{1}{16}$, and that the observed Watson–Crick pairs occur equiprobably, $p_{AU} = p_{CG} = p_{GC} = p_{UA} = \frac{1}{4}$).

Considering only primary sequence conservation, the R17 consensus therefore conveys 6 bits of information. We expect to find a match to it by chance every 64 (2^6) nucleotides. Adding the seven base pairs to the consensus description adds 14 bits of information, bringing the information content up to 20 bits, and reducing the chance of finding a spurious match to once in every million (2^{20}) nucleotides. If we search for NNN NNN NRN NAN YAN NNN NNN in the genome of the related bacteriophage MS2 (GENBANK MS2CG; the R17 genome is not in the database), we find 38 matches in the 3569 bp genome, 37 of which are spurious. If we repeat the search while requiring the seven base pairs, we find just a single match at the authentic coat protein binding site.

The above search was done with an RNA pattern-matching program similar to the program RNAMOT [Gautheret, Major & Cedergren 1990]. The program searches for deterministic (non-stochastic) motifs but with secondary structure constraints as extra terms. It works fine for small, well-defined patterns but is somewhat insensitive and problematic for finding matches to less well conserved structures. Currently, the prevailing wisdom for more sensitive, more statistically based RNA database searches is that one must write a carefully customised program for each RNA structure of interest [Dandekar & Hentze 1995]. Several such programs exist for finding transfer RNA genes [Fichant & Burks 1991; Pavesi *et al.* 1994; Lowe & Eddy 1997], and one exists for finding catalytic group I introns [Lisacek, Diaz & Michel 1994]. However, as the number of different interesting RNAs grows, this is an increasingly unsatisfactory state of affairs.

Inferring structure by comparative sequence analysis

The same base-pair induced sequence constraints that make database searching hard make consensus RNA secondary structure prediction relatively easy – relative to protein structure prediction, at least. In a structurally correct multiple alignment of RNAs, conserved base pairs are often revealed by the presence of frequent correlated compensatory mutations. Despite being a theoretical structure prediction method, RNA secondary structure prediction by this process of *comparative sequence analysis* is considered to be the most reliable means of determining an RNA secondary structure, short of solving a three-dimensional crystal or NMR structure. The accepted consensus structures of most well-studied RNAs have been derived by comparative analysis [Woese & Pace 1993] (Figure 10.5).

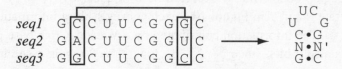

Figure 10.5 *Comparative sequence analysis recognises that the two boxed positions in this example of a multiple alignment (left) are covarying to maintain Watson–Crick complementarity. This covariation implies a base pair, leading to a consensus secondary structure prediction (right).*

Comparative analysis is a painstaking art. Inferring the correct structure by comparative analysis requires knowing a structurally correct multiple alignment, but inferring a structurally correct multiple alignment requires knowing the correct structure. A structure is 'solved' by an iterative refinement process of guessing the structure based on the current best guess of the multiple alignment, then realigning based on the new guess at the structure. The sequences to be compared must be sufficiently similar that they can be initially aligned by primary sequence identity alone to start the process, but they must be sufficiently dissimilar that a number of covarying substitutions can be detected.

A quantitative measure of pairwise sequence covariation comes from information theory [Chiu & Kolodziejczak 1991; Gutell *et al.* 1992]. The mutual information M_{ij} between two aligned columns i and j is given by

$$M_{ij} = \sum_{x_i, x_j} f_{x_i x_j} \log_2 \frac{f_{x_i x_j}}{f_{x_i} f_{x_j}}. \tag{10.1}$$

f_{x_i} is the frequency of one of the four bases (A, C, G, U) observed in column i. $f_{x_i x_j}$ is the joint (pairwise) frequency of one of the sixteen possible base pairs observed in columns i and j. M_{ij} measures how much the joint frequency distribution deviates from the distribution that is expected if the two columns vary independently. For the four-letter RNA alphabet, M_{ij} varies between 0 and 2 bits. M_{ij} is maximal if i and j individually appear completely random ($f_i = f_j = 0.25$), but i and j are perfectly correlated, for instance in a Watson–Crick base pair.

Intuitively, M_{ij} tells us how much information we get about the identity of the residue in one position if we are told the identity of the residue in the other position. In the case of a base pair with no sequence constraints, we get 2 bits of information: for instance, if we are told that i is a G, our uncertainty about j collapses from four possibilities to just one (C) so we gain 2 bits of information. If i and j are uncorrelated, the mutual information is zero. If either i or j are highly conserved positions, we also get little or no mutual information: if a position does not vary, we do not learn anything more about it by knowing the identity of its partner.

Figure 10.6 shows a contour plot of M_{ij} values calculated from a multiple alignment of 1415 tRNA sequences. The four base-paired stems of the cloverleaf structure are readily apparent. The D and TψCG stems, which are relatively highly conserved in primary sequence, are somewhat less apparent than the anticodon and acceptor stems which are extremely variable in primary sequence.

Exercise

10.1   The mutual information calculation in (10.1) requires counting frequencies of all sixteen different base pairs. This has the advantage that it makes no assumptions about Watson–Crick base pairing, so mutual information can be detected between covarying non-canonical pairs like A-A and G-G pairs. On the other hand, the calculation requires a large number of aligned sequences to obtain reasonable frequencies for sixteen possibilities. Write down an alternative information theoretic measure of base-pairing correlation that considers only two classes of i, j identities instead of all sixteen: Watson–Crick and G-U pairs grouped in one class, and all other pairs grouped in the other. Compare the properties of this calculation to the M_{ij} calculation both for small numbers of sequences and in the limit of infinite data.

10.2   RNA secondary structure prediction

Suppose we wish to predict the secondary structure of a single RNA. Many plausible secondary structures can be drawn for a sequence. The number increases exponentially with sequence length. An RNA only 200 bases long has over 10^{50} possible base-paired structures. We must distinguish the biologically correct structure from all the incorrect structures. We need both a function that assigns the correct structure the highest score, and an algorithm for evaluating the scores of all possible structures.

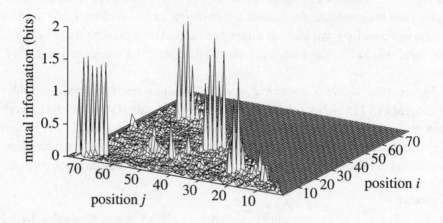

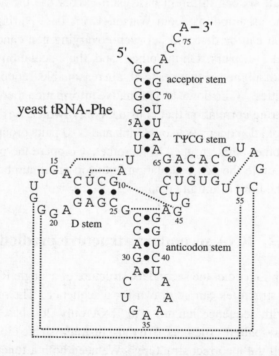

Figure 10.6 *A mutual information plot of a tRNA alignment (top) shows four strong diagonals of covarying positions, corresponding to the four stems of the tRNA cloverleaf structure (bottom; the secondary structure of yeast phenylalanine tRNA is shown). Dashed lines indicate some of the additional tertiary contacts observed in the yeast tRNA-Phe crystal structure. Some of these tertiary contacts produce correlated pairs which can be seen weakly in the mutual information plot.*

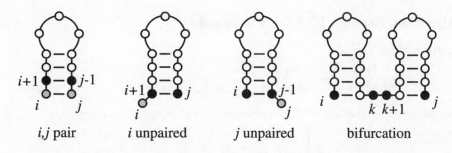

Figure 10.7 *The Nussinov algorithm looks at four ways in which the best RNA structure for a subsequence i, j can be made by adding i and/or j onto already calculated optimal structures for smaller subsequences. Pseudoknots are not considered.*

Base pair maximisation and the Nussinov folding algorithm

One approach might be to find the structure with the most base pairs. Nussinov introduced an efficient dynamic programming algorithm for this problem [Nussinov *et al.* 1978]. Although this criterion is too simplistic to give accurate structure predictions, the example is instructive because the mechanics of the Nussinov algorithm are the same as those of the more sophisticated energy minimisation folding algorithms and of probabilistic SCFG-based algorithms.

The Nussinov calculation is recursive. It calculates the best structure for small subsequences, and works its way outwards to larger and larger subsequences. The key idea of the recursive calculation is that there are only four possible ways of getting the best structure for i, j from the best structures of the smaller subsequences (Figure 10.7):

(1) add unpaired position i onto best structure for subsequence $i + 1, j$;
(2) add unpaired position j onto best structure for subsequence $i, j - 1$;
(3) add i, j pair onto best structure found for subsequence $i + 1, j - 1$;
(4) combine two optimal substructures i, k and $k + 1, j$.

More formally, the Nussinov RNA folding algorithm is as follows. We are given a sequence x of length L with symbols $x_1, \ldots, x_L$. Let $\delta(i, j) = 1$ if x_i and x_j are a complementary base pair; else $\delta(i, j) = 0$. We will recursively calculate scores $\gamma(i, j)$ which are the maximal number of base pairs that can be formed for subsequence $x_i, \ldots, x_j$.

Algorithm: Nussinov RNA folding, fill stage

Initialisation:

$$\gamma(i, i-1) = 0 \qquad \text{for } i = 2 \text{ to } L;$$
$$\gamma(i, i) = 0 \qquad \text{for } i = 1 \text{ to } L.$$

Recursion: starting with all subsequences of length 2, to length L:

$$\gamma(i,j) = \max \begin{cases} \gamma(i+1,j), \\ \gamma(i,j-1), \\ \gamma(i+1,j-1) + \delta(i,j), \\ \max_{i<k<j}\left[\gamma(i,k) + \gamma(k+1,j)\right]. \end{cases}$$
◁

Figure 10.8 shows an example of a Nussinov matrix fill in operation.

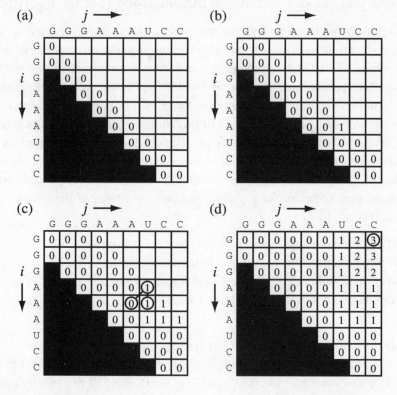

Figure 10.8 *The matrix fill stage of the Nussinov folding algorithm is shown for an example sequence* GGG AAA UCC. *(a) The initialised half-diagonal matrix. (b) The matrix after scores for subsequences of length two have been calculated. (c) An example of two different optimal substructures for the same subsequence. For the subsequence* AAAU, *either the* A *at i and the* U *at j can be paired (diagonal path) or i can be added to a substructure that already pairs the* A *at i + 1 to the* U *at j (vertical path). (d) The final matrix. The value in the upper right indicates that the maximally paired structure has three base pairs.*

The value of $\gamma(1, L)$ is the number of base pairs in the maximally base-paired structure. There are often a number of alternative structures with the same number of base pairs. To find one of these maximally base-paired structures, we trace back through the values we calculated in the dynamic programming matrix, beginning from $\gamma(1, L)$. In pseudocode, the traceback algorithm is:

Algorithm: Nussinov RNA folding, traceback stage

Initialisation: Push $(1, L)$ onto stack.

Recursion: Repeat until stack is empty:

- pop (i, j).
- if $i >= j$ continue;

  else if $\gamma(i + 1, j) = \gamma(i, j)$ push $(i + 1, j)$;

  else if $\gamma(i, j - 1) = \gamma(i, j)$ push $(i, j - 1)$;

  else if $\gamma(i + 1, j - 1) + \delta_{i,j} = \gamma(i, j)$:

  - record i, j base pair.
  - push $(i + 1, j - 1)$.

  else for $k = i + 1$ to $j - 1$: if $\gamma(i, k) + \gamma(k + 1, j) = \gamma(i, j)$:

  - push $(k + 1, j)$.
  - push (i, k).
  - break. ◁

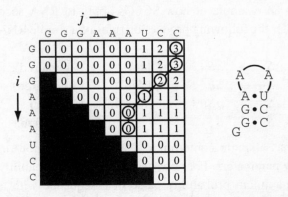

Figure 10.9 *The traceback stage of the Nussinov folding algorithm is shown for the filled matrix from Figure 10.8. An optimal traceback path is indicated with circles. The optimal structure corresponding to this path is shown at right.*

The traceback is linear in time and memory. The fill step is the limiting step as it is $O(L^2)$ in memory and $O(L^3)$ in time. An example traceback is shown in Figure 10.9. The traceback in Figure 10.9 is unbranched, so the need for the pushdown stack in the traceback algorithm is not apparent. The pushdown stack becomes important when bifurcated structures are traced back. The stack remembers one side of the the bifurcation while the other side is traced back, reminiscent of the push-down automata in Chapter 9.

Exercises

10.2   Find two more optimal structures with three base pairs besides the one in Figure 10.9. Modify the traceback algorithm so it finds one of your structures instead of the one obtained in Figure 10.9.

10.3   As we have given it, the Nussinov algorithm can produce nonsensical 'base pairs' between adjacent complementary residues (for example, one of the possible structures in the preceding exercise contains such an AU base pair). Modify the Nussinov folding algorithm so that hairpin loops must have a minimum length of h. Give the new recursion equations for the fill and traceback.

10.4   Show that the Nussinov folding algorithm can be trivially extended to find a maximally *scoring* structure where a base pair between residues a and b gets a score $s(a,b)$. (For instance, we might set $s(G,C) = 3$ and $s(A,U) = 2$ to better reflect the increased thermodynamic stability of GC pairs.)

An SCFG version of the Nussinov algorithm

The Nussinov algorithm is fundamentally similar to the SCFG algorithms in Chapter 9. As an example of how SCFGs apply to RNA secondary structure analysis, consider the following production rules of a simple RNA folding SCFG:

$$
\begin{aligned}
S &\rightarrow aS \mid cS \mid gS \mid uS &&(i \text{ unpaired}), \\
S &\rightarrow Sa \mid Sc \mid Sg \mid Su &&(j \text{ unpaired}), \\
S &\rightarrow aSu \mid cSg \mid gSc \mid uSa &&(i, j \text{ pair}), \\
S &\rightarrow SS &&\text{bifurcation.}
\end{aligned}
\tag{10.2}
$$

The SCFG has a single nonterminal S and 13 production rules with associated probability parameters. For now, assume that the probability parameters are known. The maximum probability parse of a sequence with this SCFG is an assignment of sequence positions to productions. Because the productions correspond to secondary structure elements (base pairs and single-stranded bases), the maximum probability parse is equivalent to the maximum probability secondary structure. If base pair productions have relatively high probability, the SCFG will favour parses which tend to maximise the number of base pairs in the structure.

Although the production rules for the SCFG are not in Chomsky normal form, a CYK parsing algorithm is readily written that finds the maximum probability secondary structure. Alternatively, we could convert the SCFG to Chomsky normal form and apply the algorithms in Chapter 9. Although the Chomsky normal form approach is attractive in its generality, specific algorithms for specific

SCFGs are typically more efficient. The adapted CYK algorithm is as follows. Let the probability parameters of the SCFG productions be denoted by $p(aS)$, $p(aSu)$, etc.

Algorithm: CYK for Nussinov-style RNA SCFG

Initialisation:

$$\gamma(i, i-1) = -\infty \qquad \text{for } i = 2 \text{ to } L;$$

$$\gamma(i, i) = \max \begin{cases} \log p(x_i S) \\ \log p(S x_i) \end{cases} \qquad \text{for } i = 1 \text{ to } L.$$

Recursion: for $i = 1$ to $L - 1$, $j = i + 1$ to L:

$$\gamma(i, j) = \max \begin{cases} \gamma(i+1, j) + \log p(x_i S); \\ \gamma(i, j-1) + \log p(S x_j); \\ \gamma(i+1, j-1) + \log p(x_i S x_j); \\ \max_{i < k < j} \gamma(i, k) + \gamma(k+1, j) + \log p(SS). \end{cases} \qquad \triangleleft$$

When this is done, $\gamma(1, L)$ is the log likelihood $\log P(x, \hat{\pi} | \theta)$ of the optimal structure $\hat{\pi}$ given the SCFG model θ. The traceback to find the structure corresponding to that best score is either performed analogously to the traceback in the Nussinov algorithm, or by keeping additional traceback pointers in the fill stage analogous to the CYK algorithm description in Chapter 9.

The principal difference between this and the original Nussinov algorithm is that the SCFG description is a probabilistic model. We gain access to several well-principled options for optimising the parameters of the model. We can set the SCFG's parameters by subjective estimation of the relevant probabilities, or by estimating parameters by counting state transitions in known RNA structures and converting the counts to probabilities. We can even learn probabilities from example RNAs of *unknown* structure using expectation maximisation (EM) and inside–outside training to iteratively infer both the structures and the parameters (i.e. the structures are the hidden data in the EM algorithm). Once we have written down the SCFG as a full probabilistic model of the RNA folding problem, we can 'turn the crank', applying all the probabilistic machinery we have learned in previous chapters almost by rote.

Like the Nussinov algorithm, this small SCFG is a good starting example but it is too simple to be an accurate RNA folder. It does not consider important structural features like preferences for certain loop lengths nor preferences for certain nearest neighbours in the structure caused by stacking interactions between neighbouring base pairs in a stem.

Exercises

10.5   Write down a traceback algorithm for determining the best RNA sec-
        ondary structure after the above algorithm has completed.

10.6   Devise an SCFG which uses different nonterminals to model bulge loops,
        hairpin loops, multifurcation loops and single strands.

Energy minimisation and the Zuker folding algorithm

RNA folding is dictated by biophysics rather than by counting and maximising
the number of base pairs. The most sophisticated secondary structure prediction
method for single RNAs is the Zuker algorithm, an energy minimisation algorithm which assumes that the correct structure is the one with the lowest equilibrium free energy (ΔG) [Zuker & Stiegler 1981; Zuker 1989a].

The ΔG of an RNA secondary structure is approximated as the sum of individual contributions from loops, base pairs and other secondary structure elements.
An important difference from the simpler Nussinov calculation is that the energies of stems are calculated by adding *stacking* contributions for the interface between neighbouring base pairs instead of individual contributions for each pair.
In other words, the energy of a stem of n base pairs is the sum of $n - 1$ base
stacking terms instead of n base pair terms. This produces a better fit to experimentally observed ΔG values for RNA structures but it complicates the dynamic
programming algorithm. Tables of ΔG parameters for RNA structure prediction
have been fitted to the results of experimental thermodynamic studies of small
model RNAs [Freier *et al.* 1986; Turner *et al.* 1987]. They include parameters for stacking, hairpin loop lengths, bulge loop lengths, interior loop lengths,
multi-branch loop lengths, single dangling nucleotides and terminal mismatches
on stems.

An example of the prediction of the ΔG of an RNA structure is given in Figure 10.10. Single base bulges are assumed not to disrupt stacking in the stem, so
a stacking term is included in the example in the figure. Longer bulges, which are
assumed to disrupt stacking, get no added stacking term. The hairpin loop energy
is the sum of two terms: a loop destabilisation energy dependent only on the loop
length, and a terminal mismatch energy dependent on the closing base pair and
the first and last bases of the stem. The energies used in Figure 10.10 are from
the older 'Freier rules' [Freier *et al.* 1986] at 37°C.[1]

The minimum energy structure can be calculated recursively by a dynamic programming algorithm (assuming no pseudoknots), very similar to how the maximum base-paired structure was calculated above. The principal difference is that
because of the stacking parameters, two matrices (called V and W) are kept in-

[1] Currently the most up-to-date parameters are available on the Web from
  http://www.ibc.wustl.edu/~zuker/rna/energy/.

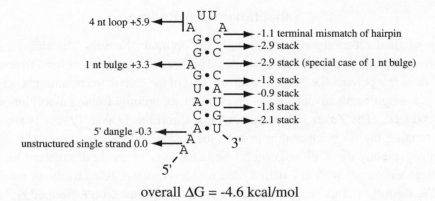

overall ΔG = -4.6 kcal/mol

Figure 10.10 *An example ΔG calculation for an RNA stem loop (the wild type R17 coat protein binding site).*

stead of one. $W(i, j)$ is the energy of the best structure on i, j. $V(i, j)$ is the energy of the best structure on i, j given that i, j are paired. The algorithm can then keep track of stacking interactions by adding new base pairs only onto the V matrix. Conceptually this two-state calculation is very similar to the use of extra insert states in pairwise dynamic programming alignment with affine gap costs (Chapter 2) to keep track of insert extensions. For a complete description of the Zuker algorithm, see Zuker & Stiegler [1981].

We could write down a SCFG that followed similar rules. The simplest stacking production rule would be, for instance, $cVg \to cgVcg$ for producing a GC pair in a stem after (stacked on) a CG, using V as a base pair generating nonterminal (as in the Zuker V matrix). With the CG terminals on the left as context for the production of the GC, this is technically a context-sensitive production, so we can't use such rules as the basis for a SCFG. However, we can convert to context-free productions by using four different nonterminals $V^{au}, V^{cg}, V^{gc}, V^{ua}$, and using right-hand sides of the form $\to gV^{gc}c$ to produce a G-C pair, for instance – the nonterminal identity V^{gc} 'remembers' that a G-C pair was just generated. (In other words, all we are doing is making the model a higher order Markov process.) The probability of a production $V^{cg} \to gV^{gc}c$, for instance, would be the probability of a C-G pair stacked on a G-C pair.[2] Other details of the Zuker algorithm and its two matrices V and W could be incorporated similarly into an analogous full probabilistic model with two nonterminals V and W (expanded for nearest neighbour context). CYK and inside–outside algorithms for an SCFG version of the Zuker algorithm have the same algorithmic complexity as the Zuker algorithm itself.

---

[2] Since only one nonterminal is possible for a given x_i, x_j pair and the other three have zero probability, the four nonterminals behave as one for the purposes of memory and time complexity in parsing algorithms.

Suboptimal RNA folding

The original Zuker algorithm finds only the optimal structure. The biologically correct structure is often not the calculated optimal structure, but rather a structure within a few percent (i.e. within the error bars) of the calculated minimum energy. It was a significant advance when an efficient suboptimal folding algorithm was introduced. The Zuker *suboptimal* folding algorithm [Zuker 1989b] is similar to running the CYK algorithm in both the inside and outside directions. One matrix (exactly the CYK algorithm) finds the ΔG of the best structure for all subsequences i, j with i, j paired, and a second matrix (effectively an outside CYK algorithm) finds the best structure for the sequence with i, j paired and the subsequence $i + 1, j - 1$ excluded.[3] The sum of the two numbers for a given i, j is the ΔG of the optimal structure that uses the pair i, j. The suboptimal folding algorithm then samples a base pair i, j 'randomly' according to its ΔG, then traces back in both the inside and outside matrices to find the optimal structure that uses that base pair. (It is therefore more correct to say that the algorithm samples *one base pair* suboptimally. The rest of the structure is the optimal structure given that base pair.)

SCFG versions of RNA folding algorithms can also sample structures according to their likelihood by a probabilistic traceback of the inside matrix, analogous to the way in which suboptimal profile HMM alignments were sampled from a forward matrix in Chapter 6.

Base pair confidence estimates

Partition function calculations for calculating the probabilities of particular base pairs or structures were introduced for energy minimisation folding algorithms by McCaskill [1990]. The McCaskill algorithm converts ΔGs to probabilities using the Gibbs–Boltzmann equation and sums probabilities of all structures instead of choosing the single minimum energy structure. The sum of the probabilities of all structures containing a base pair i, j divided by the sum over all structures is interpreted as a confidence estimate in the pair i, j.

From the SCFG viewpoint, the McCaskill algorithm is fundamentally an inside–outside algorithm, compared to the Zuker algorithm which is fundamentally a CYK algorithm. The estimation of base pair confidences for an SCFG is conceptually similar to the estimation of pairwise alignment confidences that we described for pair HMMs in Chapter 4.

---

[3] Zuker actually doubles the sequence, treats it as circular, and calculates the energy of the best structure on $j, \dots, L/1, \dots, i$. For circular RNAs, this gives the same result as the outside algorithm. For linear RNAs, the Zuker algorithm must handle the non-existent junction between the $3'$ and $5'$ end as a special case. The outside algorithm might be less complicated to implement.

Exercises

10.7   Write down the inside algorithm, outside algorithm, and inside–outside re-estimation equations for the Nussinov-style RNA folding SCFG in equation (10.2).

10.8   By analogy to profile HMM suboptimal alignment sampling, give an algorithm for sampling structures probabilistically from your inside matrix.

10.9   Show how to use your inside and outside variables to calculate the probability that positions i, j are base-paired, summed over all structures. The functional form of the answer will be analogous to your inside–outside re-estimation equations.

10.3   Covariance models: SCFG-based RNA profiles

Suppose we have a family of related RNAs – transfer RNAs or group I catalytic introns, perhaps – which share a common consensus secondary structure as well as some primary sequence motifs, and we want to search a sequence database for homologous RNAs. In Chapter 5, we used HMM-based profiles to model the consensus of protein and DNA sequence families, but we showed in Chapter 9 that HMMs are primary structure models that cannot deal effectively with RNA secondary structure constraints. In this section, we describe SCFG-based RNA structure profiles called 'covariance models' (CMs) which are the SCFG analogue of profile HMMs. Whereas profile HMMs specify a repetitive linear HMM architecture well suited for modelling multiple sequence alignments, CMs specify a repetitive tree-like SCFG architecture suited for modelling consensus RNA secondary structures.

Although we follow here the 'covariance model' approach developed in Eddy & Durbin [1994], these same general ideas and algorithms are shared by comparable SCFG-based RNA models independently developed at the same time by Sakakibara and coworkers [1994].

CMs are detailed and fairly complex probabilistic models. We first set the stage by looking in an intuitive way at more simple models of small RNA alignments.

SCFG models of ungapped RNA alignments

Figure 10.11 shows an example RNA consensus structure and an ungapped multiple alignment of an RNA family that fit the consensus. To describe this multiple alignment with an SCFG-based model, we need several different types of nonterminals to generate different types of secondary structure and sequence elements.

Base-paired columns are modelled by *pairwise* emitting nonterminals that generate both bases in the pair. Single-stranded columns are modelled by *leftwise*

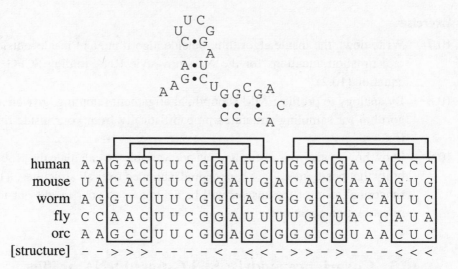

Figure 10.11 *The consensus structure of an example RNA family with no insertions or deletions is shown at the top. Five example sequences from different organisms that adopt the same structure are shown in the multiple alignment below the structure. Base-paired positions in the alignment are boxed and base-paired partners are connected by lines. The last line in the alignment is a structure consensus representation in the format we use for annotating RNA structural alignments [Konings & Hogeweg 1989].*

emitting nonterminals wherever possible. For bulges and interior loops on the 3' side of a stem, *rightwise* emitting nonterminals are sometimes needed. *Bifurcation* nonterminals are used to split into multiple stems and multi-branch loops. We define a special *start* nonterminal that acts as the initial nonterminal and as the immediate children produced from a bifurcation.[4] We also define a special *end* nonterminal that generates ϵ with probability 1 and terminates a derivation. The production rules for these states are summarised as follows, using W as a generic nonterminal to represent any of the six states:

$$
\begin{aligned}
P &\rightarrow aWb && \text{pairwise (16 pair emission probabilities),} \\
L &\rightarrow aW && \text{leftwise (4 singlet emission probabilities),} \\
R &\rightarrow Wa && \text{rightwise (4 singlet emission probabilities),} \\
B &\rightarrow SS && \text{bifurcation (probability 1),} \\
S &\rightarrow W && \text{start (probability 1),} \\
E &\rightarrow \epsilon && \text{end (probability 1).}
\end{aligned}
$$

The structure in Figure 10.11 can now be reduced to an SCFG, as shown in detail below. For clarity, only one of the possible productions is shown for each

---

[4] It is not necessary to bifurcate to start nonterminals, but this will simplify a number of subsequent algorithms. One reason for this is that we could sever the bifurcation-start connection and treat each branch of the structure as an independent SCFG model of an independent RNA domain.

nonterminal (the production corresponding to the one used in the structure and the human sequence in Figure 10.11). Pairwise productions would have 16 total productions and production probabilities for 16 possible pairs; leftwise and rightwise productions would have 4.

		Stem 1			**Stem 2**
$S_1 \rightarrow L_2 \ldots$	$S_5 \rightarrow P_6$		$S_{15} \rightarrow L_{16}$	
$L_2 \rightarrow aL_3 \ldots$	$P_6 \rightarrow gP_7c \ldots$		$L_{16} \rightarrow uP_{17} \ldots$	
$L_3 \rightarrow aB_4 \ldots$	$P_7 \rightarrow aR_8u \ldots$		$P_{17} \rightarrow gP_{18}c \ldots$	
$B_4 \rightarrow S_5S_{15}$	$R_8 \rightarrow P_9a \ldots$		$P_{18} \rightarrow gL_{19}c \ldots$	
	$P_9 \rightarrow cL_{10}g \ldots$		$L_{19} \rightarrow cP_{20} \ldots$	
	$L_{10} \rightarrow uL_{11} \ldots$		$P_{20} \rightarrow gL_{21}c \ldots$	
	$L_{11} \rightarrow uL_{12} \ldots$		$L_{21} \rightarrow aL_{22} \ldots$	
	$L_{12} \rightarrow cL_{13} \ldots$		$L_{22} \rightarrow cL_{23} \ldots$	
	$L_{13} \rightarrow gE_{14} \ldots$		$L_{23} \rightarrow aE_{24} \ldots$	
	$E_{14} \rightarrow \epsilon$		$E_{24} \rightarrow \epsilon$	

The model has an important property: its nonterminals are connected in a tree. The structure of the SCFG tree exactly mirrors the structure of the RNA and the structure of its parse trees. (This is not a necessary property of SCFGs; it was not shared by the RNA folding SCFG we saw earlier, for instance.) This allows us to adopt a convenient graphical representation of the SCFG that intuitively and compactly reflects the structure of the RNA family being modelled. It is clear from the grammar above that even simple RNA SCFGs can be very tedious to write in production rule form. A graphical representation of the same SCFG is shown in Figure 10.12.

The model has a total of 24 nonterminals, modelling a 24 nucleotide RNA alignment. The numbers do not have to be exactly the same but the number of nonterminals in the model will scale roughly linearly with the length of the alignment. One nonterminal is needed for each pair, one nonterminal for each single-stranded nucleotide, and an assortment of B, S and E nonterminals complete the model.

There is one important difference between the list of production rules and the graphical model in Figure 10.12. SCFGs typically emit symbols on transitions (i.e. $W_1 \rightarrow aW_2b$, emitting symbols and moving to a new nonterminal simultaneously). We can also choose to separate transition from emission and emit symbols on states independently of the preceding state transition. This distinction between emit-on-transition (Mealy machines) and emit-on-state (Moore machines) was discussed in Chapter 9. Moore machines have been favoured for most of the HMMs we have described, including profile HMMs. In covariance models, as an

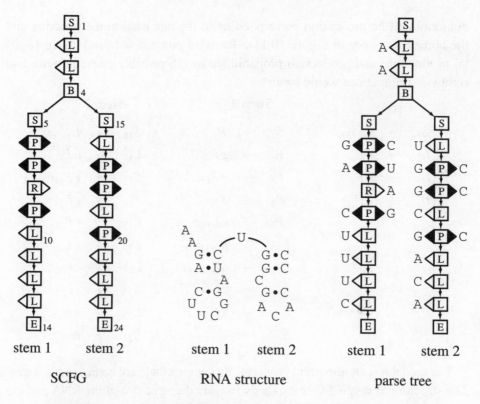

Figure 10.12 *A graphical representation of the ungapped RNA SCFG example is shown on the left. Boxes labelled P represent 16 pairwise production probabilities; boxes labelled L and R represent 4 leftwise and rightwise
production rules, respectively; boxes labelled S, B and E represent start, bifurcation and end nonterminals, respectively. The RNA consensus structure
is redrawn (middle) to correspond more closely to the tree structure of the
SCFG. A parse tree is shown for this structure (right) in which the RNA
nucleotides are assigned to states in the SCFG.*

SCFG-based extension of the profile HMM ideas in Chapter 5, we also use an
emit-on-state formalism. Likewise, we will usually refer to CM nonterminals as
'states'. Thus, our ungapped SCFG in Figure 10.12 has 16 emission probabilities
per pairwise state and 4 emission probabilities per leftwise or rightwise state, and
all the state transition probabilities are 1 (there are no alternative paths through
the ungapped model). The transition probabilities will become more interesting
when we develop models which allow insertion and deletion.

There is some ambiguity in mapping the RNA structural alignment to the
SCFG. Hairpin loops, for instance, could instead be generated right-to-left by
rightwise states. We arbitrarily chose leftwise states where possible, leftwise
states being the most similar to our previous treatment of profile HMMs.

The model in Figure 10.12 is a reasonable model of the RNA family in Figure 10.11 as long as no insertions and deletions occur. The model is comparable to an HMM composed only of match states without insert/delete states. We could use it as a full probabilistic model for RNA database searching.[5] However, we would probably miss many homologous structures; allowing insertions and deletions is important for modelling most real RNA structures. We turn now to covariance models, which expand the ungapped SCFG example into a model which tolerates insertions and deletions much like a match-state-only HMM (i.e. ungapped weight matrix) was expanded by insert and delete states into a profile HMM.

Exercise

10.10   Rewrite the list of production rules from the ungapped RNA model such that symbols are emitted independent of the previous state like an HMM. This is the formal stochastic transformational grammar that corresponds to the graphical SCFG representation in Figure 10.12.

Design of covariance models

The design goals for a CM are straightforward, but meeting these goals is not. First, a CM is built around a consensus RNA structure tree exactly like the tree we discussed above for ungapped RNA models. Secondly, a CM allows an insertion or deletion of any length at any position in the alignment. CMs use the same strategy that profile HMMs use for dealing with insertions and deletions relative to the consensus. Recall that profile HMMs repetitively use a set of three states (match, insert, delete) to model each position in a multiple alignment. A profile HMM can be thought of as a stereotyped expansion of an ungapped consensus model: every match state in the ungapped model is expanded to a match, delete and insert state in the profile HMM. Similarly, a CM expands an ungapped consensus model into a stereotyped pattern of individual states. We refer to the repetitive unit of model structure as a *node*. Profile HMMs are a linear string of a single type of three-state node. CMs are a branched tree of nodes, where there are different types of nodes with different numbers of states.

Leftwise nodes for single-stranded consensus positions expand like HMM nodes into match, insert and delete states, as do rightwise singlet nodes. Thus, leftwise nodes become a triplet of states ML, IL and D, and rightwise nodes become a triplet of states MR, IR and D.

When pairwise nodes expand, they have several insertion and deletion possibilities that they must take into account. A deletion may remove both bases in the base pair or solely the 5′ or 3′ partner, leaving the remaining unpaired partner

---

[5]   If we used this SCFG for RNA analysis, using the inside–outside and CYK algorithms for alignment would be overkill. With no gaps, only a single alignment is possible. Ungapped RNA motifs can be scored against a probabilistic model in $O(L)$ time.

as a bulge. Insertions in the base-paired stem may occur on the 5′ side of the pair, the 3′ side, or both. CMs expand a pairwise node into six states: an MP state, a D state (for complete deletion of the base pair), ML and MR states (for a single-base deletion that removes the 3′ or 5′ base, respectively), and IL and IR states that allow insertions on the 5′ or 3′ side of the pair, respectively.

The root start node is expanded to a start state S and insert states for either the 5′ or 3′ side, IL and IR. The left child start node under a bifurcation is expanded just to a single S state. The right child start node under a bifurcation is expanded to an S state and an insert-left IL state. This arrangement of the insert states assures that an insertion in any position is unambiguously assigned.[6]

Bifurcation nodes and end nodes in the consensus tree simply become B and E states in the CM.

States are then connected by state transitions. As with profile HMMs, states connect to all insert states in the current node and all non-insert states in the next node. Insert states have a state transition to themselves to allow insertions of more than one base. In pairwise nodes, IL connects to IR but not vice versa, so that insertions are unambiguously assigned to a single path through the model. This connectivity of state transitions is summarised graphically in Figure 10.13.

The complete CM is a directed graph of states, organised according to an underlying consensus tree. The 'main line' of the CM is exactly the consensus tree, but the CM also allows paths through alternative states that handle deletions and insertions.

This is only one possible design for an RNA structure profile. Insertions in stems are often base-paired instead of single-stranded (i.e. stem lengths vary). Therefore we could choose to include a pairwise-insertion (IP) state in pairwise nodes. We could remove the ML and MR states from the pairwise node and instead model deletions of a single base in a pair (which leave a bulge) as a complete deletion followed by an insertion of the bulge with the IL or IR state. A sophisticated design might even try to model the fact that long insertions in RNA are often structured. Given infinite computing resources, we could imagine replacing each insert state with a generalised SCFG RNA folding model like the one we described in the first section.

Construction of a CM from an RNA alignment

Given an RNA sequence alignment, annotation of the consensus secondary structure, and annotation of which columns should be considered to be insertions and which should be considered to be consensus columns, a CM can be precisely defined and readily constructed. Using the structure annotation on the non-insert

---

[6] Another possible arrangement would be S, IR for the left child and S for the right child; we use the chosen arrangement because of the decision to default to profile HMM-like leftwise generation when we have a choice.

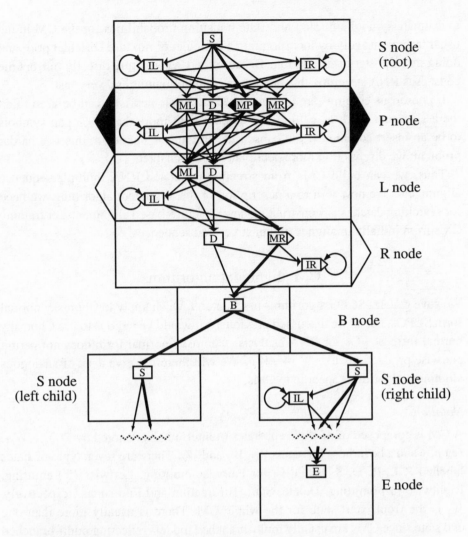

Figure 10.13 *The states of a CM (small boxes) are grouped according to the nodes of a consensus RNA structure tree (large boxes). The example here is strictly hypothetical, solely designed to put all eight types of CM nodes in the same picture. State transitions are indicated by arrows. The 'main line' consensus path is indicated by thick arrows. This main line is exactly the consensus tree itself. Jagged lines at the bottom indicate more nodes in the model that are not shown.*

columns of the alignment, a consensus structure tree is first constructed. The nodes of that tree are then filled with CM states, and states are connected by state transitions as discussed above. Based on this assignment of alignment columns to tree nodes, individual symbols can be assigned to individual CM states and each sequence in the alignment can be assigned a unique CM parse tree. Emission and transition events in these parse trees are counted. These observed counts are used to estimate symbol emission and state transition probabilities for the CM in the usual fashion, usually by incorporating a Dirichlet or mixture Dirichlet prior and doing mean posterior estimation. Because the CM structure directly mirrors the consensus RNA structure, this procedure is unambiguous and very fast.

If consensus columns are not annotated, a simple heuristic can be used to do the assignment, such as calling any column with more than 50% gap symbols to be an insert column. If the consensus structure is unknown, this is a harder problem; we discuss it in more detail later in the chapter.

Thus, we can build CMs from structure-annotated RNA multiple sequence alignments. We now turn to a description of the alignment algorithms we need for searching databases, constructing new structure-based alignments, or training CMs from initially unaligned and unstructured sequences.

CM alignment algorithms

We gave general SCFG algorithms in Chapter 9 which apply to Chomsky normal form SCFGs, but as we discussed previously, it would be tedious to use Chomsky normal form SCFGs for RNA analysis. Chomsky normal form does not permit pairwise productions like $W \rightarrow cWg$. We will therefore give a set of analogous alignment algorithms specific to CMs.

Notation

A CM is composed of M different states (nonterminals) denoted by $W_1, \ldots, W_M$. Let v, y and z be indices for states W_v, W_y and W_z. There are seven types of states labelled P, L, R, D, S, B and E, for Pairwise emitting, Leftwise (5′) emitting, Rightwise (3′) emitting, Delete, Start, Bifurcation and End states, respectively. W_1 is the (root) start state for the whole CM. There is usually more than one end state since CMs are usually multi-branched models reflecting multi-branched RNA secondary structures. The seven state types are associated with symbol emission and state transition probabilities as shown in Table 10.1.

We define numbers Δ_v^L and Δ_v^R which are the number of symbols emitted to the left and right by the state v. This simplifies the description of the algorithms. We used a similar simplifying notation in the description of the Sankoff–Cedergren N-dimensional dynamic programming algorithm in Chapter 6.

For notational and implementational convenience, each state in a CM also carries additional pieces of information. Let s_v be the *state type*, taking its value from P, L, R, D, S, B, or E, indicating one of the seven possible forms of production rule. Let $\mathcal{C}_v$ be the *children* of the state, represented by the list of one or more indices y for the states W_y that W_v can make a state transition to. Let $\mathcal{P}_v$ be the *parents* of the state, represented by the list of one or more indices y for the states W_y that make a state transition to W_v.

Bifurcation (B) states are handled specially in a CM. A bifurcation state W_v

State (s_v)	Production	Δ_v^L	Δ_v^R	Emission	Transition
P	$W_v \to x_i W_y x_j$	1	1	$e_v(x_i, x_j)$	$t_v(y)$
L	$W_v \to x_i W_y$	1	0	$e_v(x_i)$	$t_v(y)$
R	$W_v \to W_y x_j$	0	1	$e_v(x_j)$	$t_v(y)$
D	$W_v \to W_y$	0	0	1	$t_v(y)$
S	$W_v \to W_y$	0	0	1	$t_v(y)$
B	$W_v \to W_y W_z$	0	0	1	1
E	$W_v \to \epsilon$	0	0	1	1

Table 10.1. *The seven state types of a covariance model.*

always transits with probability 1 to two S states W_y and W_z, for only one choice of y and z. The children list $\mathcal{C}_v$ for a B state is a pair (y, z) for the two S children. The parent list $\mathcal{P}_y$ and $\mathcal{P}_z$ for both S state children is $\{v\}$, as only W_v transits to these states. This single choice for a bifurcation transition is very unlike Chomsky normal form, which is almost entirely described by a probabilistic choice among all y and z for bifurcation rules $W_v \to W_y W_z$. In RNA models, bifurcation states are only needed to describe multi-branch loops or multiple stems that occur in the structure. The bulk of the model consists of P, L and R states. The restriction on bifurcations greatly reduces the computational complexity of CM algorithms.

Start (S) and Delete (D) states are treated identically in alignment algorithms. The only difference between them is structural. Start states only occur as the immediate children of bifurcations or as the root state W_1. Delete states occur within P, L and R nodes of the CM.

There are three additional restrictions on CMs. First, each state may only use one type of production rule (s_v refers to both state and production type). Secondly, as in profile HMMs, states are not fully interconnected; the number of connected states in $\mathcal{C}_v$ is a constant that does not depend on the number of states M. This further reduces the complexity of the alignment algorithms compared to Chomsky normal form SCFGs. Lastly, we impose a final important restriction. States are numbered such that $y > v$ for all $y \in \mathcal{C}_v$, except for insert states, where $y \geq v$ for all $y \in \mathcal{C}_v$. This condition is important for the non-emitting states (S, D, B), guaranteeing that there are no non-emitting cycles. Similar restrictions on delete states were described for HMMs (Chapter 3).

Let us now walk through the important algorithms for manipulating RNA covariance models.

Scoring: inside algorithm

We are given an observed RNA sequence x, composed of L individual symbols $x_1, \ldots, x_i, \ldots, x_j, \ldots, x_L$. Consider first the *scoring* problem of calculating the likelihood $P(x|\theta)$ of the sequence given a covariance model θ, summed over all possible structures for x. This probability is calculated with the inside algorithm.

The inside algorithm recursively fills a three-dimensional dynamic programming matrix with values $\alpha_v(i,j)$. $\alpha_v(i,j)$ is the summed probability of all parse subtrees rooted at state v for the subsequence $x_i,\ldots,x_j$. $\alpha_v(i+1,i)$ is the probability for null subsequences of length zero; it must be included as a boundary condition because of the presence of non-emitting D, S and B states. For notational convenience, we will use $e_v(x_i,x_j)$ for all emission probabilities: for L states $e_v(x_i,x_j) = e_v(x_i)$, for R states $e_v(x_i,x_j) = e_v(x_j)$, and for non-emitting states $e_v(x_i,x_j) = 1$.

Algorithm: Inside for CMs

Initialisation: for $j = 0$ to L, $v = M$ to 1 :
$$\alpha_v(j+1,j) = \begin{cases} s_v = \text{E}: & 1; \\ s_v \in \text{S,D}: & \sum_{y\in\mathcal{C}_v} t_v(y)\alpha_y(j+1,j); \\ s_v = \text{B}: & \alpha_y(j+1,j)\alpha_z(j+1,j); \\ s_v \in \text{P,L,R}: & 0. \end{cases}$$

Recursion: for $j = 1$ to L, $i = j$ to 1, $v = M$ to 1 :
$$\alpha_v(i,j) = \begin{cases} s_v = \text{E}: & 0; \\ s_v = \text{P}, j = i: & 0; \\ s_v = \text{B}: & \sum_{k=i-1}^{j} \alpha_y(i,k)\alpha_z(k+1,j); \\ \text{otherwise}: & \\ \quad e_v(x_i,x_j)\sum_{y\in\mathcal{C}_v} t_v(y)\alpha_y(i+\Delta_v^{\text{L}},j-\Delta_v^{\text{R}}). \end{cases}$$

$\triangleleft$

When complete, the probability $P(x|\theta)$ is in $\alpha_1(1,L)$. If there are b bifurcation states and a other states ($M = a+b$), the order of complexity of the algorithm is $O(L^2M)$ in memory and $O(aML^2 + bML^3)$ in time.

Outside algorithm for CMs

For the next section on inside–outside parameter estimation, we need the outside algorithm. The outside algorithm calculates values $\beta_v(i,j)$ which are the probability of all parse trees rooted at state v that generate the complete sequence x *excluding* the subsequence $x_i,\ldots,x_j$. The outside algorithm requires calculating inside α terms first.[7] After initialising all the cells of the three-dimensional dynamic programming matrix to zero, the outside algorithm is:

[7] Actually only α_v for $s_v = S$ need to be kept from the inside pass. This is useful if memory is limiting.

Algorithm: Outside for CMs

Initialisation:

$$\beta_1(1, L) = 1.$$

Recursion: for $i = 1$ to $L+1$, $j = L$ to $i-1$, $v = 2$ to M :

$$\beta_v(i,j) = \begin{cases} \text{for } s_v = S, \mathcal{P}_v = y, \mathcal{C}_y = \{v, z\}: \\ \displaystyle\sum_{k=j}^{L} \beta_y(i,k)\alpha_z(j+1,k); \\ \text{for } s_v = S, \mathcal{P}_v = y, \mathcal{C}_y = \{z, v\}: \\ \displaystyle\sum_{k=1}^{i} \beta_y(k,j)\alpha_z(k,i-1); \\ \text{for } s_v \in \text{P,L,R,D,B,E}: \\ \displaystyle\sum_{y\in\mathcal{P}_v} e_y(x_{i-\Delta_y^L}, x_{j+\Delta_y^R})t_y(v)\beta_y(i-\Delta_y^L, j+\Delta_y^R). \end{cases}$$

$\triangleleft$

The memory and time complexities of the outside algorithm are identical to those of the inside algorithm.

Inside–outside expectation maximisation for CM parameters

If the structure of the model is known (i.e. the consensus RNA structure for the family is known) but the probability parameters are unknown, the probability parameters can be estimated from *unaligned* example RNAs using an expectation maximisation algorithm called the inside–outside algorithm. In practice, this algorithm would rarely be used. Almost all RNA consensus structures are derived from a multiple sequence alignment by comparative sequence analysis, and we described previously how a structure-annotated multiple sequence alignment can be immediately turned into a parameterised CM. However, we give the CM version of the inside–outside algorithm for completeness' sake. We can imagine situations in which a consensus structure is arrived at by other means. Also, we might not wish to assume that the multiple alignment is entirely correct, in which case we might not want to directly use the count data from the alignment, making the inside–outside algorithm more appropriate.

The probability of using state v at i, j in a derivation of a sequence x is $\frac{1}{P(x|\theta)}\alpha_v(i,j)\beta_v(i,j)$. By summing these terms in various directions, we can obtain a number of useful probabilities or expected counts, including the expected counts we need for EM re-estimation of the emission and transition probability parameters, just as we did for Chomsky normal form SCFGs.

The expected number of times that state v is used for a single sequence x is

$$c(v \text{ used}) = \frac{1}{P(x|\theta)} \sum_{i=1}^{L+1} \sum_{j=i-1}^{L} \alpha_v(i,j)\beta_v(i,j).$$

The expected number of times that a state transition $t_v(y)$ is used for a single sequence x is

$$c(v \to y \text{ used}) = \frac{1}{P(x|\theta)} \sum_{i=1}^{L+1} \sum_{j=i-1}^{L} \beta_v(i,j)e_v(x_i,x_j)t_v(y)\alpha_y(i+\Delta_v^L, j-\Delta_v^R).$$

When N independent observed sequences $x^1,\ldots,x^h,\ldots,x^N$ are used for training instead of a single observed sequence x, as is usually the case for training a model from a family of RNA sequences, the expected counts are summed over the individual sequences using inside and outside variables α^h and β^h calculated for each sequence:

$$c(v \to y \text{ used}) = \sum_{h=1}^{N} \frac{1}{P(x^h|\theta)} \sum_{i=1}^{L+1} \sum_{j=i-1}^{L} \beta_v^h(i,j)e_v(x_i^h,x_j^h)t_v(y)\alpha_y^h(i+\Delta_v^L, j-\Delta_v^R).$$

Thus, the inside–outside EM re-estimation equation for a CM transition probability from state v to state y given N training sequences $x^1,\ldots,x^N$ is[8]

$$\hat{t}_v(y) = \frac{\sum_{h=1}^{N} \frac{1}{P(x^h|\theta)} \sum_{i=1}^{L+1} \sum_{j=i-1}^{L} \beta_v^h(i,j)e_v(x_i^h,x_j^h)t_v(y)\alpha_y^h(i+\Delta_v^L, j-\Delta_v^R)}{\sum_{h=1}^{N} \frac{1}{P(x^h|\theta)} \sum_{i=1}^{L+1} \sum_{j=i-1}^{L} \alpha_v^h(i,j)\beta_v^h(i,j)}.$$

Similar arguments lead to inside–outside re-estimation equations for CM emission probabilities for state v, where the expression $\delta()$ is 1 if the condition in the parentheses is true, and 0 if the condition is false:

for $s_v = \text{P}$:

$$\hat{e}_v(a,b) = \frac{\sum_{h=1}^{N} \frac{1}{P(x^h|\theta)} \sum_{i=1}^{L} \sum_{j=i}^{L} \delta(x_i^h,x_j^h = a,b)\beta_v^h(i,j)\alpha_v^h(i,j)}{\sum_{h=1}^{N} \frac{1}{P(x^h|\theta)} \sum_{i=1}^{L} \sum_{j=i}^{L} \beta_v^h(i,j)\alpha_v^h(i,j)};$$

for $s_v = \text{L}$:

$$\hat{e}_v(a) = \frac{\sum_{h=1}^{N} \frac{1}{P(x^h|\theta)} \sum_{i=1}^{L} \sum_{j=i}^{L} \delta(x_i^h = a)\beta_v^h(i,j)\alpha_v^h(i,j)}{\sum_{h=1}^{N} \frac{1}{P(x^h|\theta)} \sum_{i=1}^{L} \sum_{j=i}^{L} \beta_v^h(i,j)\alpha_v^h(i,j)};$$

for $s_v = \text{R}$:

$$\hat{e}_v(a) = \frac{\sum_{h=1}^{N} \frac{1}{P(x^h|\theta)} \sum_{i=1}^{L} \sum_{j=i}^{L} \delta(x_j^h = a)\beta_v^h(i,j)\alpha_v^h(i,j)}{\sum_{h=1}^{N} \frac{1}{P(x^h|\theta)} \sum_{i=1}^{L} \sum_{j=i}^{L} \beta_v^h(i,j)\alpha_v^h(i,j)}.$$

The inside–outside product can also be used to estimate other quantities of

---

[8] Transition probabilities for B states are 1 by definition, so we do not need to re-estimate them.

interest. For instance, the probability that x_i and x_j are base-paired in a single sequence x by any pairwise production is

$$P(x_i, x_j \text{ paired}) \quad = \quad \frac{1}{P(x|\theta)} \sum_{v|s_v=P} \alpha_v(i,j)\beta_v(i,j).$$

Database searching: the CYK algorithm

Suppose we are given a very long sequence (a complete genome, for instance) and our task is to find one or more subsequences that match the RNA model. The algorithms we have given are well suited for global alignment, but ill suited to local alignment of one or more subsequences to a CM. Clearly we don't want the time and memory requirements of a database search algorithm to scale as the square or cube of the database sequence's length L. By limiting the length of the longest aligned subsequence to a constant D and by employing a transformation of the dynamic programming matrix coordinate system, we can implement an efficient CYK (or inside, or outside) algorithm for sequence database searching. The dynamic programming matrix is indexed by v, j, d instead of v, i, j, where d is the length of the subsequence $i, \ldots, j$ ($d = j - i + 1$) and $d \leq D$. Figure 10.14 shows how this altered coordinate system makes it straightforward to iteratively calculate a row of scores of the best alignments for subsequences of lengths $0, \ldots, D$ ending at sequence position j.

A standard CYK algorithm for SCFG alignment returns the log of the probability $P(S, \hat{\pi}|\theta)$ of the sequence S and the best parse $\hat{\pi}$ given the model θ. This score is strongly a function of the length of the aligned sequence, potentially making it difficult to choose the best matching subsequence among overlapping subsequences of different lengths in a database search. As discussed for HMMs, a nice solution to this problem is to calculate log-odds scores relative to a 'null' model of random sequences. If the random model is an independent identically distributed model in which the likelihood of the sequence under the null hypothesis is the product of individual residue frequencies f_a, then, analogous to HMM alignment scoring, log probability emission terms can be replaced by log-odds base pair or singlet nucleotide scores to make the CYK algorithm yield log-odds scores directly.[9] In the algorithm below, we use the notation $\log \hat{e}$ to indicate a log-odds emission score instead of a log probability $\log e$:

$$\begin{aligned}
\text{for } s_v = \text{P}: \quad \log \hat{e}_v(a,b) &= \log(e_v(a,b)/f_a f_b); \\
\text{for } s_v = \text{L}: \quad \log \hat{e}_v(a,b) &= \log(e_v(a)/f_a); \\
\text{for } s_v = \text{R}: \quad \log \hat{e}_v(a,b) &= \log(e_v(b)/f_b).
\end{aligned}$$

The CYK database search algorithm is as follows:

---

[9] To be a full probabilistic model, the random model would also have to specify a length distribution, but this term can usually be ignored.

Algorithm: CYK for CM database search

Initialisation: for $j = 0$ to L, $v = M$ to 1 :

$$\gamma_v(j,0) = \begin{cases} \text{for } s_v = \text{E} : \\ \quad 0; \\ \text{for } s_v \in \text{D,S} : \\ \quad \max_{y \in \mathcal{C}_v}\left[\gamma_y(j,0) + \log t_v(y)\right]; \\ \text{for } s_v = \text{B}, \mathcal{C}_v = (y,z) : \\ \quad \gamma_y(j,0) + \gamma_z(j,0); \\ \text{otherwise} : \\ \quad -\infty. \end{cases}$$

Recursion: for $j = 1$ to L, $d = 1$ to D (and $d \le j$), $v = M$ to 1 :

$$\gamma_v(j,d) = \begin{cases} \text{for } s_v = \text{E} : \\ \quad -\infty; \\ \text{for } s_v = \text{P and } d < 2 : \\ \quad -\infty; \\ \text{for } s_v = \text{B}, \mathcal{C}_v = (y,z) : \\ \quad \max_{0 \le k \le d}\left[\gamma_y(j-k,d-k) + \gamma_z(j,k)\right]; \\ \text{otherwise} : \\ \quad \max_{y \in \mathcal{C}_v}\left[\gamma_y(j - \Delta_v^{\text{R}}, d - \Delta_v^{\text{L}} - \Delta_v^{\text{R}}) + \log t_v(y)\right] \\ \quad + \log \hat{e}_v(x_i, x_j). \end{cases}$$ ◁

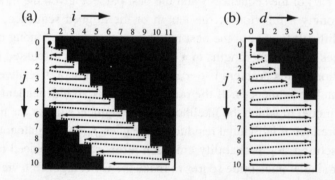

Figure 10.14 *(a) One level of a standard CYK dynamic programming matrix is shown for a database sequence of $L = 10$, indexed by start position i and end position j. (There are M different levels like this in the three-dimensional matrix, one per state in the model.) The parts of the matrix that need to be calculated if the maximum matching subsequence length is limited to $D = 5$ are shown in white. The order of calculation of the matrix cells is shown with arrows for a search algorithm that sweeps across the database sequence (i.e. increasing j). (b) An alternative coordinate system for the same CYK calculation indexed by end position j and subsequence length d, where $d = j - i + 1$. It is easier to implement a smoothly scanning CYK database search algorithm with memory requirements that are independent of L in this coordinate system since the matrix in (b) is $D \times L$ rather than $L \times L$.*

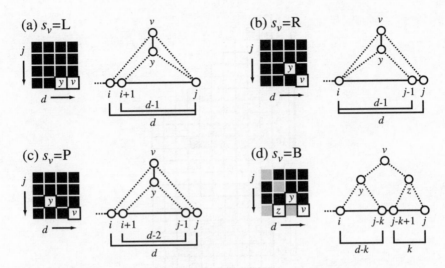

Figure 10.15 *The steps of the CYK database searching algorithm recursion are shown for four different state types. Only one level of the three-dimensional dynamic programming matrix is shown. For example, in the upper left, the value of the cell marked v, $\gamma_v(j,d)$, depends on one or more possible cells marked y, $\gamma_y(j,d-1)$, for the different states y that state v connects to. This is shown in a different way to the right of the dynamic programming matrix, by showing that if v generates a single residue leftwise, then the parse subtree rooted at state v for the subsequence of length d that ends at j is constructed by adding to subtrees for $y,j,d-1$. The calculations for R (upper right) and P (lower left) states are analogous. The calculation when v is a bifurcation (B) state depends on choosing the best bifurcation point. One such point is shown by two cells marked y and z in the dynamic programming matrix; the set of all other such connected cells is shown in grey.*

In Figure 10.15, the key steps of the recursion in this algorithm are illustrated graphically.

A few further implementation details are important. Instead of initialising all L rows, it is better to commingle the initialisation and iteration steps and move along one row j at a time, first initialising $\gamma_v(j,0)$ and then calculating $\gamma_v(j,d)$ for $d = 1,\ldots,D$. (Since the initialisation calculations for subsequences of length 0 are independent of the sequence, $\gamma_v(0,0)$ only needs to be calculated once for all v and these values can be copied for initialising subsequent $\gamma_v(j,0)$.) In the above algorithm and Figure 10.15, it is apparent that all the scores in row j are dependent only on scores in rows j and $j-1$, except for bifurcation (B) states. Bifurcation states are in turn only dependent on start state scores for the previous D rows, since B states bifurcate to two S states in the definition of a CM structure. Thus only $D+1$ rows of S state scores and two rows of scores for other states need to be stored in memory.

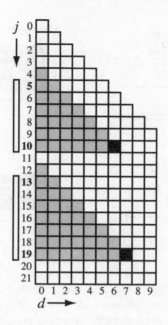

Figure 10.16 *A small example showing two non-overlapping hits in a CYK database search matrix. The two hits (from positions 5 to 10 and 13 to 19) are shown schematically as boxes to the left of the matrix. The high-scoring matrix entries $\gamma_0(j,d)$ are shown in black. The volumes of the matrix which must be in memory to reconstruct the complete alignments are shown in grey. The grey triangle also indicates how the coordinates of a cell (j,d) determine the start point i of each match even if the complete matrix is not kept in memory. Only one level of the three-dimensional matrix is shown for simplicity.*

The scores $\gamma_0(j,d)$ in row j are the log-odds scores of complete alignments to the model (i.e. for a parse tree starting from the root state, $v = 0$) ending at position j. As shown in Figure 10.16, the start point of the match i can be calculated from d: $i = j - d + 1$. That is, obtaining the start point of the alignment does not require a traceback of the dynamic programming matrix. After finding a high-scoring $\gamma_0(j,d)$, a CYK search algorithm can immediately report not just the score but also the start position i and end position j of the subsequence that gives this high score.

An implementation might choose to report all hits above a certain score threshold. However, a high-scoring alignment has 'shadows', alignments with minor differences in start and end point which will also score well. It is better to report non-overlapping hits above a threshold. A simple score post-processing algorithm which is compatible with scanning unlimited amounts of sequence with a constant memory requirement is as follows. After calculating a row j, the best score $\gamma_0(j,d)$ for a d in row j is determined. If it is greater than the reporting threshold, it is stored in a list; if it overlaps with a previous hit in the list, the lower-scoring hit is discarded; any hit in the list whose end point j is less than the current minimum start point, $j - D$, is reported as a non-overlapping match.

The time complexity of the CYK search algorithm is $O(M_a LD + M_b LD^2)$ for a model of M_a non-bifurcation states and M_b bifurcation states, a database of length L residues, and a maximum match size of D residues. The memory complexity is $O(M_a D + M_b D^2)$. Computation time scales linearly with increasing database size, and the memory required is independent of the database size.

Exercise

10.11   The same alternative matrix coordinate system can be applied to the inside or outside algorithms. Compared to CYK, the inside algorithm has the advantage that it sums over the probabilities of all possible structures and alignments for the subsequences, yet it is no more computationally complex than the CYK version. Give the inside algorithm for searching for local subsequence matches of no greater than length D.

Structural alignment: CYK algorithm with traceback

Since most of the matrix is discarded in the interest of memory efficiency, tracebacks cannot be done as we have described the search algorithm above. Therefore, alignments cannot be recovered; only scores and start and end positions can be recovered. At the expense of memory, the algorithm is readily modified to trace back and recover the optimum SCFG parse tree for a matching subsequence. Because the structure of a CM reflects the consensus secondary structure of an RNA, a CM parse tree represents both an optimal alignment to the model and an optimal secondary structure prediction. Assignments to P (pairwise) states in the parse tree indicate predicted base pairs.

CYK tracebacks can be implemented either with a second matrix of traceback pointers or by reconstructing the score calculations, as discussed for other dynamic programming algorithms in Chapter 2. All $D + 1$ rows of either traceback pointers or scores must be in memory to guarantee that a hit can be traced back. If the overlap-processing algorithm above is used, then $2D + 1$ complete rows must be kept in memory, since it takes D additional rows before a hit is determined to be non-overlapping with a later hit. If both scores and alignments are desired, it is reasonably efficient to implement both a local CYK search algorithm and a global CYK alignment algorithm with tracebacks, and to simply do two passes: first a local search pass to find the matching subsequences, and a second pass to optimally align each of these subsequences one at a time to the model in a global alignment mode with tracebacks.

The traceback starts from $\gamma_0(j, d)$ for a high-scoring subsequence of length d ending at j and works back. For a global rather than local alignment with respect to the sequence, the traceback starts from $\gamma_0(L, L)$. The process is fundamentally the same as that used to recover HMM dynamic programming tracebacks or

SCFG tracebacks from the simpler RNA models earlier in the chapter, so we will
not give full details.

Exercise

10.12   Modify the CYK algorithm so that it keeps traceback information in each
        cell to assist in recovering the optimal parse tree. What is the minimum
        information that needs to be kept for tracing back from a bifurcation
        state? What is the minimum information that needs to be kept for tracing
        back from any other state?

'Automated' comparative sequence analysis using CMs

Suppose we are given a set of *unaligned* RNA sequences and the consensus secondary structure is unknown. Combined multiple RNA alignment and consensus
secondary structure prediction is the domain of comparative sequence analysis,
which remains largely a manual process. We have given the inside–outside training algorithm, but it presupposes that we already know the structure of the model
and hence the consensus structure of the RNA family. We have also given algorithms for constructing a CM from a multiple alignment without necessarily
knowing the consensus structure. We now describe how these two ideas can be
combined into an automated comparative sequence analysis algorithm.

   The basic idea is to iterate between two steps: (a) build an optimal (or nearly
optimal) CM structure given the current alignment; and (b) build an optimal multiple alignment given the current CM.

   For (a), several approaches are possible for finding a consensus structure given
an RNA multiple alignment. A heuristic approach directly inspired by comparative sequence analysis methods was used by Eddy & Durbin [1994]. Mutual
information terms $M_{i,j}$ are first calculated for all pairs of aligned columns. A
dynamic programming folding algorithm (essentially the Nussinov algorithm) is
then used to find the structure tree which maximises the sum of the $M_{i,j}$ terms.
The fill stage of this algorithm for finding the maximum sum $S_{i,j}$ is as follows:

$$S_{i,j} = \max \begin{cases} S_{i+1,j} & \text{column } i \text{ unpaired;} \\ S_{i,j-1} & \text{column } j \text{ unpaired;} \\ S_{i+1,j-1} + M_{i,j} & \text{columns } i, j \text{ paired;} \\ \max_{i<k<j} S_{i,k} + S_{k+1,j} & \text{bifurcation.} \end{cases}$$

A traceback of this matrix then yields the consensus structure tree, which is
expanded to a CM. The advantage of this method is that by using mutual information terms, only base pairs which are best supported by comparative analysis
are paired. One disadvantage of the method is that it overpairs, because $M_{i,j} \geq 0$
and thus $S_{i,j}$ usually increases slightly even from adding a spurious pair. A sec-

ond disadvantage is that because it looks only for covariation, highly conserved structures or domains with little sequence variation may be mis-predicted.

The consensus structure is then imposed on each aligned sequence to find individual parse trees, transition and emission counts are collected from these parse trees, and counts are converted to probability parameters in the usual fashion.

There also exists a rigorous CM construction algorithm for taking an unannotated RNA multiple sequence alignment and simultaneously deriving an optimal (maximum *a posteriori*) CM structure and its parameters [S. R. Eddy, unpublished]. This algorithm is an extension of the MAP construction algorithm for profile HMMs described in Chapter 5.

For (b), constructing an optimal multiple alignment given the current model, we apply the CYK alignment algorithm to each sequence. As with HMMs, a multiple alignment of the sequences is implied by the set of their individual alignments to a common model.

This is akin to an EM algorithm except that the CYK algorithm is used in place of inside–outside at the expectation step, and a model construction algorithm (which simultaneously re-estimates both the structure and the parameters of the model, rather than just the parameters) is used at the maximisation step.

The training algorithm works almost exactly like a human comparative sequence analyst works. In some cases, it even produces the same answer. Starting from a guess at the alignment of initially unaligned sequences (either random or perhaps a sequence-based alignment), the algorithm makes an initial guess at the structure, realigns all the sequences according to that guess, and makes a new guess at the structure, iteratively converging on a consistent solution. It works well for this on ideal datasets, such as an example of 100 transfer RNA sequences used in Eddy & Durbin [1994].

However, few real datasets are so ideal. To work, the algorithm requires a large number of small RNA sequences; there must be sufficient primary sequence divergence that covariations reveal the majority of the base pairs; the consensus secondary structure must be highly conserved; and the sequences must be globally instead of just locally alignable. Moreover, the algorithm is prone to local optima, but is already so compute-intensive that the more demanding simulated annealing methods described for profile HMMs (Chapter 5) are unattractive. Although these are all just technical limitations to be overcome, it must be said that the automated comparative analysis aspect of SCFG-based analysis methods remains mostly of theoretical rather than practical interest. The most practical application of covariance model methods is in database searching for homologous RNA structures, as described in the next section.

An example of a practical application of CM algorithms

In order to get a better picture of how all this theory applies to a practical sequence analysis problem, it is worth looking at a real application of covariance models.

The largest gene family in most genomes is not a protein gene family but a family of structural RNA genes, transfer RNAs (tRNAs). For example, there are 274 tRNA genes in the yeast *Saccharomyces cerevisiae*, and about 1500 different tRNA genes in the human genome. A number of programs have been developed which find tRNA genes in genomic sequences. These tRNA detection programs are usually part of any large-scale genome annotation project. The false positive rate of these carefully hand-tuned programs is generally on the order of 0.2–0.4 false predictions per megabase of DNA [Fichant & Burks 1991; Pavesi *et al.* 1994]. This false positive rate is acceptable for small genomes like that of yeast (14 Mb), but in the 3000 Mb human genome, such a program would produce around a thousand false positives, meaning that a large fraction of the tRNA gene predictions would be wrong.

Transfer RNAs are an ideal candidate for testing covariance model methods. They are short (usually 75–95 residues); they have relatively little shared primary sequence identity but a highly conserved 'cloverleaf' secondary structure; and thousands of tRNA sequences are available for training a statistical model.

A multiple alignment consisting of 1415 tRNA sequences from a variety of organisms, including organellar and viral tRNAs, was the starting point for constructing a tRNA covariance model [Steinberg, Misch & Sprinzl 1993]. Thirty-eight sequences included short introns in the anticodon loop, so that the trained model would be 'aware' that many eukaryotic tRNA genes have short introns. A number of longer introns (usually catalytic group I and group II introns) in various positions were excluded from the alignment, since they were judged to be too long to be worth searching for with this algorithm.

A model was constructed directly from this alignment using annotation of the accepted consensus secondary structure. Computation time for model construction was negligible. The resulting model had 285 states. These were grouped into 72 nodes in the consensus model: 3 bifurcation, 28 pairwise, 33 leftwise, 1 rightwise and 7 starts. The 28 pairwise nodes correspond to 28 consensus base pairs in the four helices of the tRNA cloverleaf and the variable arm helix.

An implementation of the CYK database scanning algorithm (COVELS in the COVE program suite) searched DNA at 20 residues/second on an SGI Indigo2 R4400 workstation when restricted to a maximum match length of $D = 150$ [Eddy & Durbin 1994]. The search matrix required about 500K of RAM. The CPU time for the tRNA search was therefore much more limiting than the memory requirement. A parallelised implementation of the algorithm on a MasPar multiprocessor accelerated the search to about 2000 residues/second, but the use of specialised hardware was not deemed practical.

A hybrid program called TRNASCAN-SE was therefore written which used two different existing tRNA detection programs as fast pre-filters [Lowe & Eddy 1997]. Candidate tRNAs proposed by one or both of these programs were matched in a second step against the tRNA CM, and statistically significant hits ($\geq$ 20 bits of log-odds score) were reported as tRNAs. Table 10.2 summarises the performance of this hybrid program compared to other tRNA detection programs and to covariance models alone.

Program	Speed (bp/sec)	True pos. (%)	False pos. (per Mb)
TRNASCAN 1.3 [Fichant & Burks 1991]	400	95.1	0.37
POL3SCAN [Pavesi *et al.* 1994]	373 000	88.8	0.23
CM alone [Eddy & Durbin 1994]	20	99.8	<0.002
TRNASCAN-SE [Lowe & Eddy 1997]	30 000	99.5	<0.000 07

Table 10.2. *Comparison of tRNA gene detection methods.*

The (mostly) automatic CM based approach of TRNASCAN-SE allowed greatly improved specificity and slightly greater sensitivity than the two manually tuned methods. The expected number of false positive tRNAs in the 3000 Mb human genome was decreased from over a thousand to less than one.

In addition to sensitivity and specificity, another advantage of the CM approach is its generality – a model can be constructed of any RNA sequence alignment. In TRNASCAN-SE, for instance, a specialised additional model was easily built from an alignment of selenocysteine tRNAs (which differ from most tRNAs in several respects), enabling the program to detect these variant genes.

The principal disadvantages of CM methods for RNA sequence analysis are the high memory and CPU time requirements. In TRNASCAN-SE, the time problem was overcome by using other existing programs as fast pre-filters. However, this approach is not general, since such programs are not available for most other RNA families. Better algorithms and/or better computers will be needed to apply CM methods to larger RNAs.

10.4 Further reading

A variety of other algorithms have been applied to the single sequence RNA structure prediction problem besides dynamic programming algorithms, including genetic algorithms [Shapiro & Wu 1996; van Batenburg, Gultyaev & Pleij 1995], constraint satisfaction algorithms, and Monte Carlo sampling algorithms [Abrahams *et al.* 1990; Gultyaev 1991]. Many of these algorithms attempt to

predict pseudoknots in addition to canonical secondary structure. RNA pseudoknots continue to be a nettling issue in algorithm development for RNA structure
modelling. In general, none of these algorithms guarantees finding an optimal
pseudoknotted structure. A notable exception is the graph theoretic maximum
weighted matching algorithm of Cary & Stormo [1995]. Brown & Wilson [1995]
describe an SCFG-based approach to pseudoknot modelling using an intersection
of separate SCFG models of the pseudoknot and the rest of the structure.

The accuracy of single sequence secondary structure prediction algorithms
has been systematically compared to the results of manual comparative analysis
[Fields & Gutell 1996; Konings & Gutell 1995]. The utility of representing and
comparing RNA structures using trees has been recognised for some time [Mar-
galit *et al.* 1989; Shapiro & Zhang 1990]. There is an interesting literature on
'RNA structure space', using theoretical approaches and computer modelling to
address questions of how secondary structure constrains the evolution and function of RNAs [Schuster *et al.* 1994; Schuster 1995].

SCFG-based methods for modelling RNA structure families and multiple alignments were introduced by Eddy & Durbin [1994] and by Sakakibara, Haussler
and coworkers at UC Santa Cruz [Sakakibara *et al.* 1994]. Comparable algorithms were developed by Lefebvre [1995; 1996]. Corpet & Michot [1994] described a non-probabilistic RNA structure alignment algorithm with some similarities to the SCFG algorithms.

11

Background on probability

To make our book more self-contained, we have included a last chapter that gathers together the probabilistic ideas and methods we use. The various sections of this chapter are fairly independent, and can be dipped into as the reader wishes. Some parts are more mathematically technical than the rest of the book.

11.1 Probability distributions

We introduce here various probability distributions used throughout the book. When the outcomes we wish to assign probabilities to belong to a finite set X, a probability distribution is simply an assignment of a probability p_x to each outcome x in X. For instance, the probability distribution of outcomes of rolling a fair die would be $p_x = 1/6$ for the six outcomes $x = 1, \ldots, 6$.

If we have a continuous variable x, like the weight of an object, then the probability that that variable takes a specific value, e.g. that the weight is *exactly* 1 pound, is zero. But the probability that x takes a value in some interval, $P(x_0 \leq x \leq x_1)$ say, can be well defined and positive. As the width of the interval tends to zero, we may be able to write $P(x - \delta x/2 \leq x \leq x + \delta x/2) = f(x)\delta x$, where $f(x)$ is a function called a *probability density*, or just *density*. The probability of an interval can then be derived by integration: $P(x_0 \leq x \leq x_1) = \int_{x_0}^{x_1} f(x)dx$. A density must satisfy $f(x) \geq 0$, for all x, and $\int_{-\infty}^{\infty} f(x)dx = 1$. But note that we can have $f(x) > 1$. For instance, the density $f(x) = 10$ for $0 \leq x \leq 0.1$ and $f(x) = 0$ elsewhere is well defined.

The binomial distribution

The first distribution we consider is perhaps the simplest and most familiar: the *binomial distribution*. It is defined on a finite set consisting of all the possible results of N tries of an experiment with a binary outcome, '0' or '1'. If p is the probability of getting a '1' and $1 - p$ that of getting a '0', the probability that k out of the N tries yield a '1' is

$$P(k \text{ '1's out of } N) = \binom{N}{k} p^k (1 - p)^{N-k}, \tag{11.1}$$

where $\binom{N}{k}$ denotes the number of ways of choosing k objects from N, that is $N!/((N-k)!k!)$, and the factorial function is defined for non-negative integers as $n! = n(n-1)\cdots 1$, and $0! = 1$.

The mean m and variance σ^2 of any distribution P are defined by $m = \sum k P(k)$ and $\sigma^2 = \sum (k-m)^2 P(k)$. The positive square root of the variance, σ, is called the standard deviation. For the binomial distribution

$$m = \sum_{k=1}^{N} k \binom{N}{k} p^k (1-p)^{N-k}$$

and

$$\sigma^2 = \sum_{k=1}^{N} (k-m)^2 \binom{N}{k} p^k (1-p)^{N-k}$$

We can show (Exercise 11.1) that $m = Np$ and $\sigma^2 = Np(1-p)$.

Exercise

11.1   Calculate the mean and variance of the binomial distribution. (Hint: To find m, differentiate the binomial expansion $(p+q)^N = \sum_0^N \binom{N}{k} p^k q^{N-k}$ with respect to p and set $q = 1 - p$. For the variance, carry out two differentiations with respect to p.)

The Gaussian distribution

Consider next what happens as we let $N \to \infty$. Both the mean and the variance increase linearly with N, but we can rescale to give fixed mean and variance, defining the new variable u by $u = (k-m)/\sigma = (k-Np)/\sqrt{Np(1-p)}$. It is a classic result [Keeping 1995] that, in the limit of a large number of events, a binomial distribution becomes a Gaussian (see Figure 11.1), and with the rescaling the density is

$$f(u) = \frac{1}{\sqrt{2\pi}} \exp(-u^2/2). \tag{11.2}$$

This can be regarded as a special case of the central limit theorem, which states that the distribution of a sum of N independent random variables, normalised to the same mean and variance, tends to a Gaussian as $N \to \infty$. If a single variable takes values '0' or '1' with probabilities $1-p$ and p, respectively, the distribution of the sum of N copies of this is $P(k) = P(X_1 + \ldots + X_N \le k)$, and is precisely the binomial considered above.

The multinomial distribution

The generalisation of the binomial distribution to the case where the experiments have K independent outcomes with probabilities θ_i, $i = 1, \ldots, K$, is the *multino-*

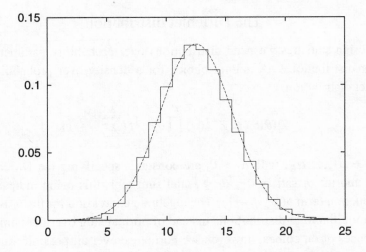

Figure 11.1 *The limit for large N of a binomial tends to a Gaussian. In this case N = 40 and p = 1/4 in (11.1).*

mial distribution. The probability of getting n_i occurrences of outcome i is given by

$$P(n|\theta) = M^{-1}(n) \prod_{i=1}^{K} \theta_i^{n_i}. \qquad (11.3)$$

Here we condition the probability on the parameters θ of the distribution, which is a natural thing to do in a Bayesian framework, because then the parameters are themselves random variables. In a classical statistics framework the probability of n could, for instance, have been denoted by $P_\theta(n)$. The normalising constant only depends on the total number of outcomes observed, $\sum_k n_k$. For fixed $\sum_k n_k$ it is

$$M(n) = \frac{n_1! \cdot n_2! \cdots n_K!}{(\sum_k n_k)!} = \frac{\prod_i n_i!}{\sum_k n_k}. \qquad (11.4)$$

For $K = 2$ the multinomial distribution reduces to the binomial distribution.

Example: Rolling a die

The outcome of rolling a die N times is described by a multinomial. The probabilities of each of the six outcomes are called $\theta_1, \ldots, \theta_6$. For a fair die where $\theta_1 = \ldots = \theta_6 = 1/6$ the probability of rolling it a dozen times and getting each outcome twice is

$$\frac{12!}{2!^6} \left(\frac{1}{6}\right)^{12} = 3.4 \times 10^{-3}.$$

□

The Dirichlet distribution

In Bayesian statistics we need distributions over probability parameters to use as prior distributions. A natural choice for a density over probabilities is the Dirichlet distribution:

$$\mathcal{D}(\theta|\alpha) = Z^{-1}(\alpha) \prod_{i=1}^{K} \theta_i^{\alpha_i - 1} \delta(\sum_{i=1}^{K} \theta_i - 1). \qquad (11.5)$$

Here $\alpha = \alpha_1, \ldots, \alpha_K$, with $\alpha_i > 0$, are constants specifying the Dirichlet distribution, and the θ_i satisfy $0 \le \theta_i \le 1$ and sum to 1, this being indicated by the delta function term $\delta(\sum_i \theta_i - 1)$. The algebraic expression for the θ_i is the same as for a multinomial distribution. Instead of normalising over the numbers n_i of occurrences of outcomes, however, we normalise over all possible values of the θ_i. To put this another way, the multinomial is a distribution over its exponents n_i, whereas the Dirichlet is a distribution over the numbers θ_i that are exponentiated. The two distributions are said to be conjugate distributions [Casella & Berger 1990], and their close formal relationship leads to a harmonious interplay in many estimation problems.

The normalising factor Z for the Dirichlet defined in (11.5) can be expressed in terms of the gamma function: [Berger 1985]

$$Z(\alpha) = \int \prod_{i=1}^{K} \theta_i^{\alpha_i - 1} \delta(\sum_i \theta_i - 1) d\theta = \frac{\prod_i \Gamma(\alpha_i)}{\Gamma(\sum_i \alpha_i)}. \qquad (11.6)$$

The gamma function is a generalisation of the factorial function to real values. For integers $\Gamma(n) = (n-1)!$. For any positive real number x,

$$\Gamma(x+1) = x\Gamma(x). \qquad (11.7)$$

It can be shown that the mean of the Dirichlet distribution is equal to the normalised parameters, i.e. the mean of θ_i is $\alpha_i / \sum_k \alpha_k$. For instance, the three distributions shown in Figure 11.2 all have the same mean $(1/8, 1/4, 5/8)$, even though the αs for the top right figure are 10 times larger than those for the top left. Note that larger αs produce a tighter distribution. Note also that when some $\alpha_i < 1$ the distribution is peaked at zero for the corresponding θ_i, as shown in the bottom left figure.

For two variables ($K = 2$) the Dirichlet distribution reduces to the more widely known beta distribution, and the normalising constant is the beta function.

Example: The dice factory

Consider again our example from Chapters 1 and 3 of a probabilistic model of a possibly loaded die with probability parameters $\theta = \theta_1, \ldots, \theta_6$. Sampling probability vectors θ from a Dirichlet parameterised by $\alpha = \alpha_1, \ldots, \alpha_6$ is like a 'dice factory' that produces different dice with different θ [MacKay & Peto 1995].

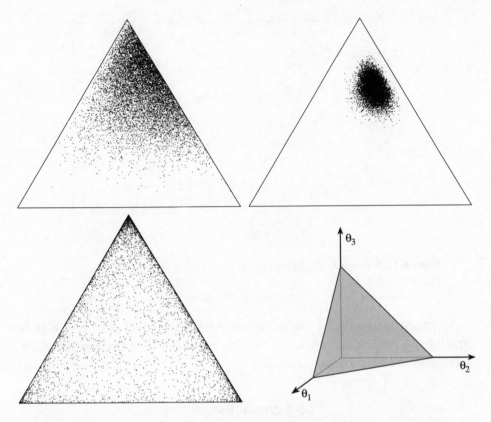

Figure 11.2 *Examples of three-dimensional Dirichlet distributions, each shown by sampling* 10 000 *points, i.e. by choosing points* θ *with probability* $\mathcal{D}(\theta|\alpha)$. *The values of* α *used are* $(1, 2, 5)$ *in the top left figure,* $(10, 20, 50)$ *top right and* $(0.1, 0.2, 0.5)$ *bottom left. The probabilities* θ *are displayed as the slice through* $3D$ *space* $(\theta_1, \theta_2, \theta_3)$ *where* $\sum \theta_i = 1$*; see the bottom right figure. A point* $(\theta_1, \theta_2, \theta_3)$ *is mapped to* $((\theta_2 - \theta_1)/\sqrt{3}, \theta_3)$ *in the plane.*

Suppose dice factory A has all six α_i set to 10, and dice factory B has all α_i set to 2. On average, both factories produce fair dice; the average of θ_i is $\frac{1}{6}$ in both cases. But if we find a loaded die with $\theta_6 = 0.5, \theta_1 = \ldots = \theta_5 = 0.1$, it is much more likely to have been produced by dice factory B:

$$\mathcal{D}(\theta|\alpha_A) = \frac{\Gamma(60)}{(\Gamma(10))^6}(0.1)^{5(10-1)}(0.5)^{10-1} = 0.119,$$

$$\mathcal{D}(\theta|\alpha_B) = \frac{\Gamma(12)}{(\Gamma(2))^6}(0.1)^{5(2-1)}(0.5)^{2-1} = 199.6.$$

The factory with the higher α parameters produces a tighter distribution in favour of fair dice. The sum $\sum \alpha_i$ is inversely proportional to the variance of the Dirichlet. (Don't be alarmed by the Dirichlet density having a value of 199.6; recall that the values of continuous probability densities at any point may be greater than one.)

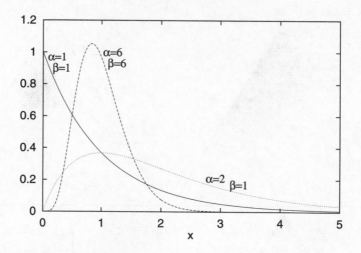

Figure 11.3 *Gamma distributions $g(x, \alpha, \beta)$ for $\alpha = \beta = 1.0$, $\alpha = \beta = 6.0$ and $\alpha = 2.0, \beta = 1.0$.*

A factory that produced almost perfectly fair dice would have very high but equal α_i. A factory that produced variably unreliable dice that are still fair on average would have low but equal α_i. □

The gamma distribution

The *gamma distribution* $g(x, \alpha, \beta)$ is given by

$$g(x, \alpha, \beta) = \frac{e^{-\beta x} x^{\alpha-1} \beta^{\alpha}}{\Gamma(\alpha)},$$

and is defined for $0 < x, \alpha, \beta < \infty$. Its mean is α/β and variance α/β^2. β is simply a scale parameter.

The gamma distribution is conjugate to the Poisson, $f(n) = e^{-p} p^n / n$, which gives the probability of seeing n events over some interval, when there is a probability p of an individual event occurring in that interval. Since the number of events in an interval is a rate, the gamma distribution is appropriate for modelling probabilities of rates, just as the Dirichlet is appropriate as a prior for emission probabilities when its conjugate, the multinomial, is used to assign probabilities to counts (p. 319). The gamma distribution has been used to model the rate of evolution at different sites in DNA sequences (p. 215).

The extreme value distribution

Suppose we take N samples from the density $g(x)$. The probability that the largest amongst them is less than x is $G(x)^N$, where $G(x) = \int_{-\infty}^{x} g(u) du$. The density for the largest value of the set of N is given by differentiating this with respect to x, giving $Ng(x)G(x)^{N-1}$. The limit for large N of $Ng(x)G(x)^{N-1}$ is called the *extreme value* density (EVD) for $g(x)$. It has a wide variety of practical uses, from modelling the breaking-point of a chain (which is determined by the weakest link), to assessing the significance of the maximum score from a set of alignments (see Chapter 2).

Let us compute the EVD when $g(x)$ is the exponential density $g(x) = \alpha e^{-\alpha x}$. Integrating gives $G(x) = 1 - e^{-\alpha x}$. Choosing y so that $e^{-\alpha y} = 1/N$, and writing $z = x - y$, we find

$$
\begin{aligned}
Ng(x)G(x)^{N-1} &= N\alpha e^{-\alpha x}(1 - e^{-\alpha x})^{N-1} = \alpha e^{-\alpha z}(1 - e^{-\alpha z}/N)^{N-1} \\
&\to \alpha e^{-\alpha z}\exp(-e^{-\alpha z}) \quad \text{for} \quad N \to \infty,
\end{aligned}
$$

where we used the well-known limit $(1 - X/N)^N \to e^{-X}$ for $N \to \infty$.[1] The cumulative probability (the probability that the extreme value is $\leq x$) is $\exp(-e^{-\alpha z})$, and is called a *Gumbel* distribution [Gumbel 1958]. The above density often gives a good approximation to the distribution of extreme values for moderate values of N. With the exponential density, Figure 11.4 shows that the maximum of a sample of size 10 gives a close approximation to the EVD.

It is a surprising fact that the Gumbel distribution is the EVD for a variety of underlying densities $g(x)$; it holds when $g(x)$ is a Gaussian too, for instance. More generally, an EVD must have the form $\exp(-f(a_N x + b_N))$, where a_N and b_N are constants depending on N and $f(x)$ is either an exponential e^{-x} or $|x|^{-\lambda}$ for some positive constant λ (see Waterman [1995] for a more precise statement of this theorem).

11.2 Entropy

Some of the terminology used in the book is borrowed from information theory (see e.g. Cover & Thomas [1991]). Information theory has strong connections to probabilistic modelling.

An *entropy* is a measure of the average uncertainty of an outcome. Given a random variable X with probabilities $P(x_i)$ for discrete set of K events $x_1,\ldots,x_K$, the Shannon entropy is defined by

$$
H(X) = -\sum_i P(x_i)\log P(x_i). \tag{11.8}
$$

In this definition, $P(x_i)\log P(x_i)$ is taken to be zero if $P(x_i) = 0$. Normally we assume that $\log$ is the natural logarithm (sometimes written ln). However, it is common to use the logarithm base 2 (called $\log_2$), in which case the unit of

---

[1] There is one delicate point in the above argument. We have to take care that $e^{-\alpha z}$ cannot grow rapidly with N, and so invalidate the limit $(1 - e^{-\alpha z}/N)^N \to \exp(-e^{-\alpha z})$. To be more precise, one has to show that the probability of large values of $e^{-\alpha z}$ according to the distribution $Ng(x)G(x)^{N-1}$ becomes vanishingly small.

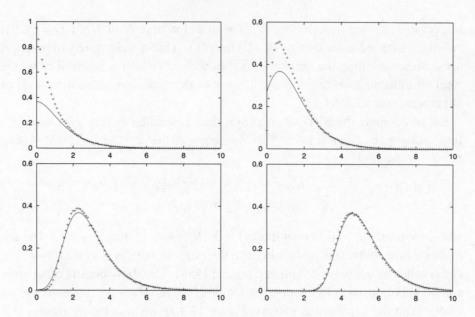

Figure 11.4 *Approximations to the extreme value distribution obtained by sampling N points from the distribution e^{-x} on $0 \leq x < \infty$, and then taking the maximum. From the top left to bottom right, $N = 1, 2, 10, 100$.*

entropy is a 'bit'. All logarithms are proportional, e.g. $\log_2(x) = \log_e(x)/\log_e(2)$, so theoretically it does not matter which logarithm is used. Often we talk about the entropy of the probability distribution P, $H(P)$, instead of $H(X)$.

The entropy is maximised when all the $P(x_i)$ are equal ($P(x_i) = 1/K$) and we are maximally uncertain about the outcome of a random sample. The maximum is the $-\sum_i \frac{1}{K} \log \frac{1}{K} = \log K$. If we are certain of the outcome of a sample from the distribution, i.e. $P(x_k) = 1$ for one k and the other $P(x_i) = 0$, the entropy is zero.

Entropy also arises as the expected score of the sequences generated by certain probabilistic models when the score is defined to be the log probability. Suppose, for instance, that the probability of residue a in some position in a sequence is p_a. Then there is a probability p_a of score $\log p_a$, and the expected score is $\sum_a p_a \log p_a$, namely the negative entropy. The same is true (see Exercise 11.2) when the model defines the probabilities at a set of independent sites.

If you are told the outcome of an event, the uncertainty is reduced from H to zero, because you have gained information. Therefore entropy is often equated with information. This can be confusing; it leads to the quite counterintuitive view that the more random something is (the higher the entropy), the more information it has. It is not confusing if we think of information as a difference in entropy. More generally *information content* or just *information* is a measure of a reduction in uncertainty after some 'message' is received; hence, the difference between the entropy before and the entropy after the message:

$$I(X) = H_{\text{before}} - H_{\text{after}}. \tag{11.9}$$

The uncertainty is not always reduced to zero; there may be noise on the communications channel, for instance, and we may remain somewhat uncertain of the outcome, in which case H_{after} is positive and the information is less than the original entropy.

In information theory it is often assumed that the probability distributions are known exactly. In many applications, however, the true distributions are not known, and therefore entropies are calculated from the frequencies of events rather than the true distributions; see Examples below.

Example: Entropy of random DNA

If each symbol (A, C, G, or T) of a DNA sequence occurs equiprobably ($p_a = 1/4$) then the entropy per DNA symbol is $-\sum_a p_a \log_2 p_a = 2$ bits.

We can think of the entropy as the number of binary yes/no questions needed to discover the outcome. For example, for random DNA, we need two questions: 'purine or pyrimidine?' followed by 'A or G?' if the answer is 'purine', and 'C or T?' otherwise. ☐

Example: Information content of a conserved position

Information content can be used to measure the degree of conservation at a site in a DNA or protein sequence alignment. Say we expect a DNA sequence to be random ($p_a = 0.25$; $H_{\text{before}} = 2$ bits), but we observe that a particular position in a number of related sequences is always an A or a G with $p_A = 0.7$ and $p_G = 0.3$. Thus $H_{\text{after}} = -0.7 \log_2 0.7 - 0.3 \log_2 0.3 = 0.88$ bits. The information content of this position is said to be $2 - 0.88 = 1.12$ bits. The more conserved the position, the higher the information content.

Notice, however, that the information content can be negative if the observed distribution has a higher entropy (is more 'random') than expected. For finding unusual patterns it is therefore better to measure the difference between the distributions by the relative entropy described below. ☐

Exercise

11.2　Assume a model in which $p_i(a)$ is the probability of amino acid a occurring in the ith position of a sequence of length l. The amino acids are considered independent. What is the probability $P(x)$ of a particular sequence $x = x_1, \ldots, x_l$? Show that the average of the log of the probability is the negative entropy $\sum P(x) \log P(x)$, where the sum is over all possible sequences x of length l.

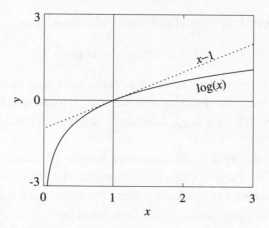

Figure 11.5 *Proof that the relative entropy (11.10) is always positive or zero if $P(x_i) = Q(x_i)$ for all i. From this graph is can be seen that $\log(x) \leq x - 1$ with equality only if $x = 1$. It follows that $-H(P\|Q) = \sum_i P(x_i)\log(Q(x_i)/P(x_i)) \leq \sum_i P(x_i)(Q(x_i)/P(x_i) - 1) = 0$, with equality holding only if, for each i, $Q(x_i) = P(x_i)$.*

Relative entropy and mutual information

We return to the definition of different types of entropy. For two distributions P and Q the *relative entropy* (also known as the Kullback–Leibler 'distance') is defined by

$$H(P\|Q) = \sum_i P(x_i)\log\frac{P(x_i)}{Q(x_i)}. \tag{11.10}$$

Information content and relative entropy are the same if the Q is a uniform 'background distribution' ($Q(x_i) = \frac{1}{K}$) that represents a completely naive initial state for H_{before}. The two terms are sometimes used interchangeably.

Relative entropy has the property that it is always greater than or equal to zero. It is easy to show that $H(P\|Q) \geq 0$ with equality if and only if $P(x_i) = Q(x_i)$ for all i (see Figure 11.5). It is often useful to think of the relative entropy $H(P\|Q)$ as a distance between the probability distributions P and Q. However, it is not symmetric, $H(P\|Q) \neq H(Q\|P)$, and it does not fulfil the formal requirements of a proper mathematical distance measure.

The relative entropy often arises as the expected score in models where the score is defined as the *log-odds*, i.e. $P(\text{data}|M)/P(\text{data}|R)$, where M is the model, and R is a null model. If p_a is the probability of residue a in some position in a sequence according to M, and q_a its probability according to R, then the score for residue a is $\log(p_a/q_a)$, and the expected score is $\sum_a p_a \log(p_a/q_a)$, which is the relative entropy.

Another important entropy measure is the *mutual information*. Two random variables X and Y are independent if $P(X,Y) = P(X)P(Y)$. It is interesting to know *how* independent they are, and that can be measured by the relative entropy 'distance' between the distributions $P(X, Y)$ and $P(X)P(Y)$,

$$M(X; Y) = \sum_{i,j} P(x_i, y_j) \log \frac{P(x_i, y_j)}{P(x_i)P(y_j)}, \qquad (11.11)$$

where the possible values for X and Y are $\{x_i\}$ and $\{y_j\}$. This is the mutual information. $M(X; Y)$ can be interpreted as the amount of information that we acquire about outcome X when we are told outcome Y.

The mutual information is maximal when X and Y always covary. If for instance all pairs except AT, TA, GC, and CG have probability zero for two positions i and j in some aligned DNA sequences, there is maximal covariation. For this situation we will always have $P(x_i, y_j) = P(x_i) = P(y_j)$ or $P(x_i, y_j) = 0$, and therefore $M = -\sum_i P(x_i) \log P(x_i)$. This is the entropy of X (or Y), so it is maximal for a uniform distribution, and the maximum is $\log K$ (assuming that X and Y have the same number, K, of possible outcomes). The maximum mutual information for DNA sequences is therefore $\log_2 4 = 2$ bits.

In Figure 10.6 the mutual information (calculated from frequencies) between every pair of columns in an RNA alignment is shown.

Example: Acceptor sites

Relative entropy is useful for finding unusual patterns in biological sequences. To illustrate this we extracted 757 acceptor sites from a database with human genes. The acceptor site is the splice site at the 3' end of the intron where the intron is spliced out to make the messenger RNA. The last two bases of the intron are almost always AG, and in this dataset they all are. We only took acceptor sites of introns occurring between two codons, i.e. not splicing in the middle of a codon. We extracted 30 bases upstream of the splice site and 20 bases downstream. In Figure 11.6 you see a small arbitrary sample of the sequences.

At each position i the frequency $p_i(a)$ of the four nucleotides was found, and the relative entropy $\sum_a p_i(a) \log_2[p_i(a)/q_a]$ calculated, where q_a is the overall distribution of the four nucleotides in the sequences. We plot this in Figure 11.6. At the AG consensus the relative entropy is very high (equal to $-\log_2(q_A)$ and $-\log_2(q_G)$ respectively). There is an interesting structure in the relative entropy upstream of the site with a minimum just two bases before the AG. There is a weak periodic signal (barely visible) of the relative entropy in the coding region, which is due to the different base composition in the three reading frames. See Brunak, Engelbrecht & Knudsen [1991] and Hebsgaard *et al.* [1996] for more discussion of information in splice sites, and Schneider & Stephens [1990] for colourful ways of displaying various entropy measures.

To see if the neighbouring positions are independent, the mutual information between the columns was calculated. For two neigbouring columns (say i and $i + 1$) the frequency of pairs $p_i(a, b)$ was found by counting how many times

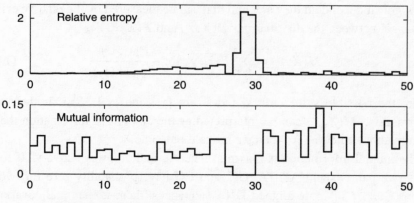

Figure 11.6 *Plots of relative entropy and mutual information for acceptor sites. Below is shown a sample of the sequences. Note the peak in relative entropy and dip in mutual information at the conserved* AG.

a occurred in column i and *b* occurred in column $i + 1$. From this the mutual information $\sum_{a,b} p_i(a,b) \log_2[p_i(a,b)/p_i(a)p_{i+1}(b)]$ was calculated, and is also plotted in Figure 11.6.

Notice that the mutual information is zero at the AG consensus: knowing that the first is A conveys no information about the next position, because it is always a G. The mutual information around the acceptor site is much less than the maximum of 2 bits, but it is non-zero, and it shows that there are correlations between neighbouring positions. This is true in most DNA. A clear periodic pattern is seen for the coding region, showing that the nucleotides are dependent in the three reading frames.   □

Exercises

11.3   Prove the above assertion about the equivalence of information content and relative entropy when q is uniform.

11.4   Show that $M(X;Y) = M(Y;X)$.

11.5   Show that $M(X;Y) = H(X) + H(Y) - H(Y,X)$, where $H(Y,X)$ is the entropy of the joint distribution $P(X,Y)$.

11.3   Inference

Probabilistic models are the main focus of this book. A model can be anything from a simple distribution to a complex stochastic grammar with many implicit probability distributions. Once the type of model is chosen, the parameters of the model have to be *inferred* from data. For instance, we may model the outcome of rolling a die with a multinomial distribution. Suppose the number of observations yielding i is n_i $(i = 1,\ldots,6)$. We do not know if it is a fair die, so we need to estimate the parameters of the multinomial distribution, i.e. the probability θ_i of getting i in a throw of the die. Here, we consider the different strategies that might be used for inference in general. For more background, see Ripley [1996] and MacKay [1992].

Maximum likelihood

Let us suppose, then, that we wish to infer parameters $\theta = \{\theta_i\}$ for a model M from a set of data D. The most obvious strategy is to maximise $P(D|\theta, M)$ over all possible θ. This is called the *maximum likelihood* criterion. Formally we write

$$\theta^{\mathrm{ML}} = \underset{\theta}{\mathrm{argmax}}\, P(D|\theta, M). \tag{11.12}$$

Generally speaking, when we treat $P(x|y)$ as a function of x we refer to it as a probability; when we treat it as a function of y we call it a likelihood. Note that a likelihood is not a probability distribution or density, but simply a function of the variable y.

Maximum likelihood has some desirable properties. For instance, it is *consistent*, in the sense that the parameter value θ_0 used to generate the dataset will also, in the limit of a large amount of data, be the value that maximises the likelihood. To see this, suppose there are K observable outcomes $\omega_1,\ldots,\omega_K$ of the model M (e.g. the 4^n possible assignments of nucleotides at a site in an aligned set of sequences). Then the frequency $n_i/\sum n_k$ of occurrences of ω_i will tend to $P(\omega_i|\theta_0, M)$ as the amount of data increases (see Exercise 11.6). Hence the log likelihood for parameter θ, given by $\sum_i (n_i/\sum n_k) \log P(\omega_i|\theta, M)$

tends to $\sum_i P(\omega_i|\theta_0, M) \log P(\omega_i|\theta, M)$. The positivity of relative entropy implies that $\sum_i P(\omega_i|\theta_0, M) \log P(\omega_i|\theta_0, M) \geq \sum_i P(\omega_i|\theta_0, M) \log P(\omega_i|\theta, M)$, for all θ. Thus the likelihood is maximised by θ_0.

A drawback of maximum likelihood is that it can give poor results when the data are scanty; we would be wiser then to rely on more prior knowledge. Consider the dice example and assume we we want to estimate the multinomial parameters from, say, three different rolls of the dice. It is shown on p. 319 that the maximum likelihood estimate of θ_i is $n_i/\sum n_k$, i.e. it is 0 for at least three of the parameters. This is obviously a bad estimate for most dice, and we would like a way to incorporate the prior knowledge that we expect all the parameters to be quite close to $1/6$.

Exercise

11.6   The *weak law of large numbers* says that the mean of a sample of size N differs from the true mean by an amount d or more with probability $\sigma^2/(Nd^2)$, where σ^2 is the variance of the distribution. Show that this implies that $n_i/\sum n_k$ tends to $P(\omega_i)$ as $\sum n_k \to \infty$, where n_i is the frequency of occurrence of ω_i.

The posterior probability distribution

The way to introduce prior knowledge is to use Bayes' theorem. Suppose there is a probability distribution over the parameters θ. Conditioning throughout on M gives the following version of Bayes' theorem:

$$P(\theta|D, M) = \frac{P(D|\theta, M) P(\theta|M)}{P(D|M)}. \tag{11.13}$$

The prior $P(\theta|M)$ has to be chosen in some reasonable manner, and that is the art of Bayesian estimation. This freedom to choose a prior has made Bayesian statistics controversial at times, but we believe it is a very convenient framework for incorporating prior (biological) knowledge into statistical estimation.

$P(\theta|D, M)$ is the posterior probability for the parameters, given the data and the model. The posterior can be used for inference in various ways. We can sample from it (see Section 11.4), and thereby locate regions of high probability for the model parameters. In Section 8.4 we show how this can be done for probabilistic models of phylogeny. If we want a specific set of parameter values for the model, we might be guided by analogy with ML and choose the *maximum a posteriori* probability (MAP) estimate,

$$\theta^{\mathrm{MAP}} = \underset{\theta}{\mathrm{argmax}}\, P(D|\theta, M) P(\theta|M). \tag{11.14}$$

Note that we ignore the data prior $P(D|M)$, because it does not depend on the parameters θ and thus the maximum point θ^{MAP} is independent of it. Another possibility is to take the *posterior mean* estimator (PME), which chooses the average of all parameter sets weighted by the posterior:

$$\theta^{\text{PME}} = \int \theta \, P(\theta|n) d\theta. \qquad (11.15)$$

The integral is over all valid probability vectors, i.e. all those that sum to one. In the following we will derive the PME for a multinomial distribution with a certain prior.

Both MAP and PME estimators are considered a little suspicious, because a non-linear transformation of the parameters usually changes the result. In technical terms they are not *equivariant* [Ripley 1996]. To see what's going on, we need to consider the effects of change of variables on densities.

Change of variables

Given a density $f(x)$, suppose there is a change of variable $x = \phi(y)$. Then we can define a density $g(y)$ by $g(y) = f(\phi(y))|\phi'(y)|$. The derivative of ϕ, $\phi'(y)$, is there because the interval δx corresponds to an interval $\delta y \phi'(y)$ under the transform ϕ, so the amount of the f density that is swept out under ϕ is proportional to this derivative; taking the derivative's absolute value ensures that the density is positive. This definition produces a correctly normalised density because $\int g(y)dy = \int f(\phi(y))|\phi'(y)|\,dy = \int f(x)dx = 1$, f being a density. We write the transformation rule formally as

$$g(y) = f(\phi(y))|\phi'(y)|. \qquad (11.16)$$

The function $f(\phi(y))$ clearly has the same maximum as $f(x)$. When we multiply by $|\phi'(y)|$, however, this maximum may shift (see Exercise 11.7). Now, the posterior $P(\theta|D,M)$ is a density, so the peak chosen by MAP can likewise change under a transformation. A similar argument shows that the PME can change under a coordinate transformation.

In contrast, the likelihood $P(D|\theta,M)$ does not transform like a density; it is simply a function of θ and a change of coordinates leaves the peak unchanged, just as the peak of $f(\phi(y))$ remains the same as that of $f(x)$ [Edwards 1992].

Exercise

11.7   Let $f(x) = 2(1-x)$ be a density on $[0,1]$. Show how this transforms to a density on y under $x = y^2$. Show that the peak and the PME of the density both shift under this transformation.

11.4 Sampling

Given probabilities $P(x_i)$ defined on the members x_i of a finite set X, to *sample* from this set means to pick elements x_i randomly with probability $P(x_i)$.

The basic practical tool for sampling is a function derived from a computer's pseudo-random number generator (i.e. the function called rand[], or something similar), that picks numbers randomly from the interval $[0, 1]$ with the uniform density. Let us call this function rand$[0, 1]$. Using it, we can choose elements x_i with frequency $P(x_i)$. We set $y = \text{rand}[0, 1]$, and then choose our element x_i by finding that i for which $P(x_1) + \ldots + P(x_{i-1}) < \text{rand}[0, 1] < P(x_1) + \ldots + P(x_{i-1}) + P(x_i)$. Clearly, the probability of rand[] lying in this range is $P(x_i)$, so x_i is picked with the correct probability.

It is actually not easy to produce random numbers with a computer. The standard function for pseudo-random numbers is usually very primitive, and will not be good enough for some applications. For example, the standard rand[] function on many UNIX computers returns an integer between 0 and $2^{15} - 1$, and one would expect to obtain 'random' bits (0 or 1) with this function by taking the value returned modulo 2. However, this gives a sequence where 0 and 1 alternate, which is clearly not random at all. On most systems there are other (and better) functions to choose from. See for instance Press *et al.* [1992] for a discussion of random number generators.

Sampling by transformation from a uniform distribution

The concept of sampling applies also to densities: Given a density f, to sample from it is to pick elements x from the space on which f is defined so that the probability of picking a point in an arbitrarily small region δR round the point x is $f(x)\delta R$. Sampling of densities can be accomplished by using pseudo-random numbers that sample from the uniform density on $[0, 1]$, and applying a change of variables that changes the density appropriately.

The theory of this goes as follows: Suppose we are given a density $f(x)$, and a map $x = \phi(y)$. From (11.16) we know that $g(y) = f(\phi(y))\phi'(y)$. If f is uniform, we have $g(y) = \phi'(y)$, so ϕ can be obtained by integration, $\phi(y) = \int_b^y g(u)du$, where b is some suitable lower bound. However, we want to pick points in x using a good pseudo-random number generator, and then map them to y. For this, we require the inverse function to ϕ, namely $y = \phi^{-1}(x)$.

Suppose for instance that we want to sample from a Gaussian. We define the cumulative Gaussian map $\phi(y) = \int_{-\infty}^y e^{-u^2/2}/\sqrt{2\pi}\, du$, and let $y = \phi^{-1}(x)$. We could make a look-up table to evaluate the inverse cumulative Gaussian function, but this is rather clumsy, and some other approach may be more convenient (e.g. Exercise 11.10).

The transformation method also applies more generally to functions of K variables, but then (11.16) must be replaced by

$$g(y_1,\ldots,y_K) = f(\phi_1(y_1,\ldots,y_K),\ldots,\phi_K(y_1,\ldots,y_K))|J(\phi)|, \qquad (11.17)$$

where $J(\phi)$ is the Jacobian, whose (i,j)-th entry is $\partial\phi_i/\partial y_j$ [Feller 1971].

Exercises

11.8   Show that the function $g(y) = \alpha^\lambda \lambda y^{\lambda-1}/(\alpha^\lambda + y^\lambda)^2$ is a density on $0 \le y < \infty$. Show that picking x uniformly from $(0,1)$ and mapping x to $y = \alpha(\frac{x}{1-x})^{1/\lambda}$ samples from $g(y)$.

11.9   Define a mapping ϕ from the variables (x,y) to (u,w) by $x = uw$, $y = (1-u)w$. Show that $J(\phi) = w$, where J is the Jacobean.

11.10 (Calculus needed!) Suppose we pick two random numbers x and y in the range $[0,1]$ and map (x,y) to the sample point $\cos(2\pi x)\log(1/y^2)$. Prove that this samples correctly from a Gaussian. This is called the Box–Muller method [Press *et al.* 1992].

Sampling from a Dirichlet by rejection

We consider now the problem of sampling from a Dirichlet, which illustrates some important principles. Suppose first that we can sample from the gamma distribution $g(x,\alpha,1)$

$$g(x,\alpha,1) = e^{-x}x^{\alpha-1}/\Gamma(\alpha)$$

for $0 < x < \infty$ (see p. 304). If we take sampled values x_1 and x_2 from two gamma distributions with parameters α_1 and α_2, respectively, then we can define a pair (u,v) with $u+v = 1$, by setting $u = x_1/(x_1+x_2)$, $v = x_2/(x_1+x_2)$; equivalently, we can set $x_1 = uw$, $x_2 = (1-u)w$ and integrate over w. Using (11.17) and the results of Exercise 11.9, the distribution $D(u,v)$ of pairs (u,v) is given by

$$
\begin{aligned}
D(u,v) &= \frac{\int_0^\infty \delta(u+v-1)e^{-uw}(uw)^{\alpha_1-1}e^{-vw}(vw)^{\alpha_2-1}w\,dw}{\Gamma(\alpha_1)\Gamma(\alpha_2)} \\
&= \frac{u^{\alpha_1-1}v^{\alpha_2-1}\delta(u+v-1)}{\Gamma(\alpha_1)\Gamma(\alpha_2)}\int_0^\infty e^{-w}w^{\alpha_1+\alpha_2-1}\,dw \\
&= u^{\alpha_1-1}v^{\alpha_2-1}\delta(u+v-1)\frac{\Gamma(\alpha_1+\alpha_2)}{\Gamma(\alpha_1)\Gamma(\alpha_2)} \\
&= \mathcal{D}(u,v|\alpha_1,\alpha_2), \qquad (11.18)
\end{aligned}
$$

where $\mathcal{D}(u,v|\alpha_1,\alpha_2)$ is the Dirichlet distribution with parameters α_1,α_2. In other words, to sample from a Dirichlet distribution of two variables (a beta distribution), we sample from two gamma distributions, whose exponents are those of the components of the Dirichlet in question, and then normalise the sampled numbers to give probabilities. This elegant result extends to Dirichlets of any number of variables (Exercise 11.11).

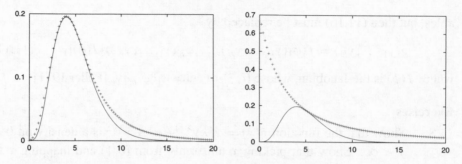

Figure 11.7 *Rejection sampling: We wish to sample from a gamma distribution $g(x,\alpha,1)$ (continuous line). It is possible to sample from the function f given by (11.19) ('+' signs), whose value always exceeds that of
the gamma distribution. Having sampled a point x from f, this point is
accepted with a probability equal to the ratio of the gamma distribution
and f at that point, i.e. with probability $g(x,\alpha,1)/f(x)$. The left figure
shows f with $\alpha = 5$, $\lambda = 3$, the right with $\alpha = 5$, $\lambda = 1$.*

We can sample from a Dirichlet, therefore, if we know how to sample from a
gamma distribution. Now we can show (Exercise 11.12) that $g(x,\alpha,1) \le f(x)$,
where

$$f(x) = \frac{4e^{-\alpha}\alpha^{\lambda+\alpha}x^{\lambda-1}}{\Gamma(\alpha)(\alpha^{\lambda}+x^{\lambda})^2}, \tag{11.19}$$

and $\lambda = \sqrt{2\alpha - 1}$. It follows that, if rand[0, 1] truly samples uniformly between
0 and 1, then $P(\text{rand}[0,1] < g(x,\alpha,1)/f(x)) = g(x,\alpha,1)/f(x)$. Thus if we first
sample from the distribution f, picking a point x with probability $f(x)$, and
accept x if rand[0, 1] $< g(x,\alpha,1)/f(x)$, then

$$P(x) = f(x)P(\text{rand}[0,1] < g(x,\alpha,1)/f(x)) = g(x,\alpha,1).$$

So this two-stage procedure enables us to sample from the gamma distribution.
It remains only to show how to sample from f. But Exercise 11.8 shows that
choosing u from [0, 1] by rand[0, 1] and defining $x = \alpha(u/1-u)^{1/\lambda}$ is equivalent
to sampling from f. For more details of the material in this section, and also for
the appropriate procedure in the case where $0 < \alpha < 1$, see Law & Kelton [1991].
Figure 11.2 was generated using this method.

This is an example of *rejection sampling*, the distribution g being obtained
by 'trimming down' from the distribution f, which is analytically tractable and
always larger than g. This only works well if $f(x)$ is a good approximation to
$g(x,\alpha,1)$; if it is not, the rejection rate will be high. The function f gives a
good approximation to $g(x,\alpha,1)$ over the range where both functions are large,
i.e. where they will be most frequently sampled from. The choice of λ is in
fact optimal for this purpose. For instance, with $\alpha = 5$ and $\lambda = \sqrt{2\alpha - 1} = 3$,

only 14% of points are rejected (Figure 11.7, left figure), whereas with $\lambda = 1$ (Figure 11.7, right figure), 65% are rejected.

Exercises

11.11   Show that (11.18) can be extended to the case of K gamma distributions, i.e. that sampling from $g(x, \alpha_i, 1)$, for $i = 1, \ldots, K$, then averaging, is equivalent to sampling from the Dirichlet $D(\theta_1, \ldots, \theta_K | \alpha_1, \ldots, \alpha_K)$. (Hint: Show that the Jacobian of the map $x_i = u_i w$, for $i \leq K - 1$, and $x_K = (1 - \sum u_i)w$ is equal to w^{K-1}.)

11.12   Prove that $g(x, \alpha, 1) \leq f(x)$, for all x and $\alpha > 1$ and $1 \geq \lambda \leq \sqrt{2\alpha - 1}$, where $f(x)$ is defined by (11.19). What happens when $\lambda > \sqrt{2\alpha - 1}$?

Sampling with the Metropolis algorithm

We often want to sample from a probabilistic model, where the analytic methods that underlie the transformation method or rejection sampling are not available. One possible approach then is to use a Markov chain defined on the space X of outcomes [Neal 1996]. We assume here that X is finite, although the ideas carry over to continuous variables and densities.

Given a point x, a chain specifies a probability $\tau(y|x)$ for the transition $x \to y$ to a point y. If we can sample from the distribution $\tau(y|x)$, i.e. given x can pick a y with probability $\tau(y|x)$, then we can generate a sequence $\{y_i\}$ where each y_i is picked by sampling from the distribution $\tau(y|y_{i-1})$.

Suppose now that we can find a τ satisfying

$$P(x)\tau(y|x) = P(y)\tau(x|y). \tag{11.20}$$

This is called the condition of *detailed balance*. It turns out that detailed balance implies

$$\frac{1}{N} \lim_{N \to \infty} C(y_i = x) = P(x), \tag{11.21}$$

for all points x, where $C(y_i = x)$ is the number of times $y_i = x$ in the sequence of length N. We can therefore approximate P as closely as we like by taking long enough sequences of $\{y_i\}$ sampled using τ. This statement needs to be qualified: Clearly, the chain needs to be able to reach every point y from any other point x; in other words, there must be a sequence of transitions that can go from x to y, for any x and y.

If we have a transition process τ that satisfies (11.20), therefore, the sequences it generates will sample P correctly. But can we find such a process? A method that achieves this is the Metropolis algorithm. It has two parts:

(1) A symmetric *proposal* mechanism. Given a point x, this selects a point y with probability $F(y|x)$. *Symmetry* means that $F(y|x) = F(x|y)$.

(2) An *acceptance* mechanism that accepts the proposed y with probability $\min(1, P(y)/P(x))$. In other words, a point y with larger posterior probability than the current x is always accepted, and one with lower probability is accepted randomly with probability $P(y)/P(x)$.

To see that this satisfies (11.20) note that, for $x \neq y$,

$$
\begin{aligned}
P(x)\tau(y|x) &= P(x)F(y|x)\min(1, P(y)/P(x)) \\
&= F(y|x)\min(P(x), P(y)) \\
&= F(x|y)\min(P(y), P(x)) \\
&= P(y)\tau(x|y).
\end{aligned}
$$

Here we used the symmetry of the proposal mechanism to replace $F(y|x)$ in the second line by $F(x|y)$ in the third.

Gibbs sampling

When we have a probabilistic model of many variables, it may often be possible to sample from the distribution obtained by keeping all variables fixed except one, i.e. the conditional distribution. Gibbs sampling exploits this idea. It works by choosing points from the conditional distribution $P(x_i|x_1, \ldots, x_{i-1}, x_{i+1}, \ldots, x_N)$ for each i, cycling repeatedly through $i = 1, \ldots, N$.

To show that this samples correctly from P, it is enough to prove detailed balance. This means that

$$
\begin{aligned}
&P(x_1, \ldots, x_n)P(\tilde{x}_i|x_1, \ldots, x_{i-1}, x_{i+1}, \ldots, x_n) \\
&= P(x_1, \ldots, x_{i-1}, \tilde{x}_i, x_{i+1}, \ldots, x_n)P(x_i|x_1, \ldots, x_{i-1}, x_{i+1}, \ldots, x_n).
\end{aligned}
$$

But we can rewrite this as

$$
\begin{aligned}
&P(x_1, \ldots, x_n)P(x_1, \ldots, x_{i-1}, \tilde{x}_i, x_{i+1}, \ldots x_n)/P(x_1, \ldots, x_{i-1}, x_{i+1}, \ldots, x_n) \\
&= P(x_1, \ldots, x_{i-1}, \tilde{x}_i, x_{i+1}, \ldots, x_n)P(x_1, \ldots, x_n)/P(x_1, \ldots, x_{i-1}, x_{i+1}, \ldots, x_n),
\end{aligned}
$$

which makes the equality obvious. Provided that the process doesn't get stuck in some subset of the parameter space, i.e. provided it is ergodic, Gibbs sampling will inevitably converge to P.

The kind of situation in which Gibbs sampling can get stuck is where there are two pieces of density which do not overlap along any of the coordinate directions, e.g. in the 2D case where half the density lies in the region $[0, 1] \times [0, 1]$ and the other half in the region $[2, 3] \times [2, 3]$. Note that if there were even a small overlap, e.g. if half the density were uniform on $[0, 1] \times [0, 1]$ and the other half uniform on $[0.99, 1.99] \times [0.99, 1.99]$, then sampling would pass between the two regions, albeit making the transition between regions quite infrequently.

Exercise

11.13   What is the expected number of samples within one region, in the preceding example, before a cross-over occurs into the other?

11.5   Estimation of probabilities from counts

Above we used the example of rolling a die. We needed to estimate the parameters of a multinomial from data: rolls of the die. The same abstract situation occurs frequently in sequence analysis, but with the number of rolls n_i with outcome i now meaning something different. For instance, n_i might be the number of times amino acid i occurs in a column of a multiple alignment.

Assume that the observations can be expressed as counts n_i for outcome i ($i = 1, \ldots, K$) and we want to estimate the probabilities θ_i for the underlying multinomial distribution. If we have plenty of data, it is natural to use the observed frequencies, $\theta_i = n_i/N$, as the estimated probabilities. Here $N = \sum_i n_i$. This is the maximum likelihood solution, θ_i^{ML}. The proof that this is so goes as follows.

We want to show that $P(n|\theta^{\text{ML}}) > P(n|\theta)$ for any $\theta \neq \theta^{\text{ML}}$. This is equivalent to showing that $\log[P(n|\theta^{\text{ML}})/P(n|\theta)] > 0$, if we only consider probability parameters yielding a non-zero probability. Using equations (11.3) and the definition of θ^{ML}, this becomes

$$\log \frac{P(n|\theta^{\text{ML}})}{P(n|\theta)} = \log \frac{\prod_i (\theta_i^{\text{ML}})^{n_i}}{\prod_i \theta_i^{n_i}}$$

$$= \sum_i n_i \log \frac{\theta_i^{\text{ML}}}{\theta_i}$$

$$= N \sum_i \theta_i^{\text{ML}} \log \frac{\theta_i^{\text{ML}}}{\theta_i} > 0.$$

The last inequality follows from the fact that the relative entropy (11.10) is always positive except when the two distributions are identical. This concludes the proof.

If data are scarce, it is not so clear what is the best estimate. If, for instance, we only have a total of two counts both on the same residue, the maximum likelihood estimate would give zero probability to all other residues. In this case, we would like to assign some probability to the other residues and not rely entirely on so few observations. Since there are no more observations, these probabilities must be determined from *prior knowledge*. This can be done via Bayesian statistics, and we will now derive the posterior mean estimator for θ.

As the prior we choose the Dirichlet distribution (11.5) with parameters α. We can then calculate the posterior (11.13) for the multinomial distribution with observations n:

$$P(\theta|n) = \frac{P(n|\theta)\mathcal{D}(\theta|\alpha)}{P(n)}.$$

For ease of notation, we have dropped the conditioning on the model M as compared to (11.13), and consider all probabilities implicitly conditioned on the model. Inserting the multinomial distribution (11.3) for $P(n|\theta)$ and the expression (11.5) for $\mathcal{D}(\theta|\alpha)$ yields

$$P(\theta|n) = \frac{1}{P(n)Z(\alpha)M(n)} \prod_i \theta_i^{n_i+\alpha_i-1} = \frac{Z(n+\alpha)}{P(n)Z(\alpha)M(n)} \mathcal{D}(\theta|n+\alpha).$$

In the last step $\prod_i \theta_i^{n_i+\alpha_i-1}$ was recognised as being proportional to the Dirichlet distribution with parameters $n+\alpha$. Here $n+\alpha$ means the set of parameters $\{n_i+\alpha_i\}$ (vector addition). Fortunately we do not have to get involved with gamma functions in order to finish the calculation, because we know that both $P(\theta|n)$ and $\mathcal{D}(\theta|n+\alpha)$ are properly normalised probability distributions over θ. This means that all the prefactors must cancel and

$$P(\theta|n) = \mathcal{D}(\theta|n+\alpha). \tag{11.22}$$

We see that the posterior is itself a Dirichlet distribution like the prior, but of course with different parameters. The observation that the above prefactor is one gives us a little corollary, which will be useful later:

$$P(n) = \frac{Z(n+\alpha)}{Z(\alpha)M(n)}. \tag{11.23}$$

Now, we only need to perform an integral in order to find the posterior mean estimator. From the definition (11.15),

$$\theta_i^{\mathrm{PME}} = \int \theta_i \mathcal{D}(\theta|n+\alpha) d\theta = Z^{-1}(n+\alpha) \int \theta_i \prod_k \theta_k^{n_k+\alpha_k-1} d\theta. \tag{11.24}$$

We can bring θ_i inside the product giving $\theta_i^{n_i+\alpha_i}$ as the ith term. Then we see that the integral is exactly of the form (11.6). We can therefore write

$$\begin{aligned} \theta_i^{\mathrm{PME}} &= \frac{Z(n+\alpha+\delta_i)}{Z(n+\alpha)} \\ &= \frac{n_i+\alpha_i}{N+A}, \end{aligned} \tag{11.25}$$

where $A = \sum_i \alpha_i$, and δ_i is a vector whose ith component is one and all its other components zero. Here we have used the property (11.7) of the gamma function, i.e. $\Gamma(x+1) = x\Gamma(x)$; this allows us to cancel all terms except $n_i+\alpha_i$ in the numerator and $N+A$ in the denominator.

This result should be compared to the ML estimate θ^{ML}. If we think of the αs as extra observations added to the real ones, this is precisely the ML estimate!

The αs are like *pseudocounts* added to the real counts. This makes the Dirichlet regulariser very intuitive, and we can in a sense forget all about Bayesian statistics and think in terms of pseudocounts. It is fairly obvious how to use these pseudocounts: if it is known *a priori* that a certain residue, say number i, is very common, we should give it a high pseudocount α_i, and if residue j is generally rare, we should give it a low pseudocount.

It is important to note the self-regulating property of the pseudocount regulariser. If there are many observations, i.e. the ns are much larger than the αs, then the estimate is essentially equal to the ML estimate. On the other hand, if there are very few observations, the regulariser would dominate and give an estimate close to the normalised αs, $\theta_i \simeq \alpha_i/A$. So typically we would choose the αs so that they are equal to the overall distribution of residues after normalisation.

Mixtures of Dirichlets

It is not easy to express all the prior knowledge about proteins in a single Dirichlet distribution; to achieve that it is natural to use several different Dirichlet distributions. We might for instance have a Dirichlet well suited to exposed amino acids, one for buried ones and so forth. In statistical terms this can be expressed as a *mixture* distribution. Assume we have m Dirichlet distributions characterised by parameter vectors $\alpha^1, \ldots, \alpha^m$. A mixture prior expresses the idea that any probability vector θ belongs to one of the components of the mixture $\mathcal{D}(\theta|\alpha^k)$ with a probability q_k. Formally:

$$P(\theta|\alpha^1,\ldots,\alpha^m) = \sum_k q_k \mathcal{D}(\theta|\alpha^k), \qquad (11.26)$$

where q_k are called the mixture coefficients. The mixture coefficients have to be positive and sum to one in order for the mixture to be a proper probability distribution. (Mixtures can be formed from any types of distributions in this way.) Whereas this probability was called $P(\theta)$ in the previous section, we are here conditioning on the αs, which was implicit before. This turns out to be convenient, because we can then use probabilities like $P(\alpha^1|n)$ below. We can then also identify q_k as the *prior* probability $q_k = P(\alpha^k)$ of each of the mixture coefficients.

For a given mixture, i.e. for fixed α parameters and mixture coefficients, it is straightforward to calculate the posterior probabilities using the results from the previous section. From the definition of conditional probabilities, we have

$$
\begin{aligned}
P(\theta|n) &= \sum_k P(\theta|\alpha^k,n)P(\alpha^k|n) \\
&= \sum_k P(\alpha^k|n)\mathcal{D}(\theta|n+\alpha^k),
\end{aligned}
$$

where we used the expression for the posterior (11.22). To compute the term

$P(\alpha^k|n)$, note that by Bayes' theorem we have

$$P(\alpha^k|n) = \frac{q_k P(n|\alpha^k)}{\sum_l q_l P(n|\alpha^l)},$$

using $q_k = P(\alpha^k)$. The probability $P(n|\alpha^k)$ is given by (11.23) (remember that $P(n)$ in the previous section was implicitly conditioned on the Dirichlet parameters, so it is $P(n|\alpha^k)$), and we get

$$P(\alpha^k|n) = \frac{q_k Z(n+\alpha^k)/Z(\alpha^k)}{\sum_l q_l Z(n+\alpha^l)/Z(\alpha^l)}. \qquad (11.27)$$

The final integration to obtain θ^{PME} can be done using the results (11.24) and (11.25) from the previous section, and yields

$$\theta_i^{\mathrm{PME}} = \sum_k P(\alpha^k|n)\frac{n_i+\alpha_i^k}{N+A}. \qquad (11.28)$$

The estimate using a mixture of Dirichlets is similar to using a single one: you just average the estimate based on each component of the mixture. However, the weight (11.27) with which they are averaged in the mixture is new. This weight is a little hard to understand intuitively, but it gives a high weight to mixture components with a high probability for the sample.

Estimating the prior

For more details of the ideas presented in the preceding section, see Brown *et al.* [1993] and Sjölander *et al.* [1996]. These authors used Dirichlet mixtures to model the distribution of column counts. They obtained the prior by estimating the mixture components and the mixture coefficients from a large dataset, i.e. a large set of count vectors.

The estimation is done as follows: The mixture defines a probability for each count vector in the database, $n^1,\ldots,n^M$,

$$P(n^l|\alpha^1,\ldots,\alpha^m;q_1,\ldots,q_m) = \int P(n^l|\theta)P(\theta|\alpha^1,\ldots,\alpha^m;q_1,\ldots,q_m)d\theta. \quad (11.29)$$

If the count vectors are considered independent, the total likelihood of the mixture is

$$P(\mathrm{data}|\mathrm{mixture}) = \prod_{l=1}^{M} P(n^l|\alpha^1,\ldots,\alpha^m;q_1,\ldots,q_m). \qquad (11.30)$$

This probability can be maximised by gradient descent or some other method of continuous optimisation.

At this point the reader is probably asking 'Why use ML estimation instead of these wonderful Bayesian approaches I just learned?' To do this you just need a prior on the parameters of the first level of priors. You can put priors on prior parameters forever. At some point you have to settle for a prior you invented or one estimated by ML or some other non-Bayesian method.

11.6 The EM algorithm

The expectation maximisation (EM) algorithm is a general algorithm for ML estimation with 'missing data' [Dempster, Laird & Rubin 1977]. The Baum–Welch algorithm for estimating hidden Markov model probabilities is a special case of the EM algorithm. For HMMs the missing data are the unknown states, since we only know the observations and not the sequence of states producing them.

Assume some statistical model is determined by parameters θ. The observed quantities are called x, and the probability of x is determined by some missing data y. For the HMM that we will treat below, θ is the set of all model parameters a and e, and y represents the path through the model. The aim is to find the model that maximises the log likelihood

$$\log P(x|\theta) = \log \sum_y P(x,y|\theta).$$

Here and in the following x means all the observations whether there is one or more sequences. To return to the notation with all sequences shown explicitly requires an extra sum over sequences in all the following formulae.

Assume now that we have a valid model, θ^t. We want to estimate a new and better model, θ^{t+1}. Using $P(x,y|\theta) = P(y|x,\theta)P(x|\theta)$, we can write the log likelihood as

$$\log P(x|\theta) = \log P(x,y|\theta) - \log P(y|x,\theta).$$

Multiplying this with $P(y|x,\theta^t)$ and summing over y yields

$$\log P(x|\theta) = \sum_y P(y|x,\theta^t)\log P(x,y|\theta) - \sum_y P(y|x,\theta^t)\log P(y|x,\theta).$$

The first term on the right we will call $Q(\theta|\theta^t)$,

$$Q(\theta|\theta^t) = \sum_y P(y|x,\theta^t)\log P(x,y|\theta). \qquad (11.31)$$

We want $\log P(x|\theta)$ to be larger than $\log P(x|\theta^t)$, so the difference should be positive. Using the two equations above we can write the difference

$$\log P(x|\theta) - \log P(x|\theta^t) =$$
$$Q(\theta|\theta^t) - Q(\theta^t|\theta^t) + \sum_y P(y|x,\theta^t)\log \frac{P(y|x,\theta^t)}{P(y|x,\theta)}.$$

The last term is the relative entropy (11.10) of $P(y|x,\theta^t)$ relative to $P(y|x,\theta)$, so it is always non-negative, so

$$\log P(x|\theta) - \log P(x|\theta^t) \geq Q(\theta|\theta^t) - Q(\theta^t|\theta^t) \qquad (11.32)$$

with equatilty only if $\theta = \theta^t$, or if $P(y|x,\theta^t) = P(y|x,\theta)$ for some other $\theta \neq \theta^t$. Choosing

$$\theta^{t+1} = \underset{\theta}{\operatorname{argmax}}\, Q(\theta|\theta^t) \qquad (11.33)$$

will always make the difference positive and thus the likelihood of the new model larger than the likelihood of θ^t. Of course, if a maximum has already been reached, $\theta^{t+1} = \theta^t$, and the likelihood will not change.

The function Q in (11.31) is an average of $\log P(x,y|\theta)$ over the distribution of y obtained with the current set of parameters θ^t. This can often be expressed analytically as a function of θ in which the constants are expectation values in the old model. This will be more concrete when we go through the derivation for HMMs shortly. The EM algorithm is usually formulated like this:

Algorithm: Expectation maximisation

E-step: Calculate the Q function (11.31).
M-step: Maximise $Q(\theta|\theta^t)$ with respect to θ.   ◁

We saw above that the likelihood increases in each iteration, so the procedure will always reach a local (or maybe global) maximum asymptotically as $t \to \infty$. For many models, such as HMMs, both of these steps can be carried out analytically. If the second step cannot be carried out exactly, we can use some numerical optimisation technique to maximise Q. In fact, it is not necessary to maximise it; it is enough to make $Q(\theta^{t+1}|\theta^t)$ larger than $Q(\theta^t|\theta^t)$. Algorithms that increase Q – without necessarily maximising it – are called generalised EM (GEM) algorithms [Dempster, Laird & Rubin 1977]. Other generalisations of the EM idea can be found in Meng & Rubin [1992] and Neal & Hinton [1993].

EM explanation of the Baum–Welch algorithm

For the HMM we shall now sketch the derivation of the EM steps which forms the Baum–Welch algorithm described in Chapter 3, p. 63. In this case we want to maximise the likelihood

$$\log P(x|\theta) = \sum_{\pi} \log P(x,\pi|\theta),$$

so the 'missing data' are the state paths π. Then Q (11.31) is given by

$$Q(\theta|\theta^t) = \sum_{\pi} P(\pi|x,\theta^t) \log P(x,\pi|\theta). \qquad (11.34)$$

For a given path each parameter in the model will appear some number of times in $P(x, \pi | \theta)$ given by the product (3.6). If it is a transition probability we will call this number $A_{kl}(\pi)$ and for the emission probabilities $E_k(b, \pi)$, i.e. $E_k(b, \pi)$ is the number of times character b is observed in state k for path π (it depends on the observation sequence, which we do not show explicitly). Then we can write (3.6) as

$$P(x, \pi | \theta) = \prod_{k=1}^{M} \prod_{b} [e_k(b)]^{E_k(b,\pi)} \prod_{k=0}^{M} \prod_{l=1}^{M} a_{kl}^{A_{kl}(\pi)},$$

where the first product is over all characters b in the alphabet. After taking the logarithm, (11.34) can now be written as

$$Q(\theta | \theta^t) = \sum_{\pi} P(\pi | x, \theta^t) \times$$

$$\left[\sum_{k=1}^{M} \sum_{b} E_k(b, \pi) \log e_k(b) + \sum_{k=0}^{M} \sum_{l=1}^{M} A_{kl}(\pi) \log a_{kl} \right]. \quad (11.35)$$

We observe that the expected values A_{kl} and $E_k(b)$ as defined in (3.20) and (3.21) on p. 64 for the Baum–Welch algorithm can be written as expectations of $A_{kl}(\pi)$ and $E_k(b, \pi)$ with respect to $P(\pi | x, \theta^t)$:

$$E_k(b) = \sum_{\pi} P(\pi | x, \theta^t) E_k(b, \pi) \quad \text{and} \quad A_{kl} = \sum_{\pi} P(\pi | x, \theta^t) A_{kl}(\pi).$$

Doing the sum over π first in (11.35) therefore gives

$$Q(\theta | \theta^t) = \sum_{k=1}^{M} \sum_{b} E_k(b) \log e_k(b) + \sum_{k=0}^{M} \sum_{l=1}^{M} A_{kl} \log a_{kl}. \quad (11.36)$$

Finally, we have to show that (3.18) maximises (11.36). Let us first look at the A term. The difference between this term for $a_{ij}^0 = \frac{A_{ij}}{\sum_k A_{ik}}$ and for any other a_{ij} is

$$\sum_{k=0}^{M} \sum_{l=1}^{M} A_{kl} \log \frac{a_{kl}^0}{a_{kl}} = \sum_{k=0}^{M} \left(\sum_{l'} A_{kl'} \right) \sum_{l=1}^{M} a_{kl}^0 \log \frac{a_{kl}^0}{a_{kl}}.$$

The last expression is the relative entropy (11.10), and thus it is larger than zero unless $a_{kl} = a_{kl}^0$. This proves that the maximum is at a_{kl}^0. Exactly the same procedure can be used for the E term.

For the HMM the E-step of the EM algorithm consists of calculating the expectations $E_k(b)$ and A_{kl}. This is done by the forward–backward procedure as described in Chapter 3. This completely determines the Q function, and the maximum is expressed directly in terms of these numbers. Therefore the M-step just consists of plugging $E_k(b)$ and A_{kl} into the re-estimation formulae for $e_k(b)$ and a_{kl} given in (3.18).

Bibliography

Abrahams, J. P., van den Berg, M., van Batenburg, E. and Pleij, C. 1990. Prediction of RNA secondary structure, including pseudoknotting, by computer simulation. *Nucleic Acids Research* 18:3035–3044.

Allison, L. and Wallace, C. S. 1993. The posterior probability distribution of alignments and its application to parameter estimation of evolutionary trees and to optimisation of muliple alignments. Technical Report TR 93/188, Monash University Computer Science.

Allison, L., Wallace, C. S. and Yee, C. N. 1992a. Finite-state models in the alignment of macromolecules. *Journal of Molecular Evolution* 35:77–89.

Allison, L., Wallace, C. S. and Yee, C. N. 1992b. Minimum message length encoding, evolutionary trees and multiple alignment. In *Hawaii International Conference on System Sciences*, volume 1, 663–674.

Altschul, S. F. 1989. Gap costs for multiple sequence alignment. *Journal of Theoretical Biology* 138:297–309.

Altschul, S. F. 1991. Amino acid substitution matrices from an information theoretic perspective. *Journal of Molecular Biology* 219:555–565.

Altschul, S. F. and Erickson, B. W. 1986. Optimal sequence alignment using affine gap costs. *Bulletin of Mathematical Biology* 48:603–616.

Altschul, S. F. and Gish, W. 1996. Local alignment statistics. *Methods in Enzymology* 266:460–480.

Altschul, S. F. and Lipman, D. J. 1989. Trees, stars, and multiple biological sequence alignment. *SIAM Journal of Applied Mathematics* 49:197–209.

Altschul, S. F., Carroll, R. J. and Lipman, D. J. 1989. Weights for data related by a tree. *Journal of Molecular Biology* 207:647–653.

Altschul, S. F., Gish, W., Miller, W., Myers, E. W. and Lipman, D. J. 1990. Basic local alignment search tool. *Journal of Molecular Biology* 215:403–410.

Altschul, S. F., Madden, T. L., Schaffer, A. A., Zhang, J., Zhang, Z., Miller, W. and Lipman, D. J. 1997. Gapped BLAST and PSI-BLAST: a new generation of protein database search programs. *Nucleic Acids Research* 25:3389–3402.

Asai, K., Hayamizu, S. and Handa, K. 1993. Prediction of protein secondary structure by the hidden Markov model. *Computer Applications in the Biosciences* 9:141–146.

Asmussen, S. 1987. *Applied Probability and Queues*. Wiley.

Atteson, K. 1997. The performance of the neighbor-joining method of phylogeny reconstruction. In Mirkin, B., McMorris, F., Roberts, F. and Rzhetsky, A., eds.,

Mathematical Hierarchies and Biology. American Mathematical Society. 133–148.

Bahl, L. R., Brown, P. F., de Souza, P. V. and Mercer, R. L. 1986. Maximum mutual information estimation of hidden Markov model parameters for speech recognition. In *Proceedings of ICASSP '86*, 49–52.

Bailey, T. L. and Elkan, C. 1994. Fitting a mixture model by expectation maximization to discover motifs in biopolymers. In Altman, R., Brutlag, D., Karp, P., Lathrop, R. and Searls, D., eds., *Proceedings of the Second International Conference on Intelligent Systems for Molecular Biology*, 28–36. AAAI Press.

Bailey, T. L. and Elkan, C. 1995. The value of prior knowledge in discovering motifs with MEME. In Rawlings, C., Clark, D., Altman, R., Hunter, L., Lengauer, T. and Wodak, S., eds., *Proceedings of the Third International Conference on Intelligent Systems for Molecular Biology*, 21–29. AAAI Press.

Bairoch, A. and Apweiler, R. 1997. The SWISS-PROT protein sequence data bank and its supplement TrEMBL. *Nucleic Acids Research* 25:31–36.

Bairoch, A., Bucher, P. and Hofmann, K. 1997. The PROSITE database, its status in 1997. *Nucleic Acids Research* 25:217–221.

Baldi, P. and Brunak, S. 1998. *Bioinformatics – The Machine Learning Approach*. MIT Press.

Baldi, P. and Chauvin, Y. 1994. Smooth on-line learning algorithms for hidden Markov models. *Neural Computation* 6:307–318.

Baldi, P. and Chauvin, Y. 1995. Protein modeling with hybrid hidden Markov model/neural network architectures. In Rawlings, C., Clark, D., Altman, R., Hunter, L., Lengauer, T. and Wodak, S., eds., *Proceedings of the Third International Conference on Intelligent Systems for Molecular Biology*, 39–47. AAAI Press.

Baldi, P., Brunak, S., Chauvin, Y. and Krogh, A. 1996. Naturally occurring nucleosome positioning signals in human exons. *Journal of Molecular Biology* 263:503–510.

Baldi, P., Chauvin, Y., Hunkapiller, T. and McClure, M. A. 1994. Hidden Markov models of biological primary sequence information. *Proceedings of the National Academy of Sciences of the USA* 91:1059–1063.

Bandelt, H.-J. and Dress, A. W. M. 1992. Split decomposition: a new and useful approach to phylogenetic analysis of distance data. *Molecular Phylogenetics and Evolution* 1:242–252.

Barton, G. J. 1993. An efficient algorithm to locate all locally optimal alignments between two sequences allowing for gaps. *Computer Applications in the Biosciences* 9:729–734.

Barton, G. J. and Sternberg, M. J. E. 1987. A strategy for the rapid multiple alignment of protein sequences. *Journal of Molecular Biology* 198:327–337.

Baserga, S. J. and Steitz, J. A. 1993. The diverse world of small ribonucleoproteins. In Gesteland, R. F. and Atkins, J. F., eds., *The RNA World*. Cold Spring Harbor Press. pp. 359–381.

Bashford, D., Chothia, C. and Lesk, A. M. 1987. Determinants of a protein fold:

unique features of the globin amino acid sequence. *Journal of Molecular Biology* 196:199–216.

Baum, L. E. 1972. An equality and associated maximization technique in statistical estimation for probabilistic functions of Markov processes. *Inequalities* 3:1–8.

Bengio, Y., De Mori, R., Flammia, G. and Kompe, R. 1992. Global optimization of a neural network–hidden Markov model hybrid. *IEEE Transactions on Neural Networks* 3:252–259.

Berger, J. O. 1985. *Statistical Decision Theory and Bayesian Analysis*. Springer-Verlag.

Berger, M. P. and Munson, P. J. 1991. A novel randomized iterative strategy for aligning multiple protein sequences. *Computer Applications in the Biosciences* 7:479–484.

Binder, K. and Heerman, D. W. 1988. *Monte Carlo Simulation in Statistical Mechanics*. Springer-Verlag.

Bird, A. 1987. CpG islands as gene markers in the vertebrate nucleus. *Trends in Genetics* 3:342–347.

Birney, E. and Durbin, R. 1997. Dynamite: a flexible code generating language for dynamic programming methods used in sequence comparison. In Gaasterland, T., Karp, P., Karplus, K., Ouzounis, C., Sander, C. and Valencia, A., eds., *Proceedings of the Fifth International Conference on Intelligent Systems for Molecular Biology*, 56–64. AAAI Press.

Bishop, M. J. and Thompson, E. A. 1986. Maximum likelihood alignment of DNA sequences. *Journal of Molecular Biology* 190:159–165.

Borodovsky, M. and McIninch, J. 1993. GENMARK: parallel gene recognition for both DNA strands. *Computers and Chemistry* 17:123–133.

Borodovsky, M. Y., Sprizhitsky, Y. A., Golovanov, E. I. and Alexandrov, A. A. 1986a. Statistical patterns in the primary structure of the functional regions of the Escherichia coli genome. I. Frequency characteristics. *Molecularnaya Biologia* 20:826–833. (English translation).

Borodovsky, M. Y., Sprizhitsky, Y. A., Golovanov, E. I. and Alexandrov, A. A. 1986b. Statistical patterns in the primary structure of the functional regions of the Escherichia Coli genome. II. Nonuniform Markov models. *Molecularnaya Biologia* 20:833–840. (English translation).

Borodovsky, M. Y., Sprizhitsky, Y. A., Golovanov, E. I. and Alexandrov, A. A. 1986c. Statistical patterns in the primary structure of the functional regions of the Escherichia Coli genome. III. Computer recognition of coding regions. *Molecularnaya Biologia* 20:1144–1150. (English translation).

Bowie, J. U., Luthy, R. and Eisenberg, D. 1991. A method to identify protein sequences that fold into a known three-dimensional structure. *Science* 253:164–170.

Box, G. E. P. and Tiao, G. C. 1992. *Bayesian Inference in Statistical Analysis*. Wiley-Interscience.

Branden, C. and Tooze, J. 1991. *Introduction to Protein Structure*. Garland.

Brendel, V., Beckmann, J. S. and Trifonov, E. N. 1986. Linguistics of nucleotide sequences: morphology and comparison of vocabularies. *Journal of Biomolecular Structure and Dynamics* 4:11–20.

Brooks, D. R. and McLennan, D. A. 1991. *Phylogeny, Ecology and Behaviour.* University of Chicago Press.

Brown, M. and Wilson, C. 1995. RNA pseudoknot modeling using intersections of stochastic context-free grammars with applications to database search. Unpublished manuscript available from http://www.cse.ucsc.edu/research/compbio/pseudoknot.html.

Brown, M., Hughey, R., Krogh, A., Mian, I. S., Sjölander, K. and Haussler, D. 1993. Using Dirichlet mixture priors to derive hidden Markov models for protein families. In Hunter, L., Searls, D. B. and Shavlik, J., eds., *Proceedings of the First International Conference on Intelligent Systems for Molecular Biology,* 47–55. AAAI Press.

Brunak, S., Engelbrecht, J. and Knudsen, S. 1991. Prediction of human mRNA donor and acceptor sites from the DNA sequence. *Journal of Molecular Biology* 220:49–65.

Bucher, P. and Hofmann, K. 1996. A sequence similarity search algorithm based on a probabilistic interpretation of an alignment scoring system. In States, D. J., Agarwal, P., Gaasterland, T., Hunter, L. and Smith, R. F., eds., *Proceedings of the Fourth International Conference on Intelligent Systems for Molecular Biology,* 44–51. AAAI Press.

Bucher, P., Karplus, K., Moeri, N. and Hofmann, K. 1996. A flexible motif search technique based on generalized profiles. *Computers and Chemistry* 20:3–24.

Buneman, P. 1971. The recovery of trees from measures of dissimilarity. In Hodson, F. R., Kendall, D. G. and Tautu, P., eds., *Mathematics in the Archaeological and Historical Sciences.* Edinburgh University Press. pp. 387–395.

Burge, C. and Karlin, S. 1997. Prediction of complete gene structures in human genomic DNA. *Journal of Molecular Biology* 268:78–94.

Camin, J. H. and Sokal, R. R. 1965. A method for deducing branching sequences in phylogeny. *Evolution* 19:311–327.

Cardon, L. R. and Stormo, G. D. 1992. Expectation maximization algorithm for identifying protein-binding sites with variable lengths from unaligned DNA fragments. *Journal of Molecular Biology* 223:159–170.

Carrillo, H. and Lipman, D. 1988. The multiple sequence alignment problem in biology. *SIAM Journal of Applied Mathematics* 48:1073–1082.

Cary, R. B. and Stormo, G. D. 1995. Graph-theoretic approach to RNA modeling using comparative data. In Rawlings, C., Clark, D., Altman, R., Hunter, L., Lengauer, T. and Wodak, S., eds., *Proceedings of the Third International Conference on Intelligent Systems for Molecular Biology,* 75–80. AAAI Press.

Casella, G. and Berger, R. L. 1990. *Statistical Inference.* Duxbury Press.

Cavender, J. A. 1978. Taxonomy with confidence. *Mathematical Biosciences* 40:271–280.

Cech, T. R. and Bass, B. L. 1986. Biological catalysis by RNA. *Annual Review of Biochemistry* 55:599–629.

Chan, S. C., Wong, A. K. C. and Chiu, D. K. Y. 1992. A survey of multiple sequence comparison methods. *Bulletin of Mathematical Biology* 54:563–598.

Chang, W. I. and Lawler, E. L. 1990. Approximate string matching in sublinear expected time. In *Proceedings of the 31st Annual IEEE Symposium on Foundations Computer Science*, 116–124. IEEE.

Chao, K. M., Hardison, R. C. and Miller, W. 1994. Recent developments in linear-space alignment methods: a survey. *Journal of Computational Biology* 1:271–291.

Chao, K. M., Pearson, W. R. and Miller, W. 1992. Aligning two sequences within a specified diagonal band. *Computer Applications in the Biosciences* 8:481–487.

Chiu, D. K. Y. and Kolodziejczak, T. 1991. Inferring consensus structure from nucleic acid sequences. *Computer Applications in the Biosciences* 7:347–352.

Chomsky, N. 1956. Three models for the description of language. *IRE Transactions Information Theory* 2:113–124.

Chomsky, N. 1959. On certain formal properties of grammars. *Information and Control* 2:137–167.

Chothia, C. and Lesk, A. M. 1986. The relation between the divergence of sequence and structure in proteins. *EMBO Journal* 5:823–826.

Churchill, G. A. 1989. Stochastic models for heterogeneous DNA sequences. *Bulletin of Mathematical Biology* 51:79–94.

Churchill, G. A. 1992. Hidden markov chains and the analysis of genome structure. *Computers and Chemistry* 16:107–115.

Claverie, J.-M. 1994. Some useful statistical properties of position-weight matrices. *Computers and Chemistry* 18:287–294.

Collado-Vides, J. 1989. A transformational-grammar approach to the study of the regulation of gene expression. *Journal of Theoretical Biology* 136:403–425.

Collado-Vides, J. 1991. A syntactic representation of units of genetic information – a syntax of units of genetic information. *Journal of Theoretical Biology* 148:401–429.

Corpet, F. and Michot, B. 1994. RNAlign program: alignment of RNA sequences using both primary and secondary structures. *Computer Applications in the Biosciences* 10:389–399.

Cover, T. M. and Thomas, J. A. 1991. *Elements of Information Theory*. John Wiley & Sons, Inc.

Cox, D. R. 1962. Further results on tests of separate families of hypotheses. *Journal of the Royal Statistical Society, B* 24:406–424.

Cox, D. R. and Miller, H. D. 1965. *The Theory of Stochastic Processes*. Chapman & Hall.

Dandekar, T. and Hentze, M. W. 1995. Finding the hairpin in the haystack: searching for RNA motifs. *Trends in Genetics* 11:45–50.

Dayhoff, M. O., Eck, R. V. and Park, C. M. 1972. In Dayhoff, M. O., ed., *Atlas of Protein Sequence and Structure*, volume 5. National Biomedical Research Foundation, Washington D.C. pp. 89–99.

Dayhoff, M. O., Schwartz, R. M. and Orcutt, B. C. 1978. A model of evolutionary change in proteins. In Dayhoff, M. O., ed., *Atlas of Protein Sequence and Structure*, volume 5, supplement 3. National Biomedical Research Foundation, Washington D.C. pp. 345–352.

Dembo, A. and Karlin, S. 1991. Strong limit theorems of empirical functionals for large exceedances of partial sums of i.i.d. variables. *Annals of Probability* 19:1737–1755.

Dempster, A. P., Laird, N. M. and Rubin, D. B. 1977. Maximum likelihood from incomplete data via the EM algorithm. *Journal of the Royal Statistical Society* B 39:1–38.

Dong, S. and Searls, D. B. 1994. Gene structure prediction by linguistic methods. *Genomics* 23:540–551.

Doolittle, R. F., Feng, D.-F., Tsang, S., Cho, G. and Little, E. 1996. Determining divergence times of the major kingdoms of living organisms with a protein clock. *Science* 271:470–477.

Eck, R. V. and Dayhoff, M. O. 1966. *Atlas of Protein Sequence and Structure*. National Biomedical Research Foundation.

Eddy, S. R. 1995. Multiple alignment using hidden Markov models. In Rawlings, C., Clark, D., Altman, R., Hunter, L., Lengauer, T. and Wodak, S., eds., *Proceedings of the Third International Conference on Intelligent Systems for Molecular Biology*, 114–120. AAAI Press.

Eddy, S. R. 1996. Hidden Markov models. *Current Opinion in Structural Biology* 6:361–365.

Eddy, S. R. and Durbin, R. 1994. RNA sequence analysis using covariance models. *Nucleic Acids Research* 22:2079–2088.

Eddy, S. R., Mitchison, G. and Durbin, R. 1995. Maximum discrimination hidden Markov models of sequence consensus. *Journal of Computational Biology* 2:9–23.

Edwards, A. W. F. 1970. Estimation of the branch points of a branching diffusion process. *Journal of the Royal Statistical Society, B* 32:155–174.

Edwards, A. W. F. 1992. *Likelihood*. Johns Hopkins Universty Press.

Edwards, A. W. F. 1996. The origin and early development of the method of minimum evolution for the reconstruction of phylogenetic trees. *Systematic Biology* 45:179–191.

Edwards, A. W. F. and Cavalli-Sforza, L. 1963. The reconstruction of evolution. *Annals of Human Genetics* 27:105.

Edwards, A. W. F. and Cavalli-Sforza, L. 1964. Reconstruction of evolutionary trees. In Heywood, V. H. and McNeill, J., eds., *Phenetic and Phylogenetic Classification*. Systematics Association Publication No. 6. pp. 67–76.

Efron, B. and Tibshirani, R. J. 1993. *An Introduction to the Bootstrap*. Chapman and Hall.

Efron, B., Halloran, E. and Holmes, S. 1996. Bootstrap confidence levels for phylogenetic trees. *Proceedings of the National Academy of Sciences of the USA* 93:13429–13434.

Feller, W. 1971. *An Introduction to Probability Theory and its Applications, Vol II.* John Wiley and Sons.

Felsenstein, J. 1973. Maximum-likelihood estimation of evolutionary trees from continuous characters. *American Journal of Human Genetics* 25:471–492.

Felsenstein, J. 1978a. Cases in which parsimony or compatibility methods will be positively misleading. *Systematic Zoology* 27:401–410.

Felsenstein, J. 1978b. The number of evolutionary trees. *Systematic Zoology* 27:27–33.

Felsenstein, J. 1981a. Evolutionary trees from DNA sequences: a maximum likelihood approach. *Journal of Molecular Evolution* 17:368–376.

Felsenstein, J. 1981b. A likelihood approach to character weighting and what it tells us about parsimony and compatibility. *Biological Journal of the Linnean Society* 16:183–196.

Felsenstein, J. 1985. Confidence limits on phylogenies: an approach using the bootstrap. *Evolution* 39:783–791.

Felsenstein, J. 1996. Inferring phylogenies from protein sequences by parsimony, distance, and likelihood methods. *Methods in Enzymology* 266:418–427.

Felsenstein, J. and Churchill, G. A. 1996. A hidden Markov model approach to variation among sites in rate of evolution. *Molecular Biology and Evolution* 13:93–104.

Feng, D.-F. and Doolittle, R. F. 1987. Progressive sequence alignment as a prerequisite to correct phylogenetic trees. *Journal of Molecular Evolution* 25:351–360.

Feng, D.-F. and Doolittle, R. F. 1996. Progressive alignment of amino acid sequences and construction of phylogenetic trees from them. *Methods in Enzymology* 266:368–382.

Fichant, G. A. and Burks, C. 1991. Identifying potential tRNA genes in genomic DNA sequences. *Journal of Molecular Biology* 220:659–671.

Fields, D. S. and Gutell, R. R. 1996. An analysis of large rRNA sequences folded by a thermodynamic method. *Folding and Design* 1:419–430.

Fitch, W. M. 1971. Toward defining the course of evolution: minimum change for a specifed tree topology. *Systematic Zoology* 20:406–416.

Fitch, W. M. and Margoliash, E. 1967a. Construction of phylogenetic trees. *Science* 155:279–284.

Fitch, W. M. and Margoliash, E. 1967b. A method for estimating the number of invariant amino acid coding positions in a gene using cytochrome c as a model case. *Biochemical Genetics* 1:65–71.

Frasconi, P. and Bengio, Y. 1994. An EM approach to grammatical inference: input/output HMMs. In *Proceedings of the 12th IAPR International Conference on Pattern Recognition*, volume 2, 289–294. IEEE Comput. Soc. Press.

Freier, S. M., Kierzek, R., Jaeger, J. A., Sugimoto, N., Caruthers, M. H., Neilson, T. and Turner, D. H. 1986. Improved free-energy parameters for predictions of RNA duplex stability. *Proceedings of the National Academy of Sciences of the USA* 83:9373–9377.

Fujiwara, Y., Asogawa, M. and Konagaya, A. 1994. Stochastic motif extraction using hidden Markov model. In Altman, R., Brutlag, D., Karp, P., Lathrop, R. and Searls, D., eds., *Proceedings of the Second International Conference on Intelligent Systems for Molecular Biology*, 121–129. AAAI Press.

Gautheret, D., Major, F. and Cedergren, R. 1990. Pattern searching/alignment with RNA primary and secondary structures: an effective descriptor for tRNA. *Computer Applications in the Biosciences* 6:325–331.

Gerstein, M. and Levitt, M. 1996. Using iterative dynamic programming to obtain accurate pairwise and multiple alignments of protein structures. In States, D. J., Agarwal, P., Gaasterland, T., Hunter, L. and Smith, R. F., eds., *Proceedings of the Fourth International Conference on Intelligent Systems for Molecular Biology*, 59–67. AAAI Press.

Gerstein, M., Sonnhammer, E. L. L. and Chothia, C. 1994. Volume changes in protein evolution. *Journal of Molecular Biology* 236:1067–1078.

Gersting, J. L. 1993. *Mathematical Structures for Computer Science*. W. H. Freeman.

Gesteland, R. F. and Atkins, J. F., eds. 1993. *The RNA World*. Cold Spring Harbor Laboratory Press.

Gilbert, W. 1986. The RNA world. *Nature* 319:618.

Gold, L., Polisky, B., Uhlenbeck, O. and Yarus, M. 1995. Diversity of oligonucleotide functions. *Annual Review of Biochemistry* 64:763–797.

Goldman, N. 1993. Statistical tests of models of DNA substitution. *Journal of Molecular Evolution* 36:182–198.

Goldman, N. and Yang, Z. 1994. A codon-based model of nucleotide substitution for protein-coding DNA sequences. *Molecular Biology and Evolution* 11:725–735.

Gonnet, G. H., Cohen, M. A. and Benner, S. A. 1992. Exhaustive matching of the entire protein sequence database. *Science* 256:1443–1445.

Gotoh, O. 1982. An improved algorithm for matching biological sequences. *Journal of Molecular Biology* 162:705–708.

Gotoh, O. 1993. Optimal alignment between groups of sequences and its application to multiple sequence alignment. *Computer Applications in the Biosciences* 9:361–370.

Gotoh, O. 1996. Significant improvement in accuracy of multiple protein alignments by iterative refinement as assessed by reference to structural alignments. *Journal of Molecular Biology* 264:823–838.

Grate, L. 1995. Automatic RNA secondary structure determination with stochastic context-free grammars. In Rawlings, C., Clark, D., Altman, R., Hunter, L., Lengauer, T. and Wodak, S., eds., *Proceedings of the Third International Conference on Intelligent Systems for Molecular Biology*, 136–144. AAAI Press.

Gribskov, M. and Veretnik, S. 1996. Identification of sequence patterns with profile analysis. *Methods in Enzymology* 266:198–212.

Gribskov, M., Lüthy, R. and Eisenberg, D. 1990. Profile analysis. *Methods in Enzymology* 183:146–159.

Gribskov, M., McLachlan, A. D. and Eisenberg, D. 1987. Profile analysis: detection of distantly related proteins. *Proceedings of the National Academy of Sciences of the USA* 84:4355–4358.

Gultyaev, A. P. 1991. The computer simulation of RNA folding involving pseudoknot formation. *Nucleic Acids Research* 19:2489–2494.

Gumbel, E. J. 1958. *Statistics of Extremes*. Columbia University Press.

Gupta, S. K., Kececioglu, J. D. and Schaffer, A. A. 1995. Improving the practical space and time efficiency of the shortest-paths approach to sum-of-pairs multiple sequence alignment. *Journal of Computational Biology* 2:459–472.

Gutell, R. R. 1993. Collection of small subunit (16S and 16S-like) ribosomal RNA structures. *Nucleic Acids Research* 21:3051–3054.

Gutell, R. R., Power, A., Hertz, G. Z., Putz, E. J. and Stormo, G. D. 1992. Identifying constraints on the higher-order structure of RNA: continued development and application of comparative sequence analysis methods. *Nucleic Acids Research* 20:5785–5795.

Hannenhalli, S., Chappey, C., Koonin, E. V. and Pevsner, P. A. 1995. Genome sequence comparison and scenarios for gene rearrangements: a test case. *Genomics* 30:299–311.

Harpaz, Y. and Chothia, C. 1994. Many of the immunoglobulin superfamily domains in cell adhesion molecules and surface receptors belong to a new structural set which is close to that containing variable domains. *Journal of Molecular Biology* 238:528–539.

Harrison, M. A. 1978. *Introduction to Formal Language Theory*. Addison-Wesley.

Hasegawa, M., Kishino, H. and Yano, T. 1985. Dating the human-ape splitting by a molecular clock of mitochondrial DNA. *Journal of Molecular Evolution* 22:160–174.

Haussler, D., Krogh, A., Mian, I. S. and Sjölander, K. 1993. Protein modeling using hidden Markov models: analysis of globins. In Mudge, T. N., Milutinovic, V. and Hunter, L., eds., *Proceedings of the Twenty-Sixth Annual Hawaii International Conference on System Sciences*, volume 1, 792–802. IEEE Computer Society Press.

Hebsgaard, S. M., Korning, P. G., Tolstrup, N., Engelbrecht, J., Rouzé, P. and Brunak, S. 1996. Splice site prediction in Arabidopsis thaliana pre-mRNA by combining local and global sequence information. *Nucleic Acids Research* 24:3439–3452.

Hein, J. 1989a. A new method that simultaneously aligns and reconstructs ancestral sequences for any number of homologous sequences, when the phylogeny is given. *Molecular Biology and Evolution* 6:649–668.

Hein, J. 1989b. A tree reconstruction method that is economical in the number of pairwise comparisons used. *Molecular Biology and Evolution* 6:669–684.

Hein, J. 1993. A heuristic method to reconstruct the history of sequences subject to recombination. *Journal of Molecular Evolution* 36:396–405.

Henderson, J., Salzberg, S. and Fasman, K. H. 1997. Finding genes in DNA with a hidden Markov model. *Journal of Computational Biology* 4:127–141.

Hendy, M. D. and Penny, D. 1989. A framework for the quantitative study of evolutionary trees. *Systematic Zoology* 38:297–309.

Henikoff, J. G. and Henikoff, S. 1996. Using substitution probabilities to improve position-specific scoring matrices. *Computer Applications in the Biosciences* 12:135–143.

Henikoff, S. and Henikoff, J. G. 1991. Automated assembly of protein blocks for database searching. *Nucleic Acids Research* 19:6565–6572.

Henikoff, S. and Henikoff, J. G. 1992. Amino acid substitution matrices from protein blocks. *Proceedings of the National Academy of Sciences of the USA* 89:10915–10919.

Henikoff, S. and Henikoff, J. G. 1994. Position-based sequence weights. *Journal of Molecular Biology* 243:574–578.

Hertz, G. Z., Hartzell III, G. W. and Stormo, G. D. 1990. Identification of consensus patterns in unaligned DNA sequences known to be functionally related. *Computer Applications in the Biosciences* 6:81–92.

Higgins, D. G. and Sharp, P. M. 1989. Fast and sensitive multiple sequence alignments on a microcomputer. *Computer Applications in the Biosciences* 5:151–153.

Higgins, D. G., Bleasby, A. J. and Fuchs, R. 1992. CLUSTAL V: improved software for multiple sequence alignment. *Computer Applications in the Biosciences* 8:189–191.

Hillis, D. M. and Bull, J. J. 1993. An empirical test of bootstrapping as a method for assessing confidence in phylogenetic analysis. *Systematic Biology* 42:182–192.

Hillis, D. M., Bull, J. J., White, M. E., Badgett, M. R. and Molineux, I. J. 1992. Experimental phylogenetics: generation of a known phylogeny. *Science* 255:589–592.

Hirosawa, M., Hoshida, M., Ishikawa, M. and Toya, T. 1993. MASCOT: multiple alignment system for protein sequences based on three-way dynamic programming. *Computer Applications in the Biosciences* 9:161–167.

Hirschberg, D. S. 1975. A linear space algorithm for computing maximal common subsequences. *Communications of the ACM* 18:341–343.

Hogeweg, P. and Hesper, B. 1984. The alignment of sets of sequences and the construction of phyletic trees: an integrated method. *Journal of Molecular Evolution* 20:175–186.

Holm, L. and Sander, C. 1993. Protein structure comparison by alignment of distance matrices. *Journal of Molecular Biology* 233:123–138.

Hopcroft, J. E. and Ullman, J. D. 1979. *Introduction to Automata Theory, Languages, and Computation*. Addison-Wesley.

Huang, X. and Zhang, J. 1996. Methods for comparing a DNA sequence with a protein sequence. *Computer Applications in the Biosciences* 12:497–506.

Hudson, R. R. 1990. Gene genealogies and the coalescent process. In Futuyma, D. and Antonovics, J., eds., *Gene Genealogies and the Coalescent Process*. Oxford University Press. pp. 1–44.

Huelsenbeck, J. P. and Rannala, B. 1997. Phylogenetic methods come of age: testing hypotheses in an evolutionary context. *Science* 276:227–232.

Hughey, R. and Krogh, A. 1996. Hidden Markov models for sequence analysis: extension and analysis of the basic method. *Computer Applications in the Biosciences* 12:95–107.

Jacob, F. 1977. Evolution and tinkering. *Science* 196:1161–1166.

Jefferys, W. H. and Berger, J. O. 1992. Ockham's razor and Bayesian analysis. *American Scientist* 80:64–72.

Juang, B. H. and Rabiner, L. R. 1991. Hidden Markov models for speech recognition. *Technometrics* 33:251–272.

Jukes, T. H. and Cantor, C. 1969. Evolution of protein molecules. In *Mammalian Protein Metabolism*. Academic Press. pp. 21–132.

Karlin, S. and Altschul, S. F. 1990. Methods for assessing the statistical significance of molecular sequence features by using general scoring schemes. *Proceedings of the National Academy of Sciences of the USA* 87:2264–2268.

Karlin, S. and Altschul, S. F. 1993. Applications and statistics for multiple high-scoring segments in molecular sequences. *Proceedings of the National Academy of Sciences of the USA* 90:5873–5877.

Karplus, K. 1995. Evaluating regularizers for estimating distributions of amino acids. In Rawlings, C., Clark, D., Altman, R., Hunter, L., Lengauer, T. and Wodak, S., eds., *Proceedings of the Third International Conference on Intelligent Systems for Molecular Biology*, 188–196. AAAI Press.

Keeping, E. S. 1995. *Introduction to Statistical Inference*. Dover Publications.

Kim, J. and Pramanik, S. 1994. An efficient method for multiple sequence alignment. In Altman, R., Brutlag, D., Karp, P., Lathrop, R. and Searls, D., eds., *Proceedings of the Second International Conference on Intelligent Systems for Molecular Biology*, 212–218. AAAI Press.

Kim, J., Pramanik, S. and Chung, M. J. 1994. Multiple sequence alignment using simulated annealing. *Computer Applications in the Biosciences* 10:419–426.

Kimura, M. 1980. A simple method for estimating evolutionary rates of base substitutions through comparative studies of necleotide sequences. *Journal of Molecular Evolution* 16:111–120.

Kimura, M. 1983. *The Neutral Theory of Molecular Evolution*. Cambridge University Press.

Kingman, J. F. C. 1982a. The coalescent. *Stochastic Processes and their Applications* 13:235–248.

Kingman, J. F. C. 1982b. On the genealogy of large populations. *Journal of Applied Probability* 19A:27–43.

Kirkpatrick, S., Gelatt, Jr., C. D. and Vecchi, M. P. 1983. Optimization by simulated annealing. *Science* 220:671–680.

Kishino, H., Miyata, T. and Hasegawa, M. 1990. Maximum likelihood inference of protein phylogeny and the origin of chloroplasts. *Journal of Molecular Evolution* 31:151–160.

Konings, D. A. M. and Gutell, R. R. 1995. A comparison of thermodynamic foldings with comparatively derived structures of 16S and 16S-like rRNAs. *RNA* 1:559–574.

Konings, D. A. M. and Hogeweg, P. 1989. Pattern analysis of RNA secondary structure: similarity and consensus of minimal-energy folding. *Journal of Molecular Biology* 207:597–614.

Krogh, A. 1994. Hidden Markov models for labeled sequences. In *Proceedings of the 12th IAPR International Conference on Pattern Recognition*, 140–144. IEEE Computer Society Press.

Krogh, A. 1997a. Gene finding: putting the parts together. In Bishop, M., ed., *Guide to Human Genome Computing*. Academic Press, 2nd edition. To appear.

Krogh, A. 1997b. Two methods for improving performance of a HMM and their application for gene finding. In Gaasterland, T., Karp, P., Karplus, K., Ouzounis, C., Sander, C. and Valencia, A., eds., *Proceedings of the Fifth International Conference on Intelligent Systems for Molecular Biology*, 179–186. AAAI Press.

Krogh, A. 1998. An introduction to hidden Markov models for biological sequences. In Salzberg, S., Searls, D. and Kasif, S., eds., *Computational Biology: Pattern Analysis and Machine Learning Methods*. Elsevier. Chapter 4. In press.

Krogh, A. and Mitchison, G. 1995. Maximum entropy weighting of aligned sequences of proteins or DNA. In Rawlings, C., Clark, D., Altman, R., Hunter, L., Lengauer, T. and Wodak, S., eds., *Proceedings of the Third International Conference on Intelligent Systems for Molecular Biology*, 215–221. AAAI Press.

Krogh, A., Mian, I. S. and Haussler, D. 1994. A hidden Markov model that finds genes in *E. coli* DNA. *Nucleic Acids Research* 22:4768–4778.

Krogh, A., Brown, M., Mian, I. S., Sjölander, K. and Haussler, D. 1994. Hidden Markov models in computational biology: applications to protein modeling. *Journal of Molecular Biology* 235:1501–1531.

Kuhner, M. K., Yamato, J. and Felsenstein, J. 1995. Estimating effective population size and mutation rate from sequence data using Metropolis–Hastings sampling. *Genetics* 140:1421–1430.

Kulp, D., Haussler, D., Reese, M. G. and Eeckman, F. H. 1996. A generalized hidden Markov model for the recognition of human genes in DNA. In States, D. J., Agarwal, P., Gaasterland, T., Hunter, L. and Smith, R. F., eds., *Proceedings of the Fourth International Conference on Intelligent Systems for Molecular Biology*, 134–142. AAAI Press.

Langley, C. H. and Fitch, W. M. 1974. An examination of the constancy of the rate of molecular evolution. *Journal of Molecular Evolution* 3:161–177.

Lari, K. and Young, S. J. 1990. The estimation of stochastic context-free grammars using the inside–outside algorithm. *Computer Speech and Language* 4:35–56.

Lari, K. and Young, S. J. 1991. Applications of stochastic context-free grammars using the inside–outside algorithm. *Computer Speech and Language* 5:237–257.

Larsen, N. and Zwieb, C. 1993. The signal recognition particle database (SRPDB). *Nucleic Acids Research* 21:3019–3020.

Law, A. M. and Kelton, W. D. 1991. *Simulation Modelling and Analysis*. McGraw-Hill.

Lawrence, C. E. and Reilly, A. A. 1990. An expectation maximization (EM) algorithm for the identification and characterization of common sites in unaligned biopolymer sequences. *Proteins* 7:41–51.

Lawrence, C. E., Altschul, S. F., Boguski, M. S., Liu, J. S., Neuwald, A. F. and Wootton, J. C. 1993. Detecting subtle sequence signals: a Gibbs sampling strategy for multiple alignment. *Science* 262:208–214.

Lefebvre, F. 1995. An optimized parsing algorithm well suited to RNA folding. In Rawlings, C., Clark, D., Altman, R., Hunter, L., Lengauer, T. and Wodak, S., eds., *Proceedings of the Third International Conference on Intelligent Systems for Molecular Biology*, 222–230. AAAI Press.

Lefebvre, F. 1996. A grammar-based unification of several alignment and folding algorithms. In States, D. J., Agarwal, P., Gaasterland, T., Hunter, L. and Smith, R. F., eds., *Proceedings of the Fourth International Conference on Intelligent Systems for Molecular Biology*, 143–154. AAAI Press.

Lindenmayer, A. 1968. Mathematical models for cellular interactions in development I. filaments with one-sided inputs. *Journal of Theoretical Biology* 18:280–299.

Lipman, D. J., Altschul, S. F. and Kececioglu, J. D. 1989. A tool for multiple sequence alignment. *Proceedings of the National Academy of Sciences of the USA* 86:4412–4415.

Lisacek, F., Diaz, Y. and Michel, F. 1994. Automatic identification of group I intron cores in genomic DNA sequences. *Journal of Molecular Biology* 235:1206–1217.

Lowe, T. M. and Eddy, S. R. 1997. tRNAscan-SE: a program for improved detection of transfer RNA genes in genomic sequence. *Nucleic Acids Research* 25:955–964.

Lukashin, A. V., Engelbrecht, J. and Brunak, S. 1992. Multiple alignment using simulated annealing: branch point definition in human mRNA splicing. *Nucleic Acids Research* 20:2511–2516.

Luthy, R., McLachlan, A. D. and Eisenberg, D. 1991. Secondary structure-based profiles: use of structure-conserving scoring tables in searching protein sequence databases for structural similarities. *Proteins* 10:229–239.

Luthy, R., Xenarios, I. and Bucher, P. 1994. Improving the sensitivity of the sequence profile method. *Protein Science* 3:139–146.

MacKay, D. J. C. 1992. Bayesian interpolation. *Neural Computation* 4:415–447.

MacKay, D. J. C. and Peto, L. 1995. A hierarchical Dirichlet language model. *Natural Language Engineering* 1:1–19.

Margalit, H., Shapiro, B. A., Oppenheim, A. B. and Maizel, J. V. 1989. Detection of common motifs in RNA secondary structures. *Nucleic Acids Research* 17:4829–4845.

Mathews, J. and Walker, R. L. 1970. *Mathematical Methods of Physics*. W. A. Benjamin.

Mau, B., Newton, M. A. and Larget, B. 1996. Bayesian phylogenetic inference via Markov chain Monte Carlo methods. Technical Report 961, Statistics Department, University of Wisconsin-Madison.

Maxwell, E. S. and Fournier, M. J. 1995. The small nucleolar RNAs. *Annual Review of Biochemistry* 64:897–934.

McCaskill, J. S. 1990. The equilibrium partition function and base pair binding probabilities for RNA secondary structure. *Biopolymers* 29:1105–1119.

McClure, M. A., Vasi, T. K. and Fitch, W. M. 1994. Comparative analysis of multiple protein-sequence alignment methods. *Journal of Molecular Evolution* 11:571–592.

McKeown, M. 1992. Alternative mRNA splicing. *Annual Review of Cell Biology* 8:133–155.

Melefors, O. and Hentze, M. W. 1993. Translational regulation by mRNA/protein interactions in eukaryotic cells: ferritin and beyond. *BioEssays* 15:85–90.

Meng, X.-L. and Rubin, D. B. 1992. Recent extensions to the EM algorithm. *Bayesian Statistics* 4:307–320.

Mevissen, H. T. and Vingron, M. 1996. Quantifying the local reliability of a sequence alignment. *Protein Engineering* 9:127–132.

Miller, W. and Myers, E. W. 1988. Sequence comparison with concave weighting functions. *Bulletin of Mathematical Biology* 50:97–120.

Mitchison, G. 1998. Probabilistic modelling of phylogeny and alignment. *Molecular Biology and Evolution* submitted.

Mitchison, G. and Durbin, R. 1995. Tree-based maximal likelihood substitution matrices and hidden Markov models. *Journal of Molecular Evolution* 41:1139–1151.

Miyazawa, S. 1994. A reliable sequence alignment method based on probabilities of residue correspondence. *Protein Engineering* 8:999–1009.

Mott, R. 1992. Maximum likelihood estimation of the statistical distribution of Smith–Waterman local sequence similarity scores. *Bulletin of Mathematical Biology* 54:59–75.

Myers, E. W. 1994. A sublinear algorithm for approximate keyword searching. *Algorithmica* 12:345–374.

Myers, E. W. and Miller, W. 1988. Optimal alignments in linear space. *Computer Applications in the Biosciences* 4:11–17.

Myers, G. 1995. Approximately matching context-free languages. *Information Processing Letters* 54:85–92.

Neal, R. M. 1996. *Bayesian Learning in Neural Networks*. Springer (Lecture Notes in Statistics).

Neal, R. M. and Hinton, G. E. 1993. A new view of the EM algorithm that justifies incremental and other variants. Preprint, Dept. of Computer Science, Univ. of Toronto, available from ftp://archive.cis.ohio-state.edu/pub/neuroprose/neal.em.ps.Z.

Needleman, S. B. and Wunsch, C. D. 1970. A general method applicable to the search for similarities in the amino acid sequence of two proteins. *Journal of Molecular Biology* 48:443–453.

Noller, H. F., Hoffarth, V. and Zimniak, L. 1992. Unusual resistance of peptidyl transferase to protein extraction procedures. *Science* 256:1416–1419.

Normandin, Y. and Morgera, S. D. 1991. An improved MMIE training algorithm for speaker-independent, small vocabulary, continuous speech recognition. In *Proceedings of ICASSP '91*, 537–540.

Nussinov, R., Pieczenik, G., Griggs, J. R. and Kleitman, D. J. 1978. Algorithms for loop matchings. *SIAM Journal of Applied Mathematics* 35:68–82.

Pavesi, A., Conterio, F., Bolchi, A., Dieci, G. and Ottonello, S. 1994. Identification of new eukaryotic tRNA genes in genomic DNA databases by a multistep weight matrix analysis of transcriptional control regions. *Nucleic Acids Research* 22:1247–1256.

Pearson, W. R. 1995. Comparison of methods for searching protein sequence databases. *Protein Science* 4:1145–1160.

Pearson, W. R. 1996. Effective protein sequence comparison. *Methods in Enzymology* 266:227–258.

Pearson, W. R. and Lipman, D. J. 1988. Improved tools for biological sequence comparison. *Proceedings of the National Academy of Sciences of the USA* 4:2444–2448.

Pearson, W. R. and Miller, W. 1992. Dynamic programming algorithms for biological sequence comparison. *Methods in Enzymology* 210:575–601.

Pedersen, A. G., Baldi, P., Brunak, S. and Chauvin, Y. 1996. Characterization of prokaryotic and eukaryotic promoters using hidden Markov models. In States, D. J., Agarwal, P., Gaasterland, T., Hunter, L. and Smith, R. F., eds., *Proceedings of the Fourth International Conference on Intelligent Systems for Molecular Biology*, 182–191. AAAI Press.

Peltz, S. W. and Jacobson, A. 1992. mRNA stability: in trans-it. *Current Opinion in Cell Biology* 4:979–983.

Pesole, G., Attimonelli, M. and Saccone, C. 1994. Linguistic approaches to the analysis of sequence information. *Trends in Biotechnology* 12:401–408.

Pietrokovski, S., Hirshon, J. and Trifonov, E. N. 1990. Linguistic measure of taxonomic and functional relatedness of nucleotide sequences. *Journal of Biomolecular Structure and Dynamics* 7:1251–1268.

Preparata, F. P. and Shamos, M. I. 1985. *Computational Geometry*. Springer-Verlag.

Press, W. H., Teukolsky, S. A., Vetterling, W. T. and Flannery, B. P. 1992. *Numerical Recipes in C*. Cambridge University Press.

Rabiner, L. R. 1989. A tutorial on hidden Markov models and selected applications in speech recognition. *Proceedings of the IEEE* 77:257–286.

Rabiner, L. R. and Juang, B. H. 1986. An introduction to hidden Markov models. *IEEE ASSP Magazine* 3:4–16.

Rabiner, L. R. and Juang, B. H. 1993. *Fundamentals of Speech Recognition*. Prentice-Hall.

Rannala, B. and Yang, Z. 1996. Probability distribution of molecular evolutionary trees: a new method of phylogenetic inference. *Journal of Molecular Evolution* 43:304–311.

Reese, M. G., Eeckman, F. H., Kulp, D. and Haussler, D. 1997. Improved splice site detection in Genie. *Journal of Computational Biology* 4:311–323.

Renals, S., Morgan, N., Bourlard, H., Cohen, M. and Franco, H. 1994. Connectionist probability estimators in hmm speech recognition. *IEEE Transactions on Speech and Audio Processing* 2:161–174.

Riis, S. K. and Krogh, A. 1997. Hidden neural networks: a framework for HMM/NN hybrids. In *Proceedings of ICASSP '97*, 3233–3236. IEEE.

Ripley, B. D. 1996. *Pattern Recognition and Neural Networks*. Cambridge University Press.

Rosenblueth, D. A., Thieffry, D., Huerta, A. M., Salgado, H. and Collado-Vides, J. 1996. Syntactic recognition of regulatory regions in *Escherichia coli*. *Computer Applications in the Biosciences* 12:415–422.

Russell, R. B. and Barton, G. J. 1992. Multiple protein sequence alignment from tertiary structure comparison: assignment of global and residue confidence levels. *Proteins* 14:309–323.

Saitou, N. 1996. Reconstruction of gene trees from sequence data. *Methods in Enzymology* 266:427–448.

Saitou, N. and Nei, M. 1987. The neighbor-joining method: a new method for reconstructing phylogenetic trees. *Molecular Biology and Evolution* 4:406–425.

Sakakibara, Y., Brown, M., Hughey, R., Mian, I. S., Sjölander, K., Underwood, R. C. and Haussler, D. 1994. Stochastic context-free grammars for tRNA modeling. *Nucleic Acids Research* 22:5112–5120.

Sankoff, D. 1975. Minimal mutation trees of sequences. *SIAM Journal of Applied Mathematics* 28:35–42.

Sankoff, D. and Cedergren, R. J. 1983. Simultaneous comparison of three or more sequences related by a tree. In Sankoff, D. and Kruskal, J. B., eds., *Time Warps, String Edits, and Macromolecules: the Theory and Practice of Sequence Comparison*. Addison-Wesley. Chapter 9, pp. 253–264.

Sankoff, D. and Kruskal, J. B. 1983. *Time Warps, String Edits, and Macromolecules: The Theory and Practice of Sequence Comparison*. Addison-Wesley.

Sankoff, D., Morel, C. and Cedergren, R. J. 1973. Evolution of 5S RNA and the nonrandomness of base replacement. *Nature New Biology* 245:232–234.

Schneider, T. D. and Stephens, R. M. 1990. Sequence logos: a new way to display consensus sequences. *Nucleic Acids Research* 18:6097–6100.

Schuster, P. 1995. How to search for RNA structures. Theoretical concepts in evolutionary biotechnology. *Journal of Biotechnology* 41:239–257.

Schuster, P., Fontana, W., Stadler, P. F. and Hofacker, I. L. 1994. From sequences to shapes and back: a case study in RNA secondary structures. *Proceedings of the Royal Society: Biological Sciences, Series B* 255:279–284.

Schwartz, R. and Chow, Y.-L. 1990. The N-best algorithm: an efficient and exact procedure for finding the n most likely hypotheses. In *Proceedings of ICASSP'90*, 81–84.

Searls, D. B. 1992. The linguistics of DNA. *American Scientist* 80:579–591.

Searls, D. B. and Murphy, K. P. 1995. Automata-theoretic models of mutation and alignment. In Rawlings, C., Clark, D., Altman, R., Hunter, L., Lengauer, T. and Wodak, S., eds., *Proceedings of the Third International Conference on Intelligent Systems for Molecular Biology*, 341–349. AAAI Press.

Shapiro, B. A. and Wu, J. C. 1996. An annealing mutation operator in the genetic algorithms for RNA folding. *Computer Applications in the Biosciences* 12:171–180.

Shapiro, B. A. and Zhang, K. 1990. Comparing multiple RNA secondary structures using tree comparisons. *Computer Applications in the Biosciences* 6:309–318.

Shimamura, M., Yasue, H., Ohshima, K., Abe, H., Kato, H., Kishiro, T., Goto, M.,

Munechika, I. and Okada, N. 1997. Molecular evidence from retroposons that whales form a clade within even-toed ungulates. *Nature* 388:666–670.

Shpaer, E. G., Robinson, M., Yee, D., Candlin, J. D., Mines, R. and Hunkapiller, T. 1996. Sensitivity and selectivity in protein similarity searches: a comparison of Smith–Waterman in hardware to BLAST and FASTA. *Genomics* 38:179–191.

Sibbald, P. R. and Argos, P. 1990. Weighting aligned protein or nucleic acid sequences to correct for unequal representation. *Journal of Molecular Biology* 216:813–818.

Sjölander, K., Karplus, K., Brown, M., Hughey, R., Krogh, A., Mian, I. S. and Haussler, D. 1996. Dirichlet mixtures: a method for improved detection of weak but significant protein sequence homology. *Computer Applications in the Biosciences* 12:327–345.

Smith, T. F. and Waterman, M. S. 1981. Identification of common molecular subsequences. *Journal of Molecular Biology* 147:195–197.

Sokal, R. R. and Michener, C. D. 1958. A statistical method for evaluating systematic relationships. *University of Kansas Scientific Bulletin* 28:1409–1438.

Sonnhammer, E. L. L., Eddy, S. R. and Durbin, R. 1997. Pfam: a comprehensive database of protein domain families based on seed alignments. *Proteins* 28:405–420.

Staden, R. 1988. Methods to define and locate patterns of motifs in sequences. *Computer Applications in the Biosciences* 4:53–60.

Steinberg, S., Misch, A. and Sprinzl, M. 1993. Compilation of tRNA sequences and sequences of tRNA genes. *Nucleic Acids Research* 21:3011–3015.

Stolcke, A. and Omohundro, S. M. 1993. Hidden Markov model induction by Bayesian model merging. In Hanson, S. J., Cowan, J. D. and Giles, C. L., eds., *Advances in Neural Information Processing Systems 5*, volume 5, 11–18. Morgan Kaufmann Publishers, Inc.

Stormo, G. D. 1990. Consensus patterns in DNA. *Methods in Enzymology* 183:211–221.

Stormo, G. D. and Hartzell III, G. W. 1989. Identifying protein–binding sites from unaligned DNA fragments. *Proceedings of the National Academy of Sciences of the USA* 86:1183–1187.

Stormo, G. D. and Haussler, D. 1996. Optimally parsing a sequence into different classes based on multiple types of evidence. In States, D. J., Agarwal, P., Gaasterland, T., Hunter, L. and Smith, R. F., eds., *Proceedings of the Fourth International Conference on Intelligent Systems for Molecular Biology*, 369–375. AAAI Press.

Studier, J. A. and Keppler, K. J. 1988. A note on the neighbour-joining algorithm of Saitou and Nei. *Molecular Biology and Evolution* 5:729–731.

Swofford, D. L. and Olsen, G. J. 1996. Phylogeny reconstruction. In Hillis, D. M. and Moritz, C., eds., *Molecular Systematics*. Sinauer Associates. pp. 407–511.

Tatusov, R. L., Altschul, S. F. and Koonin, E. V. 1994. Detection of conserved segments in proteins: iterative scanning of sequence databases with alignment blocks. *Proceedings of the National Academy of Sciences of the USA* 91:12091–12095.

Taylor, W. R. 1987. Multiple sequence alignment by a pairwise algorithm. *Computer Applications in the Biosciences* 3:81–87.

Thompson, E. A. 1975. *Human Evolutionary Trees*. Cambridge University Press.

Thompson, J. D., Higgins, D. G. and Gibson, T. J. 1994a. CLUSTAL W: improving the sensitivity of progressive multiple sequence alignment through sequence weighting, position specific gap penalties and weight matrix choice. *Nucleic Acids Research* 22:4673–4680.

Thompson, J. D., Higgins, D. G. and Gibson, T. J. 1994b. Improved sensitivity of profile searches through the use of sequence weights and gap excision. *Computer Applications in the Biosciences* 10:19–29.

Thorne, J. L., Kishino, H. and Felsenstein, J. 1992. Inching toward reality: an improved likelihood model of sequence evolution. *Methods in Enzymology* 34:3–16.

Tolstrup, N., Rouzé, P. and Brunak, S. 1997. A branch point consensus from Arabidopsis found by non-circular analysis allows for better prediction of acceptor sites. *Nucleic Acids Research* 25:3159–3164.

Tuerk, C., MacDougal, S. and Gold, L. 1992. RNA pesudoknots that inhibit human immunodeficiency virus type 1 reverse transcriptase. *Proceedings of the National Academy of Sciences of the USA* 89:6988–6992.

Turner, D. H., Sugimoto, N., Jaeger, J. A., Longfellow, C. E., Freier, S. M. and Kierzek, R. 1987. Improved parameters for prediction of RNA structure. *Cold Spring Harbor Symposia Quantitative Biology* 52:123–133.

van Batenburg, F. H. D., Gultyaev, A. P. and Pleij, C. W. A. 1995. An APL-programmed genetic algorithm for the prediction of RNA secondary structure. *Journal of Theoretical Biology* 174:269–280.

Vingron, M. 1996. Near-optimal sequence alignment. *Current Opinion in Structural Biology* 6:346–352.

Vingron, M. and Waterman, M. S. 1994. Sequence alignment and penalty choice: review of concepts, case studies and implications. *Journal of Molecular Biology* 235:1–12.

Waterman, M. S. 1995. *Introduction to Computational Biology*. Chapman & Hall.

Waterman, M. S. and Eggert, M. 1987. A new algorithm for best subsequence alignments with application to tRNA–rRNA comparisons. *Journal of Molecular Biology* 197:723–725.

Waterman, M. S. and Perlwitz, M. D. 1984. Line geometries for sequence comparisons. *Bulletin of Mathematical Biology* 46:567–577.

Watson, J. D., Hopkins, N. H., Roberts, J. W., Steitz, J. A. and Weiner, A. M. 1987. *Molecular Biology of the Gene*. Benjamin/Cummings.

Wilmanns, M. and Eisenberg, D. 1993. Three-dimensional profiles from residue-pair preferences: identification of sequences with beta/alpha-barrel fold. *Proceedings of the National Academy of Sciences of the USA* 90:1379–1383.

Witherell, G. W., Gott, J. M. and Uhlenbeck, O. C. 1991. Specific interaction between RNA phage coat proteins and RNA. *Progress in Nucleic Acid Research and Molecular Biology* 40:185–220.

Woese, C. R. and Pace, N. R. 1993. Probing RNA structure, function, and history by comparative analysis. In Gesteland, R. F. and Atkins, J. F., eds., *The RNA World*. Cold Spring Harbor Laboratory Press. pp. 91–117.

Wray, G. A., Levinto, J. S. and Shapiro, L. H. 1996. Molecular evidence for deep precambrian divergences among metazoan phyla. *Science* 274:568–573.

Wu, S. and Manber, U. 1992. Fast text searching allowing errors. *Communications of the ACM* 35:83–90.

Yada, T. and Hirosawa, M. 1996. Detection of short protein coding regions within the Cyanobacterium genome: application of the hidden Markov model. *DNA Research* 3:355–361.

Yada, T., Sazuka, T. and Hirosawa, M. 1997. Analysis of sequence patterns surrounding the translation initiation sites on Cyanobacterium genome using the hidden Markov model. *DNA Research* 4:1–7.

Yang, Z. 1993. Maximum-likelihood estimation of phylogeny from DNA sequences when substitution rates differ over sites. *Molecular Biology and Evolution* 10:1396–1401.

Yang, Z. 1994. Maximum likelihood phylogenetic estimation from DNA sequences with variable rates over sites: approximate methods. *Journal of Molecular Evolution* 39:306–314.

Zuckerkandel, E. and Pauling, L. 1962. Molecular disease, evolution and genetic heterogeneity. In Marsha, M. and Pullman, B., eds., *Horizons in Biochemistry*. Academic Press. pp. 189–225.

Zuker, M. 1989a. Computer prediction of RNA structure. *Methods in Enzymology* 180:262–288.

Zuker, M. 1989b. On finding all suboptimal foldings of an RNA molecule. *Science* 244:48–52.

Zuker, M. 1991. Suboptimal sequence alignment in molecular biology: alignment with error analysis. *Journal of Molecular Biology* 221:403–420.

Zuker, M. and Stiegler, P. 1981. Optimal computer folding of large RNA sequences using thermodynamics and auxiliary information. *Nucleic Acids Research* 9:133–148.

Author index

Page references in italics refer to the bibliography

Subject index